Creating Language

Creating Language

Integrating Evolution, Acquisition, and Processing

Morten H. Christiansen and Nick Chater

The MIT Press
Cambridge, Massachusetts
London, England

First MIT Press paperback edition, 2018

This book was set in Times and Syntax by Toppan Best-set Premedia Limited.

Library of Congress Cataloging-in-Publication Data

Names: Christiansen, Morten H., 1963– author. | Chater, Nick, author.
Title: Creating language: integrating evolution, acquisition, and processing / Christiansen, Morten H., and Nick Chater; foreword by Peter W. Culicover.
Description: Cambridge, MA: The MIT Press, [2016] | Includes bibliographical references and index.
Identifiers: LCCN 2015038404 | ISBN 9780262034319 (hardcover : alk. paper) ISBN 9780262535113 (paperback)
Subjects: LCSH: Creativity (Linguistics) | Language acquisition. | Cognition. | Psycholinguistics.
Classification: LCC P37 .C547 2016 | DDC 401/.9—dc23 LC record available at http://lccn.loc.gov/2015038404

149822760

Contents

Foreword

Peter W. Culicover

Language is something that is so familiar to us that it is second nature. Yet when we pay close attention to it, it presents us with a cascade of puzzles, paradoxes and mysteries. Why do only humans have language? How did languages come to acquire their particular properties? Why are they so much the same in some respects, yet so different in others?

Come to think of it, what's the right way to properly describe the properties of languages? How does knowledge of language get into the heads of language learners? When young children begin to acquire a language, is what they know qualitatively different from what an adult knows? And if it is different, how do they make the transition from child knowledge to adult knowledge? If languages have rules, where are those rules in the mind? What are we doing when we are speaking and understanding language, and what relationship does that have to our knowledge? And so on.

Contemporary research on language is largely concerned with general questions such as these, and the more specific questions that follow from them as they become more precisely framed. Much of the scientific discourse about the nature of language, the present book included, must be understood against the backdrop of the answers that have been offered by Noam Chomsky over the past fifty years, essentially since the publication of his *Aspects of the Theory of Syntax* (Chomsky, 1965). At the nexus of Chomsky's thesis is this (Chomsky, 1968, 79):

> We must postulate an innate structure [in the human mind] that is rich enough to account for the disparity between experience and knowledge, one that can account for the construction of the empirically justified generative grammars within the given limitations of time and access to data … this postulated innate mental structure must not be so rich and restrictive as to exclude certain known languages … [t]he factual situation is obscure enough to leave room for much difference of opinion over the true nature of this innate mental structure that makes acquisition of language possible.

"Much difference of opinion" is an understatement. There have been many proposals and counterproposals about what this "innate structure" is, and even denials that there is any such structure in the mind that is specifically devoted to language. I won't try to rehearse any of them here. But quite generally, it appears that the more specific a proposal is about what this structure consists of, the more difficult it is to understand how it might have evolved as a property of the human mind. (This may be the reason that Chomsky's thinking on the question has evolved to the point that he has proposed that the only property that characterizes the human capacity for language is recursion (Hauser, Chomsky, & Fitch, 2002).) It is this kind of difficulty that makes the challenge of answering the sorts of questions raised above so much fun, and so engaging.

Moreover, Chomsky talks of "generative grammar," the mental representation of language in the mind. I have no problem with this—I was trained as a linguist at MIT in the 1960s, I think that there is something in my (human) mind that embodies my knowledge of language, and I'm happy to call it a "generative grammar," since it enables me to deal with an arbitrary number of novel sentences (such as the ones in this Foreword) in English. But what is this something, and how does it embody my knowledge of language? It is a mystery. Many would say that my linguist's description of this knowledge does not correspond in any architectural way to the thing that embodies the knowledge. But why does the thing behave as though it has, or sort of has, such a description (a phenomenon that I call "squinting at Dali's Lincoln" in chapter 1 of Culicover [2013])?

Let me give just one concrete example of a linguistic property that raises all kinds of challenging questions about what it is we know and how we know it (and believe me, there are thousands). In English it is generally possible to question an argument by placing an interrogative expression that corresponds to it at the beginning of the sentence (*What kind of filling do you suppose Otto is going to put on that sandwich; I suppose Otto is going to put Bergkäse on his sandwich*). It is unacceptable to do that when the argument in question is inside of a relative clause (**What kind of filling did you meet a person who is going to put on that sandwich?; I met a person who is going to put Bergkäse on his sandwich*). But, curiously, in Japanese the interrogative expression does not appear at the beginning of the clause, but remains in place in the relative clause, and the result is acceptable.

Why is this? Could it be that humans have evolved an "innate structure" for language that specifically excludes the English-type questions? On the one hand we are tempted to say "yes" because it is hard to see how such knowledge could be acquired through experience. On the other hand, we are tempted to say "no," because it is a puzzle how having such knowledge would enhance survival of an individual who had it. (Not that we really know that much about the role of knowledge in evolution.) But what is the alternative?

It seems that we are faced with a paradox in this and a multitude of similar and not so similar cases. Speakers behave as if they know things that they couldn't have learned because of lack of evidence. But are we being too glib in our understanding of the words "know," "learn," and "evidence"? What exactly do I "know" when I say that the sentence with the asterisk sounds bad? Is it possible that there is a way to explain my judgment without assuming that I "know" that this is a problematic sentence or that I have "learned" that it is? Is the fact that I never experience a particular type of construction "evidence" that it is problematic? What about all of the things that I have never heard that are not problematic (like the sentences in this Foreword, for example)?

The answer to these questions, of course, is "maybe—let's think about it." On this note, another quote from Chomsky is very appropriate: "We know too little about mental structures to advance dogmatic claims …" (Chomsky, 1980, 49). Of course he's right. There are always alternatives to be entertained, and we can always try to show that what might appear to be a candidate for a property of this "innate structure" is actually the consequence of something else—something that doesn't have to do with language specifically, or something that has been incorporated into languages because it facilitates expression or communication, or has spread through social transmission.

In fact, I'll go further and say that we *must* try to show that there is an alternate explanation to any and all proposed specific aspects of "innate structure," as Morten Christiansen and Nick Chater do in this book. If they are on the right track (and I think they are about many things), they will have shed light on some mysteries about why languages have certain properties and how language gets into our heads. If they turn out in the end to be wrong about something (an unfortunate inevitability in our line of work, if we are willing to take risks), they may have lent additional credence to our understanding of what is likely to be part of the "innate structure."

There's really no way to know without working through the details, and work through the details they do.

References

Chomsky, N. (1965). *Aspects of the theory of syntax*. Cambridge, MA: MIT Press.

Chomsky, N. (1968). *Language and mind*. New York: Harcourt Brace Jovanovich.

Chomsky, N. (1980). Rules and representations. *Behavioral and Brain Sciences*, *3*, 1–15.

Culicover, P. W. (2013). *Grammar and complexity: Language at the intersection of competence and performance*. Oxford: Oxford University Press.

Hauser, M. D., Chomsky, N., & Fitch, W. T. (2002). The faculty of language: What is it, who has it, and how did it evolve? *Science*, *298*, 1569–1579.

Preface

The phenomenon of human language and language use is remarkable. The spectacular complexity and subtlety of the world's languages, and our astonishing ability to acquire and use them to convey instructions, depths of feeling, scientific theories, and so much more, may indeed be the most distinctive collective human achievement. In this book, we argue that to begin to understand this astonishing phenomenon, we must look at its origins. That is, we must consider how language is created: moment by moment, in the generation and understanding of individual utterances; year by year, as new language learners acquire the skill of generating and understanding; and generation by generation, as languages change, split and fuse, through processes of cultural evolution, from what we imagine to be the rudimentary communicative systems of our far distant ancestors to the richness and diversity of natural languages today. It turns out, as we shall see, that understanding the creation of language across the three timescales of language processing, acquisition, and evolution, throws new light on each level. By considering the cascade of processes that led language to *come to be*, we can perhaps better understand what language *is*. This book attempts to show that this can be done, and, in the process, how we can think afresh about some of the profound puzzles in the science of human language—why language has the structure it does, how we process it so effortlessly, why languages are learnable, and how languages have evolved.

This book has been a long time in the making. It results from joint work going back more than two decades, to the early 1990s, when we were beginning our research careers at the then Centre for Cognitive Science (now part of the School of Informatics) and the Department of Psychology at the University of Edinburgh. It has been a collaboration that "snapped together" almost instantly, and has been unbreakable since, despite many changes of institution and the two of us ending up on different continents. The perspective reflected in this book, viewing language through the interlinked processes by which language is created, has come into ever clearer focus over the last decade, resulting in a series of joint papers which, in reworked and adapted forms, constitute some of the core elements of this book. Integrating these ideas into a single framework has been a far greater challenge than we ever imagined, but also a hugely exciting and stimulating project. We hope some of this enthusiasm and exhilaration spills over to the reader.

We have many people to thank. We are, first and foremost, hugely grateful for the unfailing support of our families—MHC: Anita Govindjee and Sunita Christiansen; NC: Louie Fooks, Maya Fooks, and Caitlin Fooks—to whom this book is gratefully dedicated. They have both tolerated the long hours that this project has consumed, and been delightful hosts for our frequent stays in each other's homes. This book, and our work together over the last twenty years, would not have been possible without them.

A great number of people have contributed to the development of the ideas in this book. Our thanks go to the many people who have worked with us as students or post-docs, including Chris Conway, Rick Dale, Michelle Ellefson, Thomas Farmer, Anne Hsu, Hajnal Jolsvai, Ethan Jost, Gary Lupyan, Stewart McCauley, Jennifer Misyak, Padraic Monaghan, Luca Onnis, Florencia Reali, Patty Reeder, Matthew Roberts, Esther van den Bos, and Julia Ying, and to the many people whose input and ideas have helped shape this work (perhaps more than they know) in ways ranging from direct collaboration to inspiring discussions, including Inbal Arnon, Rens Bod, Andy Clark, Bill Croft, Dan Dediu, Jeff Elman, Karen Emmorey, Nick Enfield, Stanka Fitneva, Hartmut Fitz, Stefan Frank, Adele Goldberg, James Hurford, Mutsumi Imai, Simon Kirby, Stephen Levinson, Elena Lieven, Maryellen MacDonald, Jay McClelland, Karl Magnus Petersson, Martin Pickering, Michael Ramscar, Pete Richerson, Tom Schoenemann, Mark Seidenberg, Richard Shillcock, Linda Smith, Mike Tomasello, Bruce Tomblin, Paul Vitányi, and Bill Wang. Special thanks go to Peter Culicover, for reading and commenting on the manuscript, and for agreeing so generously to write a foreword to this book. We are also grateful for detailed feedback on the chapters in this book from MHC's lab group—Brandon Conley, Jess Flynn, Scott Goldberg, Ethan Jost, Yena Kang, Jordan Limperis, Stewart McCauley, Samantha Reig, Sven Wang, and Julia Ying—as well as from Inbal Arnon, Rick Dale, Thomas Farmer, Stefan Frank, Jennifer Misyak, Padraic Monaghan, and Florencia Reali. Moreover, the book has benefitted from constructive comments from cross-disciplinary audiences at three major lecture series that MHC delivered across the world: the 2009 Nijmegen Lectures at the Max Planck Institute for Psycholinguistics, Nijmegen, the Netherlands; an Ida Cordelia Beam Distinguished Visiting Professorship in 2010 at the University of Iowa; and a Visiting Professorship at the University of Hong Kong in 2012, as well as at keynotes and plenaries by both authors at numerous high-profile conferences. Finally, a big thank you to Gary Kerridge, who so efficiently put our references in proper order, to Anita Govindjee for her careful editing of the entire book, and to Phil Laughlin, our editor at MIT Press, for his enthusiastic support and for having faith in this project despite its alarmingly long gestation.

Our work has been supported by the generosity of many funding bodies over many years. In particular, the preparation of this book has been supported by the Human Frontiers of Science Program (grant RGP0177/2001-B), the Binational Science Foundation (grant number 2011107 [with Inbal Arnon]), a Charles A. Ryskamp Fellowship from the American Council of Learned Societies, the Santa Fe Institute, and the Max Planck Institute for Psycholinguistics to MHC; and by ERC grant 295917-RATIONALITY, the ESRC Network for Integrated Behavioral Science (grant number ES/K002201/1), the Leverhulme Trust (grant number RP2012-V-022), and UK Research Councils (Grant EP/K039830/1) to NC.

We are grateful, too, to the publishers who have graciously allowed us to rework a number of previous papers (often rather drastically) into chapters of this book. Chapter 2 is based on M. H. Christiansen and N. Chater, 2008, Language as shaped by the brain, *Behavioral and Brain Sciences*, *31*, 489–558, and M.H. Christiansen and R.-A Mueller, 2014, Cultural recycling of neural substrates during language evolution and development, in M. S. Gazzaniga and G. R. Mangun (Eds.), *The Cognitive Neurosciences V* (pp. 675–682), Cambridge, MA: MIT Press. Chapter 3 draws on N. Chater and M. H. Christiansen, 2010, Language acquisition meets language evolution, *Cognitive Science*, *34*, 1131–1157. Chapter 4 is based on M. H. Christiansen and N. Chater, in press, The Now-or-Never bottleneck: A fundamental constraint on language, *Behavioral and Brain Sciences*. Chapter 5 derives in part from M. H. Christiansen & P. Monaghan, in press, Division of labor in vocabulary structure: Insights from corpus analyses, *Topics in Cognitive Science*: and M. H. Christiansen, 2013, Language has evolved to depend on multiple-cue integration, in R. Botha & M. Everaert (Eds.), *The evolutionary emergence of language: Evidence and inference* (pp. 42–61), Oxford: Oxford University Press. Chapter 7 draws on M. H. Christiansen and M. C. MacDonald, 2009, A usage-based approach to recursion in sentence processing, *Language Learning, 59 (Suppl. 1)*, 126–161; and M. H. Christiansen and N. Chater, 2015, The language faculty that wasn't: A usage-based account of natural language recursion, *Frontiers in Psychology, 6,* 1182.

Above all, we hope that the approach developed here, drawing out the connections between the creation of language across multiple timescales, provides the starting point for integrating the study of language structure, processing, acquisition, and evolution. The science of language has led to some remarkable theoretical achievements, and has produced many wonderful discoveries about how language and language use work. But it can often seem that the field has broken into innumerable fragments. We hope that *Creating Language* contributes to the project of putting the language sciences back together again.

Morten H. Christiansen and Nick Chater
Ithaca, NY (US) and Coventry (UK)

I Theoretical and Empirical Foundations

1 Language Created across Multiple Timescales

Language is a process of free creation; its laws and principles are fixed, but the manner in which the principles of generation are used is free and infinitely varied. Even the interpretation and use of words involves a process of free creation.
—Noam Chomsky, 1970, *Language and Freedom*

We create language in the moment. Almost everything we say, of any significant complexity, is likely to be something we have never said before; much of what we say, no one has ever uttered before. And beyond the creation of the individual sentence, of course, we are able to organize these sentences into essays, novels, speeches, jokes, emails, recipes, newspaper articles, and academic monographs. But the real mystery is how we can freely express our stream of consciousness and thoughts in real time, using a wonderfully elaborate and complex system of words and grammatical constructions; and do this so naturally, speedily, and apparently effortlessly, that most of us are blithely unaware of our spectacular linguistic creativity.

Yet we are not born with language. Instead, each child has to develop the ability to speak and understand language from scratch—and, extraordinarily, each child must "create" its own language processing system without any explicit instruction. This would seem to make learning the "system" of human language much more challenging than tasks for which we routinely offer comprehensive university-level training, such as learning to program a computer, master a logical formalism, or carry out advanced mathematics. Nonetheless, young children are able to recreate the linguistic skills of their parents merely, it would appear, by being immersed in a population of language users—such osmosis is not, of course, observed for learning computer programming, logic, or mathematics.

We create language on a third timescale, too—not just in the moment when generating language, or through our childhoods as we acquire our mother tongue, but collectively as generations of speakers innovate, modify, and propagate the sounds, words, and grammatical constructions to create the languages of today. Languages are in continual flux: the entire Indo-European language group—including languages as diverse as Danish, Greek, Hindi, and Welsh—has diverged from a common root in less than nine thousand years (Gray & Atkinson, 2003). The emergence of the thousands of languages of the world is one of the most extraordinary achievements of human culture. How has this been possible?

In this light, it is clear that our ability to create language in the moment, during development, and across generations, poses fundamental questions for the biological and social sciences as well as the humanities: How did language evolve in response to environmental and biological forces? How is language acquired by each new generation? And how is language processed "online" in everyday social interactions? Across the language sciences, these questions are typically treated as separate topics, to be addressed more or less independently. In contrast, this book is premised on the idea that questions about evolution, acquisition, and processing are best addressed *together* because there are strong constraints working across these domains of inquiry, allowing each to shed light on the others. Specifically, we construe language as being created across multiple timescales: the timescale of seconds at which we create language by producing and interpreting utterances; the timescale of tens of years during which we create our knowledge of language in acquisition and update it across our lifespan; and the timescale of perhaps a hundred thousand years or more over which language was created in the human lineage and evolved into its current form. Figure 1.1 provides a schematic illustration of the relationship between the three timescales in the processing, acquisition, and evolution of language[1].

Our framework builds on a construction-based approach to language, in which an utterance is viewed as consisting of a set of constructions (e.g., Croft, 2001; Goldberg, 2006; O'Grady, 2005). Constructions are learned pairings between form and meaning (in a broad sense including discourse function), including morphemes or words (e.g., *-ing*, *read*), multiword sequences (e.g., *a cup of tea*), partially filled lexical patterns (e.g., *the Xer, the Yer*), and more abstract linguistic schemas (e.g., the transitive construction SUBJ V OBJ, as in *I like licorice*). The processing of language—both in comprehension and production—at the timescale of the utterance is fundamental to language creation at the other two timescales. Each processing event provides an acquisition opportunity, and language acquisition consists of a vast number of such

1. Wang (1978) also proposes looking at language across three timescales: microhistory—which overlaps with our acquisition timescale, mesohistory (centuries/millennia), and macrohistory (many millennia)—which our evolutionary timescale collapse into a single timescale.

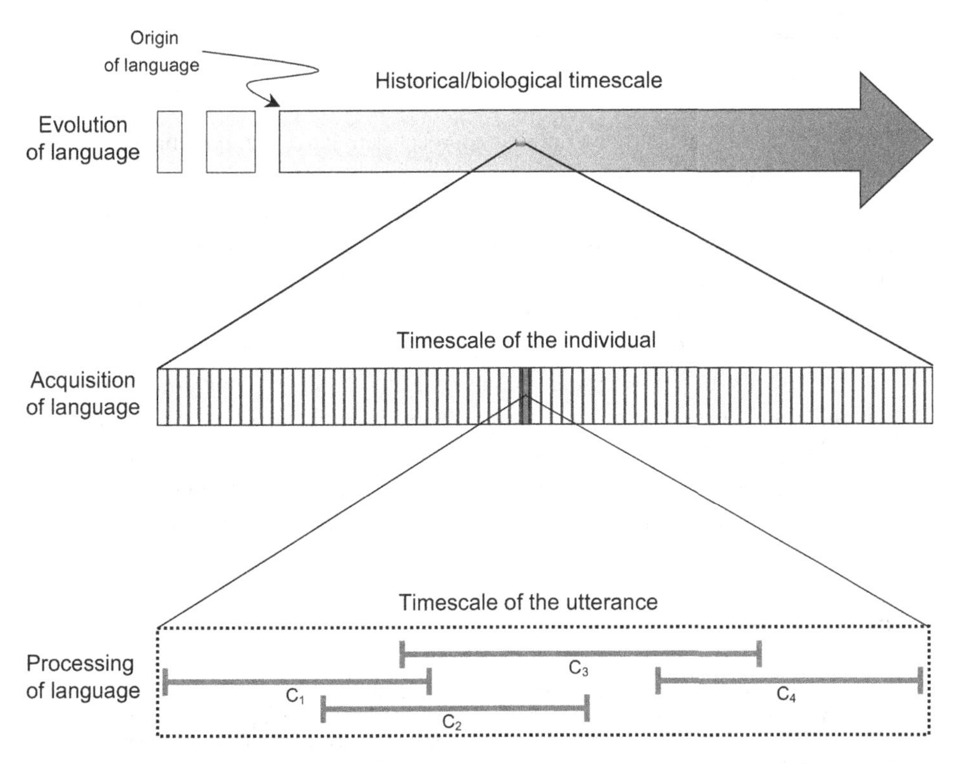

Figure 1.1
Three timescales of language creation: the timescale of the utterance during which language is processed (with C_i corresponding to constructions); the timescale of the individual lifetime during which language is acquired; and the historical/biological timescale during which language originated and evolved in our species.

processing events distributed across the lifespan of an individual. Consequently, we view language acquisition as being usage-based (e.g., O'Grady, 2005; Tomasello, 2003); that is, language acquisition in essence involves learning how to process utterances. In the same way that language acquisition consists of a multitude of processing events, the evolution of language over time has been shaped through cultural transmission by the language skills of many generations of language-learning individuals (e.g., Beckner et al., 2009; Christiansen & Chater, 2008; Hurford, 1999; Smith & Kirby, 2008; Tomasello, 2003; for a review, see Dediu et al., 2013). Thus, language creation is closely intertwined across multiple timescales, highlighting the importance of theoretical interdependencies between the processing, acquisition, and evolution of language.

1.1 Generative Approaches to Language Processing, Acquisition, and Evolution

In the mainstream generative grammar approaches to language, possible interrelations between the processing, acquisition and evolution of language are rarely explored (though see Culicover, 2013a; Culicover & Nowak, 2003; Pinker, 1994; Jackendoff, 2002). This may be a consequence of Chomsky's methodological dictums (see box 1.1) that the study of language proper should be separated from how it is used and processed (Chomsky, 1965), acquired over development (Chomsky, 1975), and how it evolved (Chomsky, 2005). Tracing exactly how processing, acquisition and evolution may be related within current generative grammar approaches is complicated by the fact that the field appears to be in a state of a flux following the introduction of the Minimalist Program (MP; Chomsky, 1995) in the mid-1990s. Current conceptions of Universal Grammar (UG) differ substantially, ranging from views

Box 1.1
Chomsky's Hidden Legacy

Noam Chomsky has had a huge impact on the study of language and helped usher in the cognitive revolution in psychology and the cognitive sciences. Over the past nearly sixty years, his evolving vision of generative grammar has had a major influence on thinking in the language sciences: directly in his own work and the work of the proponents of his theories, as well as indirectly in the research of many of his opponents. However, Chomsky also appears to have had a more subtle impact on the language sciences, affecting how language researchers tend to approach their subject of study. Over the years, he has gradually sought to isolate the study of language from considerations regarding processing, acquisition, and evolution. First, Chomsky (1965) contended that performance data from processing are not relevant for understanding the nature of language, which he viewed as the idealized linguistic competence of a speaker/hearer: "... investigation of performance will proceed only so far as understanding of underlying competence permits" (p. 10). He then proposed that "... [language] learning is 'instantaneous'" (Chomsky, 1975, p. 119), implying that acquisition is not relevant for the understanding of language proper. Finally, Chomsky (2005) made a similar claim with regard to language evolution, suggesting that "... the Great Leap that yields Merge ... was effectively instantaneous ..." (p. 12). The result of what we call *Chomsky's hidden legacy* is not only that linguistics has been separated from the other language sciences but also that the study of processing, acquisition, and evolution have become separated from one another. These methodological and theoretical divisions have become so ingrained in the language sciences that they set the theoretical ground-rules, even for many of Chomsky's critics. For example, when developing his cross-linguistic analysis of how processing constraints shape grammar, Hawkins (1994) readily accepted Chomsky's conception of the logical problem of language acquisition. Similarly, Cowie (1999), despite her fierce critique of Chomsky's nativist perspective on language acquisition, nonetheless adopted his ideas about the separation of competence and performance, with the former being central to the study of language. Thus, the impact of Chomsky's methodological dictums looms large within the language sciences, even among researchers who otherwise oppose his perspective on specific aspects of language. A key goal of this book is to present a comprehensive framework within which to re-integrate the processing, acquisition, and evolution of language, with the aim of moving beyond Chomsky's hidden legacy.

close to the Principles and Parameters Theory (PPT; Chomsky, 1981) of pre-minimalist generative grammar (e.g., Crain, Goro & Thornton, 2006; Crain & Pietroski, 2006) to a simpler vision of the components of generative grammar (Culicover & Jackendoff, 2005; Pinker & Jackendoff, 2005), to Minimalism itself, in which language acquisition is confined to learning a lexicon from which cross-linguistic variation is proposed to arise (Boeckx, 2006; Chomsky, 1995).

However, common to the great majority of current generative approaches is the separation of the universal characteristics of language acquisition (whatever these are hypothesized to be) from the processing of linguistic input/output. Figure 1.2 provides a simplified illustration of how innate mechanisms (UG) for language are related to its acquisition and processing within generative approaches. The interpretations of the various components of this schematization depend on the conception of UG in play in a particular theoretical framework. From the viewpoint of PPT, UG consists of a set of genetically-specified universal linguistic principles combined with a set of parameters to account for variations across languages (Crain et al., 2006). Information from the language environment is used during acquisition to determine the parameter settings relevant for individual languages (dotted arrow). The language environment also provides input for learning about the language periphery; that is, aspects of the language (including the lexicon) that are not encompassed by the core linguistic properties of UG and which may be learned by domain-general mechanisms. A speaker's knowledge of the core and peripheral aspects of language comes together during processing to allow the comprehension and production of utterances, isolating processing from acquisition.

The *Simpler Syntax* (SS) version of generative grammar (e.g., Culicover, 2013a; Culicover & Jackendoff, 2005; Jackendoff, 2002; Pinker & Jackendoff, 2005) combines elements from construction grammar (e.g., Goldberg, 2006) with structural principles from PPT (such as phrase structure principles [X-bar theory], agreement, and case-marking). These basic principles form part of a genetically specified UG, but are considered to be less restrictive than in PPT and function more as guides for learning. As in construction grammar, the language environment plays a crucial role in the acquisition of both words and grammatical structure, the latter being acquired as part of the lexicon in the form of constructions. A lexical entry can function as a schema that may figure in new utterances or be abstracted with other constructions to form an inheritance hierarchy of increasingly general schemas. UG pre-specifies the highest level of abstraction in the hierarchy and provides (soft) constraints on the creation of lower-level generalizations in order to maximize hierarchical coherence (indicated by the dotted line in figure 1.2). This means

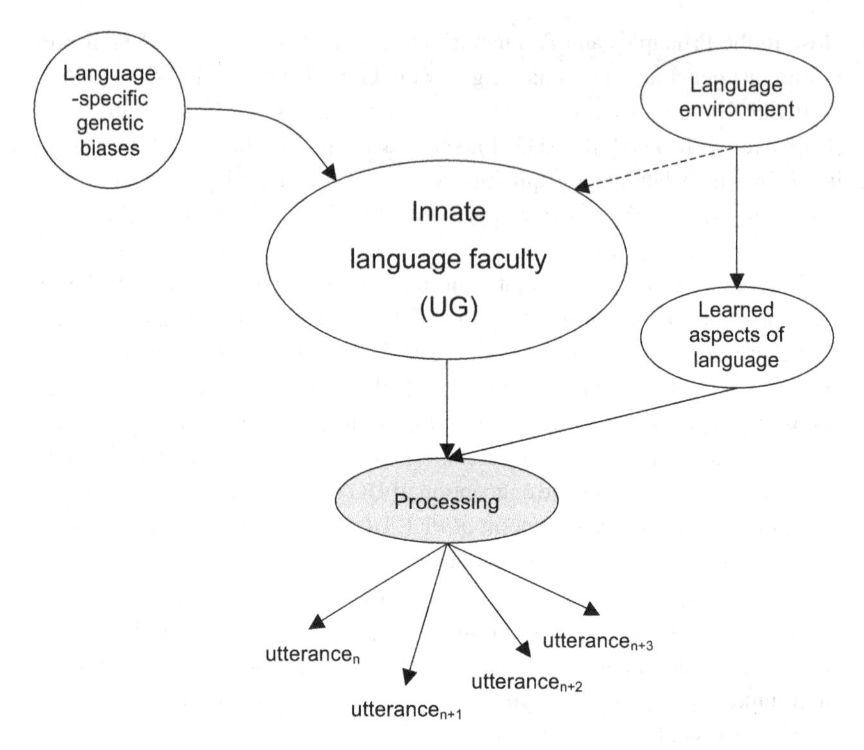

Figure 1.2
Illustration of how innate mechanisms for language are related to the acquisition and processing of linguistic structure within most current approaches to generative grammar.

that there is no sharp distinction between core and peripheral elements of language; rather, the generality of syntactic constraints is viewed as fundamentally graded (Culicover, 1999). However, the construction-based knowledge of language (including UG) is separated from a working-memory-based processing system for language production and comprehension (Jackendoff, 2007), thereby isolating processing from acquisition, as in previous generative approaches.

The MP approach to generative grammar assumes that language is a perfect system for mapping between sound and meaning (Chomsky, 1995). In a departure from earlier generative approaches, only recursion (in the form of Merge) is considered to be unique to human language ability (Hauser, Chomsky & Fitch, 2002), thus constituting the genetically specified *Innate Language Faculty* in figure 1.2. Variation among languages is now explained in terms of lexical parameterization; that is, differences between languages are no longer accounted for by a set of parameters associated with grammars (as in PPT) but instead by parameters associated with particular lexical items. The system for acquiring the lexicon is suggested to be species-specific but not special to language (Hauser et al., 2002) and thus along with peripheral features of language constitutes aspects of language that may be learned by domain-general mechanisms. Similar to other generative approaches, the innate components of language and linguistic knowledge acquired through development are considered to be separate from processing.

A common assumption of the three recent approaches to generative grammar is that processing is secondary and typically relegated to an unspecified, or loosely specified, performance system (though see Jackendoff, 2007). This separation of processing and acquisition also affects how evolution of the innate language faculty (or UG) is construed. Figure 1.3 provides a schematization of how natural selection over time would result in UGs that provide for increasingly advanced linguistic abilities by selecting genes (or combinations of genes) specific to the language faculty (curved arrows). A separate performance system (gray ovals) mediates comprehension and production (straight solid arrows), and acts as a go-between that permits the primary linguistic input to fine-tune native language knowledge in UG (dashed arrows). Finally, the language environment itself changes through processes of cultural transmission (double arrows), rather than as a consequence of the innate language faculty (though UG will constrain the space of possible changes).

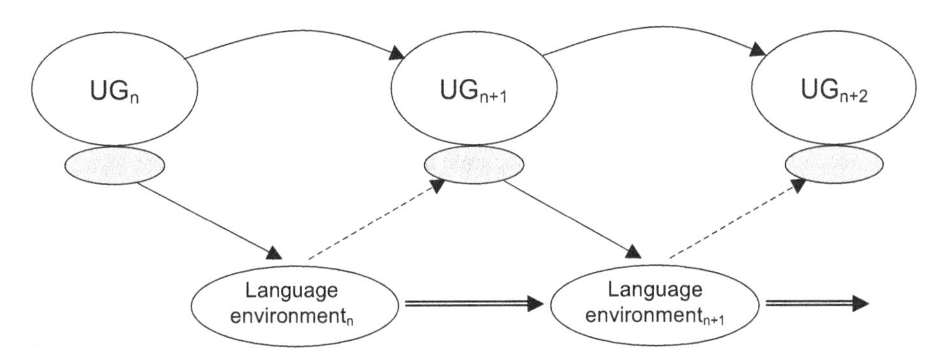

Figure 1.3
Illustration of the gradual evolution of increasingly sophisticated innate mechanisms for language (UGs; curved arrows) as a consequence of repeated cycles of learning from the language environment (dotted arrows) and the production thereof (solid straight arrows), both mediated by the performance system (grey ovals). The language environment itself also changes over time due to processes of cultural transmission (double arrows).

This figure most straightforwardly captures the gradual evolution of more sophisticated UGs within the PPT framework (Pinker, 1994; Pinker & Bloom, 1990) in which parameters are inferred from the language environment (dashed arrows) while also allowing for the periphery to evolve (double arrows). The SS perspective on generative grammar also assumes a gradual evolution of UGs (Culicover & Jackendoff, 2005). Acquisition here, too, is mediated by the performance system (although not in terms of parameters). Although the core-periphery distinction is eschewed in SS, some aspects of the language environment are likely to change due to processes of cultural transmission, language contact, etc. (double arrows). It is less clear how to fit the MP approach into an evolutionary framework. Hauser et al. (2002) suggest that the innate language faculty (in the form of recursion) may not have evolved specifically for language but instead from older domain-specific systems (e.g., for navigation) that became domain-general, allowing language to incorporate recursive structure. Nonetheless, they acknowledge that the innate language faculty could have been subject to natural selection, in which case figure 1.3 provides a possible evolutionary scenario. Specifically, the innate recursive mechanism may have been adapted to function better in its specific linguistic role as mediator between sensori-motor and conceptual-intentional systems. Such adaptation would have involved interactions with the language environment mediated by the performance system. Most of the adaptation, however, is likely to have taken place in the lexical system to provide the machinery for dealing with language variation (e.g., lexical parameterization).

1.2 An Integrated Framework for Language Processing, Acquisition, and Evolution

Generativist approaches are not the only ones to treat the processing, acquisition, and evolution of language as separate theoretical endeavors. For example, typological and usage-based approaches to language processing typically downplay issues related to the acquisition and evolution of language (e.g., Clark, 1996; Hawkins, 1994, 2004). Similarly, work on language acquisition tends to not consider questions pertaining to the processing and evolution of language (e.g., Cowie, 1999; Hirsh-Pasek & Golinkoff, 1996; O'Grady, 1997), and studies of language evolution usually pay little attention to research on language acquisition and processing (e.g., Aitchison, 2000; Botha, 2003; Burling, 2005; Corballis, 2002; Dunbar, 1998; Lieberman, 2000; Loritz, 1999; Nettle, 1999). In contrast, we argue that there are strong theoretical connections between the processing, acquisition, and evolution of language—allowing each to shed light on the others—and that key questions within each area can only be fully addressed through an integrated approach.

As noted earlier, our framework builds on the construction-based approach to language (e.g., Bybee, 2007; Goldberg, 2006; O'Grady, 2005; Tomasello, 2003). Language acquisition fundamentally involves learning how to process utterances made up of multiple constructions. Figure 1.4 illustrates the intertwined nature of processing and acquisition, both of which are part and parcel of the same underlying system (dotted oval). Language acquisition consists of innumerable attempts at processing individual utterances (gray ovals). Over time, syntactic generalizations are learned across specific constructions (C's) or combinations thereof (dotted lines). Thus, we see no sharp distinction between language acquisition and processing. Linguistic experience continues to affect processing throughout the lifespan (as evidenced by training studies with adult participants; e.g., Mitchell & Cuetos, 1991; Wells, Christiansen, Race, Acheson, & MacDonald, 2009) but has a more dramatic overall impact on processing ability in childhood due to developmental (e.g., Munakata, McClelland, Johnson, & Siegler, 1997) and linguistic (e.g., Braine & Brooks, 1995; Theakston, 2004) processes of entrenchment. It is also important to note that we do not view learning and processing as starting from a *tabula rasa* (i.e., a "clean slate" without prior constraints). Rather, there are substantial biological constraints both on how utterances are processed and on how construction-based linguistic generalizations are learned. In contrast to generative approaches, however, these constraints do not derive from a UG specific to language but from domain-general biological biases.

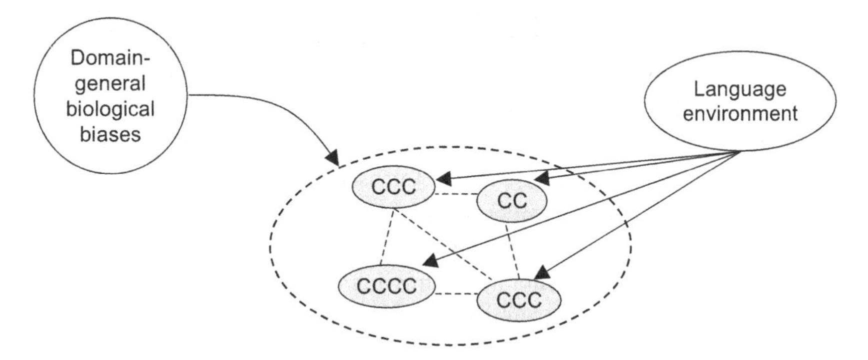

Figure 1.4
An illustration of the intertwined nature of language processing and acquisition constrained by biological domain-general biases. The language environment provides acquisition opportunities in the form of utterances, processed in terms of their component constructions (C's) across which generalizations may be learned (dotted lines).

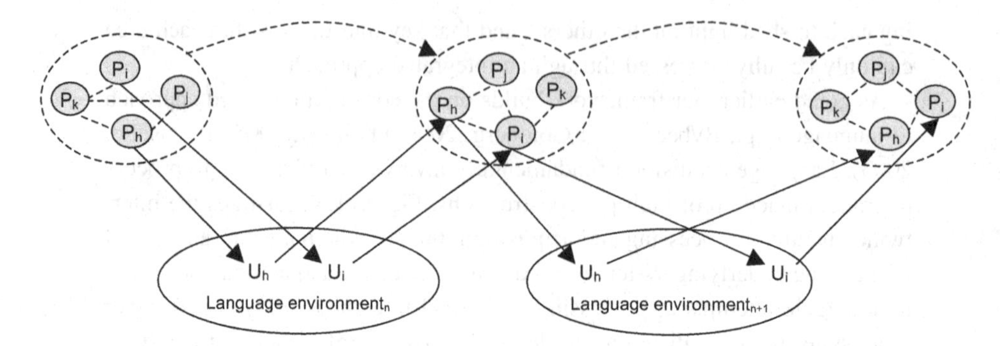

Figure 1.5
The evolution of language viewed as the selection of constructions from a speech community (dashed ovals) that better fit biological cognitive and communicative biases (curved arrows) on processing events (grey ovals). The language environment is shaped by the interaction between experience-based production and comprehension of utterances (downward arrows) that in turn are learned through processing (upward arrows).

Figure 1.5 places our view of acquisition as processing in an evolutionary context. We construe the evolution of language as the gradual selection of constructions that better fit the biological domain-general biases on learning, processing, and communication (curved dashed arrows), most of which are likely to have predated the origin of language (Christiansen & Chater, 2008). Language users within a given speech community (dashed oval) acquire their knowledge of language through innumerable processing events (P's in gray ovals) in which they attempt to comprehend language (upward arrows), including generalizations across constructions (dashed lines). Language users also generate the utterances (U's) that make up the language environment (downward arrows) from which subsequent learners acquire their native language. When viewed across generations of language users, cultural transmission—constrained by the users' learning and processing mechanisms—thus become the primary factor in language evolution (see also e.g., Beckner et al., 2009; Hurford, 1999; Smith & Kirby, 2008; Tomasello, 2003). This evolutionary account explains the close fit between the mechanisms involved in the processing and acquisition of language and the structure of language itself: language has been adapted through cultural transmission over generations of language users to fit the cognitive biases inherent in the mechanisms used for processing and acquisition.

Although the processing, acquisition, and evolution of language takes place at very different timescales, these three types of language creation are nonetheless closely intertwined, so that the study of each provides powerful theoretical

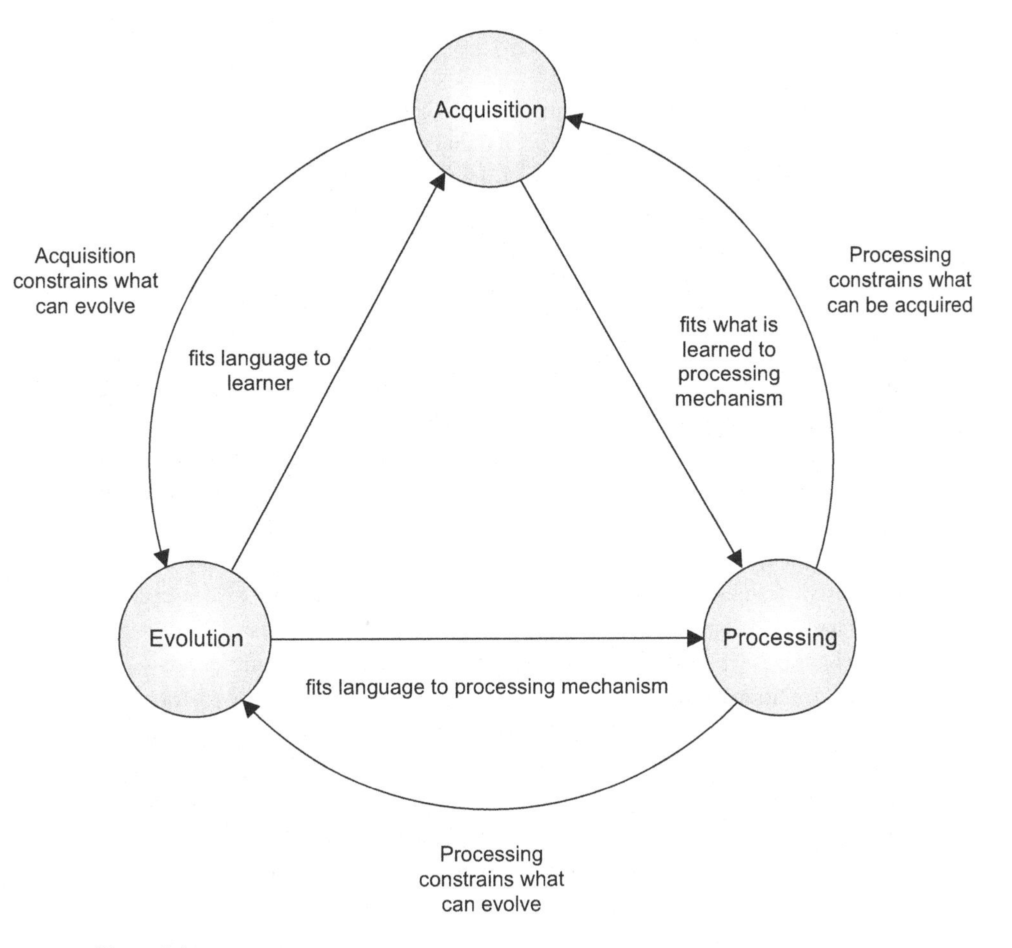

Figure 1.6
The interrelations between the evolution, acquisition, and processing of language.

constraints on the study of the others. These interrelationships are illustrated in figure 1.6. Biological biases on processing constrain the kinds of constructions that can be acquired. Domain-general constraints arising at the timescales of processing and acquisition combine to provide limitations on what can evolve over the historical timescale. Consequently, the evolution of linguistic structure through cultural transmission will fit language to the language user in terms of processing and acquisition. Moreover, developmental processes associated with acquisition will fit what is learned to the mechanisms involved in processing language.

1.3 Overview of the Book

The first part of this book presents the new theoretical framework for understanding language evolution, acquisition, and processing[2]. A growing body of research across the language sciences indicates that the apparently unified picture of language from generative approaches to language does not really provide a compelling account: the relationship between processing and acquisition is tenuous; language evolution seems miraculous. The standard defense of the generative picture against concerns of this kind is that it is "the only game in town"; there is simply no alternative framework for understanding language. But this defense is no longer valid. Important developments in a range of fields, including statistical natural language processing, corpus-based linguistics, learnability theory, functional linguistics, computational modeling, psycholinguistic experiments with children and adults, and developments in evolutionary linguistics have pointed to alternative ways of characterizing the evolution, acquisition, and processing of language. We aim to provide a comprehensive framework that brings together these diverse lines of evidence to understand how language is created across multiple timescales.

In chapter 2, we start by discussing research on language evolution, showing that the traditional notion that UG is a biological endowment of abstract linguistic constraints can be ruled out on evolutionary grounds. Instead, we argue that the fit between the mechanisms employed for language and the way in which language is acquired and used can be explained by processes of cultural evolution shaped by the human brain. Thus, drawing parallels with the reuse of pre-existing cortical brain circuits for recent human innovations such as reading and arithmetic, we suggest that language likewise has evolved by "piggybacking" on prior neural mechanisms, constrained by social and pragmatic considerations, the nature of our thought processes, perceptuo-motor factors, as well as cognitive limitations on learning, memory, and processing. Through cultural transmission involving repeated cycles of learning and use, these constraints have shaped the world's languages into the diversity of forms we can observe today.

The third chapter explores the implications of the cultural evolution of language for understanding the problem of language acquisition, which is cast in a new and much more tractable form. In essence, the child faces a problem of induction, where the objective is to *coordinate* with others (C-induction), rather than to model the structure of the natural world (N-induction). We

2. This book primarily focuses on research in the language sciences from the past few decades. For a comprehensive historical overview of early psycholinguistic work, see Levelt (2012).

suggest that, of the two, C-induction is dramatically easier. More broadly, we argue that understanding the acquisition of any cultural form, linguistic or otherwise, during development requires considering the corresponding question of how that cultural form arose through processes of cultural evolution. This perspective helps resolve the so-called "logical" problem of language acquisition—i.e., how children correctly generalize from limited input to the whole language—because the language itself has been shaped by previous generations of learners to fit the domain-general biases that children bring to bear on acquisition. Our approach also provides insight into the nature of language universals, and has far-reaching implications for evolutionary psychology.

Chapter 4 discusses how the immediacy of language processing provides a fundamental constraint on accounts of language acquisition and evolution. Language happens in the here-and-now. Because memory is fleeting, new material will rapidly obliterate previous material, creating what we call the *Now-or-Never bottleneck.* To successfully deal with the continual deluge of linguistic information, the brain must compress and recode its input into "chunks" as rapidly as possible. It must deploy all available information predictively to ensure that local linguistic ambiguities are dealt with *Right-First-Time*; once the original input is lost, there is no way to recover it. Similarly, language learning must also occur in the here-and-now. This implies that language acquisition involves learning how to process linguistic structure, rather than inducing a grammar. Incoming language incrementally gets recoded into chunks of increasing granularity, from sounds to constructions, and beyond. Importantly, several key properties of language follow naturally from this perspective, including the local nature of linguistic dependencies, the quasi-regular nature of linguistic structure, multiple levels of linguistic representation, and duality of patterning (i.e., that meaningful units are composed of smaller elements).

The second part of the book discusses some of the major implications of the theoretical approach developed in Part I for our understanding of how language works. The focus will be on providing an alternative account of specific aspects of language acquisition and processing when viewed in light of the cultural evolution of language.

The idea of language as shaped by the brain implies that linguistic structure has to be acquired largely by mechanisms that are not uniquely dedicated for this purpose. In chapter 5, we propose that language has evolved to rely on a multitude of probabilistic information sources for its acquisition, allowing language to be as expressive as possible while still being learnable by domain-general learning mechanisms. As a case study, we focus on the structure of the vocabulary and how it reveals a complex relationship between systematicity and arbitrariness in the mapping between the sound of a word and its meaning. Results from corpus analyses, computational modeling, and human experimentation reveal that systematicity may not only help the child learn early word meanings but also facilitate the acquisition of basic aspects of syntax. A probabilistic relationship exists between what a word sounds like and how it is used: nouns tend to sound like other nouns and verbs like other verbs. Importantly, these sources of phonological information, or "cues," not only play an important role in language acquisition but also affect syntactic processing in adulthood. Thus, the integration of phonological cues with other types of information is integral to the computational architecture of our language system. This integration, in turn, is one of the key factors that makes language learnable given the rich sources of information available in the input, without an innate UG.

The multiple-cue integration perspective on language acquisition highlights the rich nature of the input. In combination with the emphasis on the cultural evolution of language, this points to an experience-based account of language processing, in which exposure to language plays a crucial role in determining language ability. The sixth chapter therefore emphasizes the importance of experience for understanding language processing, focusing on the processing of relative clauses as an example. Evidence from corpus analyses, computational modeling, and psycholinguistic experimentation demonstrates that variation in relative clause processing—including differences across individuals—can be explained by variations in linguistic experience. Additional experimental data suggest that individual differences in domain-general abilities for sequence learning and memory-based chunking, in turn, may affect individuals' ability to learn from linguistic experience. We conclude that our language abilities emerge through the complex interactions between linguistic experience and multiple constraints deriving from learning and processing.

The experience-based approach to language processing suggests that our ability to generate complex syntactic structures may emerge gradually, construction by construction. This contrasts with most current approaches to linguistic structure, which suggest that our generative linguistic capacity derives from recursion as a built-in fundamental property of grammar. Chapter 7 presents a usage-based perspective on recursive sentence processing in which recursion is proposed to be an acquired linguistic processing skill. Evidence from computational modeling and behavioral experimentation underscores the role of experience in the processing of recursive structure and the importance of domain-general sequence learning as an underlying mechanism for our limited recursive abilities. We close the chapter by arguing that recursion, when viewed in light of evolution and acquisition, is best viewed as a usage-based skill.

This book presents an overarching framework for the language sciences that may be seen as ignoring long-held theoretical distinctions in mainstream (generative) linguistics: core linguistic phenomena versus peripheral aspects of language, competence versus performance, acquiring language versus learning a skill, language evolution versus language change. In the final chapter, we argue that breaking down these distinctions is a necessary part of building a unified account of the phenomenon of human language. Moreover, this approach promises to create a direct relationship between the language sciences and linguistic theory—viewing linguistic structure as processing history—as well as integrating often disconnected scientific inquiries into language processing, language acquisition, and language evolution. We conclude that an integrated approach to these three intertwined timescales of language creation may, perhaps counter-intuitively, simplify rather than complicate our understanding of the nature of language.

2 Language as Shaped by the Brain

> The formation of different languages and of distinct species, and the proofs that both have been developed through a gradual process, are curiously the same … We see variability in every tongue, and new words are continually cropping up; but as there is a limit to the powers of the memory, single words, like whole languages, gradually become extinct.
> —Charles Darwin, 1871, *The Descent of Man*

Imagine a world without language. What would be the fate of humanity, if language as we know it suddenly ceased to exist? Would we end up as the ill-fated grunting hominids in *Planet of the Apes*? Not likely, perhaps, but try to imagine the devastating consequences of language loss on a more personal level. Language is so tightly interwoven into the very fabric of our everyday lives that losing even parts of it has detrimental repercussions. Consider, for example, the problems facing someone with agrammatic aphasia as evidenced in the following speech sample from a patient explaining that he has returned to the hospital for some dental work:

> Ah … Monday … ah, Dad and Paul Haney [referring to himself by his full name] and Dad … hospital. Two … ah, doctors … and ah … thirty minutes … and yes … ah … hospital. And, er, Wednesday … nine o'clock. And er Thursday, ten o'clock… doctors. Two doctors … and ah … teeth. Yeah, … fine. (Ellis & Young, 1988: p. 242)

Now, imagine an evolutionary scenario in which our ancestors did not evolve a capability for language. In such a picture, it is obvious that our culture and society would have developed very differently. For example, instructing the young in new skills without the use of language would have been difficult, especially as those skills became more complex; passing on knowledge of the natural world, and social norms, would be extremely challenging; planning and implementing complex collective projects would be almost unimaginable. Developing new technologies, building villages, planting crops, irrigating the land, and so on, would therefore be severely impeded—it is not clear that it would have been possible to move beyond hunting and gathering.

As should be clear, language is an incredibly powerful and highly sophisticated means of communication, apparently unique to humans, allowing us to communicate about an unbounded number of objects, situations and events. Language permits us to transfer cultural information readily from generation to generation—originally, in terms of purely oral transmission of knowledge, including instructions, religious beliefs, and enlightening tales. Later, the development of writing ensured a more permanent means of storing information, making it easier to share large quantities of knowledge across time. And most recently, telecommunications networks, computers, and smartphones have allowed us to communicate rapidly over great distances (for example, via email, text messages, or social networking websites).

Language thus constitutes one of the most pervasive and complex aspects of human cognition. Yet, before children can tie their shoes or ride a bicycle, they will already have a good grasp of their native tongue. The relative ease of acquisition suggests that when the child makes a "guess" about the structure of language on the basis of apparently limited evidence, she has an uncanny tendency to guess right. This strongly suggests that there must be a close relationship between the mechanisms by which the child acquires and processes language, and the structure of language itself.

What is the origin of this close relationship between the mechanisms children use in acquisition and the structure of language? One view is that specialized brain mechanisms devoted to language acquisition have evolved over long periods of natural selection (e.g., Pinker & Bloom, 1990; Pinker & Jackendoff, 2009). A second view rejects the idea that these specialized brain mechanisms have arisen through adaptation, just as it has been argued that many biological structures are not the product of adaptation (e.g., Berwick, Friederici, Chomsky, & Bolhuis, 2013; Bickerton, 1995; Gould, 1993; Jenkins, 2000; Lightfoot, 2000), and proposes an alternative evolutionary account. Both these viewpoints put the explanatory emphasis on brain mechanisms specialized for language—and ask how they have evolved.

Here, we develop and argue for a third view, which takes the opposite starting point. It asks not, "why is the brain so well suited to learning language?" but instead, "why is language so well suited to being learned by the brain?" Following Darwin (1871), we propose that *language* has adapted through gradual processes of cultural evolution to be easy to learn, produce, and understand. From this perspective, the structure of human language must inevitably be shaped around human learning and processing biases deriving from the structure of our thought processes, perceptuo-motor factors, cognitive limitations, and pragmatic constraints. Language is easy for us to learn and use not because our brains embody knowledge of language, but because *language has adapted to our brains*. According to this view, whatever domain-general learning and processing biases people happen to have will tend to become embedded in the structure of language because it will be easier to learn to understand and produce languages, or specific linguistic forms, that fit these biases.

2.1 The Logical Problem of Language Evolution

For a period spanning three decades, Chomsky (1965, 1972, 1980b, 1986, 1988, 1993) has been the most influential advocate of the view that language requires specialized brain mechanisms, arguing that a substantial innate endowment of language-specific knowledge is necessary for language acquisition to be possible. These constraints form a Universal Grammar (UG), that is, a collection of grammatical principles that hold across all human languages. In this framework, a child's language ability gradually unfolds according to a genetic blueprint in much the same way as a chicken grows a wing (Chomsky, 1988). The staunchest proponents of this view even go as far as to claim that "doubting that there are language-specific, innate computational capacities today is a bit like being still dubious about the very existence of molecules, in spite of the awesome progress of molecular biology" (Piattelli-Palmarini, 1994: p. 335).

As discussed in chapter 1, there is considerable variation in current conceptions of the exact nature of UG, ranging from approaches close to the Principles and Parameters Theory (PPT; Chomsky, 1981) of pre-minimalist generative grammar (e.g., Crain, Goro, & Thornton, 2006; Crain & Pietroski, 2006), to the Simpler Syntax (SS) version of generative grammar (Culicover, 1999, 2013a; Culicover & Jackendoff, 2005; Jackendoff, 2002, 2010; Pinker & Jackendoff, 2005), to the Minimalist Program (MP) in which language acquisition is confined to learning a lexicon from which cross-linguistic variation is proposed to arise (e.g., Boeckx, 2006; Chomsky, 1995; Hornstein & Boeckx, 2009). Common to these three current approaches to generative grammar is the central assumption that the constraints of UG (whatever their form) are fundamentally arbitrary—i.e., not determined by functional considerations. That is, these principles cannot be explained in terms of learning, cognitive constraints, or communicative effectiveness. For example, consider the principles of binding, which have come to play a key role in generative linguistics (Chomsky, 1981). The principles of binding capture patterns of, among other things, reflexive pronouns (e.g., *himself, themselves*) and accusative pronouns (*him, them*, etc.), which appear, at first sight, to defy functional explanation. Consider examples (1)-(4), where the subscripts indicate coreference, and asterisks indicate ungrammaticality.

(1) $John_i$ sees $himself_i$
(2) *$John_i$ sees him_i
(3) $John_i$ said $he_{i/j}$ won
(4) *He_i said $John_i$ won

In (1), the pronoun *himself* must refer to John; in (2) it cannot. In (3), the pronoun *he* may refer to John or to another person; in (4), it cannot refer to John. These and many other cases indicate that an extremely rich set of patterns govern the behavior of pronouns, and these patterns appear arbitrary—it appears that numerous alternative patterns would, from a functional standpoint, serve equally well. These patterns are instantiated in PPT by the principles of binding theory (Chomsky, 1981), in SS by constraints arising from structural and/or syntax-semantics interface principles (Culicover & Jackendoff, 2005), and in MP by limitations on movement (internal merge, Hornstein, 2001). Independent of their specific formulations, the constraints on binding, while apparently universal across natural languages, are assumed to be arbitrary—and hence may be presumed to be part of the genetically encoded UG.

Supposedly arbitrary universals, such as the restrictions on binding, contrast with functional constraints on language. Whereas the former are hypothesized to derive from the internal workings of a UG-based language system, the latter originate from cognitive and pragmatic constraints related to language acquisition and use. Consider the tendency in English to place long phrases after short ones, for example, as evidenced by so-called *heavy-NP shifts*, illustrated in (5) and (6)

(5) John found $_{PP}$[under his bed] $_{NP}$[the book he had not been able to locate for over two months].
(6) John found $_{NP}$[the book he had not been able to locate for over two months] $_{PP}$[under his bed].

In (5), the long (or "heavy") direct-object noun phrase (NP), *the book he had not been able to locate for over two months*, appears at the end of the sentence, separated from its canonical postverbal position by the prepositional phrase (PP) *under his bed*. Both corpus analyses (Hawkins, 1994) and psycholinguistic sentence-production experiments (Stallings, MacDonald, & O'Seaghdha, 1998) suggest that (5) is much more acceptable than the standard (or "non-shifted") version in (6), in which the direct object NP is placed immediately following the verb.

Whereas individuals speaking head-initial languages, such as English, prefer short phrases before long phrases, speakers of head-final languages, such as Japanese, have been shown to have the opposite long-before-short preference (Yamashita & Chang, 2001). In both cases, the preferential ordering of long versus short phrases can be explained in terms of minimization of memory load and maximization of processing efficiency (Hawkins, 2004). In light of such arguments, the patterns of length-induced phrasal reordering are generally considered, within generative grammar, to be a "performance issue" related to functional constraints (e.g., making language processing more efficient), which are outside the purview of UG (although some functionally-oriented linguists have suggested that these kind of performance constraints may shape grammar itself; e.g., Hawkins, 1994, 2004; see also chapter 4). In contrast, the constraints inherent in UG are viewed as arbitrary and non-functional in the sense that they do not relate to communicative or pragmatic considerations, nor are they assumed to arise from limitations on the mechanisms involved in using or acquiring language. Indeed, some generative linguists have argued that aspects of UG can even *hinder* communication (e.g., Chomsky, 2005; Lightfoot, 2000).

If we suppose that the arbitrary principles of UG are genetically specified, then this raises the question of the evolutionary origin of this genetic endowment. As we noted above, two views have been proposed.

Adaptationists emphasize a gradual evolution of the human language faculty through *natural selection* (e.g., Briscoe, 2003; Corballis, 1992, 2003; Dunbar, 2003; Greenfield, 1991; Hurford, 1991; Jackendoff, 2002; Nowak, Komarova & Niyogi, 2001; Pinker, 1994, 2003b; Pinker & Bloom, 1990; Pinker & Jackendoff, 2005, 2009). Linguistic ability confers added reproductive fitness, leading to a selective pressure for language genes[1], which over many generations come to gradually encode increasingly elaborate grammars.

Non-adaptationists (e.g., Berwick et al., 2013; Bickerton, 1995—but see Bickerton, 2003; Chomsky, 1988, 2005, 2010; Jenkins, 2000; Lightfoot, 2000; Piattelli-Palmarini, 1989) suggest that natural selection only played a minor role in the emergence of language in humans, focusing instead on a variety of alternative possible evolutionary mechanisms by which UG could have emerged *de novo* (e.g., due to very few key mutation "events"; Chomsky, 2010; Lanyon, 2006).

1. For purposes of exposition, we use the term "language genes" as shorthand for genes that may be involved in encoding a potential UG. By using this term we do not mean to suggest that this relationship necessarily involves a one-to-one correspondence between individual genes and a specific aspect of language or cognition (see Dediu & Christiansen, in press, for discussion).

We argue that both of these views, as currently formulated, face profound theoretical difficulties resulting in a *logical problem of language evolution*[2]. This is because, on analysis, it is mysterious how proto-language—which must have been, at least initially, a cultural product likely to be highly variable both over time and geographical locations—could ultimately have become genetically fixed as a highly elaborate biological structure. That is, we will argue that there is no currently viable account of how a genetically encoded UG could have evolved. If this is right, then the brain cannot encode principles of UG—and therefore neither adaptationist nor non-adaptationist solutions are required. Instead, language has been shaped by the brain: language reflects pre-existing, and hence non-language-specific, neural constraints.

2.2 Evolution of Universal Grammar by Biological Adaptation

Adaptation provides a natural explanation for the origin of any innate biological structure. In general, natural selection favors genes that code for biological structures that increase *fitness* (in terms of expected numbers of viable offspring).[3] A biological structure contributes to fitness by fulfilling some function or purpose—the heart is assumed to pump blood, the legs to provide locomotion. Similarly, it has been conjectured that the function of UG is to support language acquisition. If so, natural selection will generally favor biological structures that fulfill their purpose well, so that, over generations, hearts will become well adapted to pumping blood, legs well adapted to locomotion, and any putative biological endowment for language acquisition will become well adapted to acquiring language.

Perhaps the most influential statement of the adaptationist viewpoint is by Pinker and Bloom (1990). They start from the position that "*natural selection is the only scientific explanation of adaptive complexity*. 'Adaptive complexity' describes any system composed of many interacting parts where the details of the parts' structure and arrangement suggest design to fulfill some function"

2. Intermediate positions, which accord some role to both non-adaptationists and adaptationist mechanisms, are of course possible. Such intermediate viewpoints inherit the logical problems that we discuss below for both types of approach, in proportion to the relative contribution presumed to be associated with each. Moreover, we note that our arguments have equal force irrespective of whether one assumes that language has a vocal (e.g., Dunbar, 2003) or manual-gestural (e.g., Corballis, 2003) origin.

3. Strictly, the appropriate measure is the more subtle *inclusive* fitness, which takes into account the reproductive potential not just of an organism, but also a weighted sum of the reproductive potentials of its kin, where the weighting is determined by the closeness of kinship (Hamilton, 1964). Mere reproduction is only relevant to long-term gene frequencies to the degree that one's offspring have a propensity to reproduce, and so down the generations.

(p. 709; their emphasis). As another example of adaptive complexity, they refer to the exquisite optical and computational sophistication of the vertebrate visual system. Pinker and Bloom note that such a complex and elaborate mechanism has an extremely low probability of occurring by chance. Whatever the influence of non-adaptational factors (see below), they argue that there must additionally have been substantial adaptation to fine-tune a system as complex as the visual system. Given that language appears comparably complex to vision, Pinker and Bloom conclude that it is also highly improbable that language is entirely the product of non-adaptationist processes. Indeed, Pinker (2003b) argues that "the default prediction from a Darwinian perspective on human psychological abilities" (p. 16) is the adaptationist view.

The scope and validity of the adaptationist viewpoint in biology is controversial (e.g., Dawkins, 1986; Gould, 2002; Gould & Lewontin, 1979; Hecht Orzak, & Sober, 2001); and some theorists have used this controversy to question adaptationist views of the origin of UG (e.g., Bickerton, 1995; Lewontin, 1998). Here, we take a different tack. We argue that, whatever the merits of adaptationist explanation in general, and as applied to vision in particular, the adaptationist account cannot extend to a putative UG.

2.2.1 Why Universal Grammar Could Not Be an Adaptation to Language

Let us suppose that a genetic encoding of universal properties of language did, as the adaptationist view holds, arise as an adaptation to the environment and, specifically, to the *linguistic* environment. This point of view seems to apply most naturally for aspects of language that have a transparent *functional* value. For example, being able to memorize a large number of arbitrary form-meaning mappings (i.e., words) seems to have great functional advantages. A biological endowment that allows for the acquisition of large vocabularies would be likely to lead to enhanced communication, and hence to be positively selected. Thus, over time, functional aspects of language might be expected to become genetically encoded across the entire population. But UG, according to Chomsky (e.g., 1980b, 1988), consists precisely of linguistic principles that appear highly abstract and arbitrary—i.e., which have no functional significance. To what extent can an adaptationist account of the evolution of a biological basis for language explain how genes could arise which somehow encode such abstract and arbitrary properties of language?

Pinker and Bloom (1990) provide an elegant approach to this question. They suggest that the constraints imposed by UG, such as the binding constraints mentioned above, can be construed as communication protocols for transmitting information over a serial channel, analogous to protocols used in computer-to-computer communication. While the general features of such protocols (e.g., the use of a small set of discrete symbols) may be functionally important, many of the specific aspects of the protocol do not matter, as long as everyone (within a given speech community) adopts the *same* protocol. Similarly, when computers communicate with one another across the Internet, a particular protocol might have features such as a 128-bit IP address, IPv6 packets, Stateless Address Auto Configuration, etc. However, there are many other settings that would be just as effective. What is important is that the computers that are to interact adopt the *same* set of settings—otherwise communication will fail. Adopting the same settings is therefore of fundamental functional importance to communication between computers, but the particular choice of settings is not. Similarly, when it comes to the specific features of UG, Pinker and Bloom suggest that "in the evolution of the language faculty, many 'arbitrary' constraints may have been selected simply because they defined parts of a standardized communicative code in the brains of some critical mass of speakers" (1990: p. 718). Thus, such arbitrary constraints on language can come to have crucial adaptive value to the language-user; genes that favor such constraints will be positively selected. Over many generations, the arbitrary constraints may then become innately specified.

We argue that this viewpoint faces three fundamental difficulties, concerning language change, the dispersion of hominid populations, and the question of *what* is genetically encoded.

Problem 1: Language change

Pinker and Bloom's (1990) analogy with communications protocols, while apt, is something of a double-edged sword. Communication protocols and other technical standards typically diverge rapidly unless there is concerted oversight and enforcement to maintain common standards. Maintaining and developing common standards is a central part of software and hardware development. And, ominously for Pinker and Bloom's conclusions, linguistic conventions also change rapidly[4]. Thus, the linguistic environment over which selectional pressures operate presents a "moving target" for natural selection. And if linguistic conventions change more rapidly than genes can follow via natural selection, then genes that encode biases for particular conventions will be eliminated because, as the language changes, the initially helpful biases will later be incorrect, and hence will *decrease* fitness. More generally, in a fast

4. Although certain key high-frequency words, such as pronouns, kinship terms, etc., tend to be somewhat resistant to change (Pagel, Atkinson, & Meade, 2007), a large number of other words and structural features of language are subject to rapid change (e.g., Dunn, Greenhill, Levinson, & Gray, 2011; Greenhill, Atkinson, Meade, & Gray, 2010).

changing environment, phenotypic flexibility to deal with various environments will typically be favored over genes that bias the phenotype narrowly toward a particular environment.

It may be tempting to suggest that the linguistic principles of UG are the very aspects of language that will *not* change, and hence that these aspects of language will provide a stable linguistic environment over which adaptation can operate. But this appeal would of course be circular, because the genetic endowment of UG is proposed to *explain* language universals, so it cannot be assumed that the language universals pre-date the emergence of the genetic basis for UG. We shall repeatedly have to steer around this *circularity trap* below.

Chater, Reali, and Christiansen (2009) illustrate the problems raised by language change in a series of computer simulations. They assume the simplest possible set-up: that (binary) linguistic principles and language "genes" stand in one-to-one correspondence. Each gene has three alleles—one biased in favor of each version of the corresponding principle, and one neutral allele[5]. Agents learn the language by trial-and-error, where their guesses are biased according to which alleles they have. The fittest agents are allowed to reproduce, and a new generation of agents is produced by sexual recombination and mutation. When the language is fixed, there is a selection pressure in favor of the "correctly" biased genes, and these rapidly come to dominate the population. This is an instance of the *Baldwin effect* (Baldwin, 1896; for discussion see Weber & Depew, 2003) in which information that is initially learned later becomes encoded in the genome. A frequently cited hypothetical example of the Baldwin effect is the development of calluses on the keels and sterna of ostriches (Waddington, 1942). The proposal is that calluses initially developed in response to abrasions where the keel and sterna touch the ground during sitting. Natural selection then favored individuals that could develop calluses more rapidly, until callus development became triggered within the embryo and could occur without environmental stimulation. Pinker and Bloom suggest that the Baldwin effect in a similar way could be the driving force behind the adaptation of UG. Natural selection will favor learners who are genetically disposed to rapidly acquire the language to which they are

5. This set-up closely resembles that used by Hinton and Nowlan (1987) in their simulations of the Baldwin effect, and to which Pinker and Bloom (1990) refer in support of their adaptationist account of language evolution. The simulations are also similar in format to other models of language evolution (e.g., Briscoe, 2003; Kirby & Hurford, 1997; Nowak, Komarova & Niyogi, 2001). Note, however, the simulations discussed here have a very different purpose from work on understanding historical language change from a UG perspective, for example, as involving successive changes in linguistic parameters (e.g., Baker, 2001; Lightfoot, 2000; Yang, 2002).

exposed. The simulations by Chater et al. confirmed that over many generations this process would lead to a genetically specified UG.

However, the simulations also revealed that when language is allowed to change (e.g., either through innovations and modifications from within the speech community, or through exogenous forces such as language contact), the effect reverses. This is because biased genes are severely selected against when they become inconsistent with the fast-changing linguistic environment, and neutral genes come to dominate the population. The selection in favor of neutral genes occurs even for low levels of language change (i.e., the effect occurs, to some degree, even if the speed of language change equals the rate of genetic mutation). But, of course, linguistic change (prior to any genetic encoding) is likely to have been much faster than genetic change. After all, in the modern era, language change has been astonishingly rapid, leading, for example, to the wide phonological and syntactic diversity of the Indo-European language group, from a common ancestral language about 9,000 years ago (Gray & Atkinson, 2003). Language in hunter-gatherer societies changes at least as rapidly. Papua New Guinea, settled within the last 50,000 years, has an estimated one-quarter of the world's languages. These are enormously linguistically diverse, and most originate in hunter-gatherer communities (Diamond, 1992)[6]. Thus, from the point of view of natural selection, it appears that language, like other cultural adaptations, changes far too rapidly to provide a stable target over which natural selection can operate. Human language learning therefore may be analogous to typical biological responses to high levels of environmental change—i.e., to develop general-purpose strategies that apply across rapidly-changing environments, rather than specializing to any particular environment. This strategy appears to have been used in biology by "generalists" such as cockroaches and rats, in contrast, for example, to pandas and koalas, which are adapted to extremely narrow environmental niches.

A potential limitation of our argument so far is that we have assumed that changes in the linguistic environment are unrelated to language genes. But perhaps many aspects of language change may arise because the language is adapting due to selection pressures from learners, and hence from their genes. One might imagine the following argument: suppose there is a slight, random,

6. Some recent theorists have proposed that a further pressure for language divergence between groups is the sociolinguistic tendency for groups to "badge" their in-group by difficult-to-fake linguistic idiosyncrasies (Baker, 2003; Boyd & Richerson, 1987; Nettle & Dunbar, 1997). Such pressures would increase the pace of language divergence, and thus exacerbate the problem of divergence for adaptationist theories of language evolution.

genetic preference for languages with feature *A* rather than *B*. Then this may influence the language spoken by the population to have feature *A*, and this may in turn select for genes that favor feature *A*. Such coevolutionary feedback might, in principle, serve to amplify small random differences into what could become, *ultimately*, rigid arbitrary language universals.

This line of reasoning, while intuitively attractive, is not confirmed by simulations. Chater et al. (2009) found that when linguistic change is genetically influenced rather than random, it turns out that, while this "amplification" can in principle occur, leading to a Baldwin effect, it does not emerge from small random genetic biases. Instead, it only occurs when language is initially strongly influenced by genes. But if "arbitrary" features of language have to be predetermined strongly by the genes from the very beginning, then this leaves little scope for subsequent operation of the Baldwin effect as envisioned by Pinker and Bloom.

Importantly, note that we are not denying the general possibility of gene-behavior coevolution, which appears to be widespread in biology (e.g., Futuyma & Slatkin, 1983). Indeed, there are key human cases of gene-behavior coevolution in the form of so-called *niche construction* where the genetically-influenced behavior of an organism affects the environment to which those genes are adapting (e.g., Odling-Smee, Laland, & Feldman, 2003; Weber & Depew, 2003). For example, the development of arable agriculture and dairying appear to have coevolved with genes for the digestion of starch (Perry et al., 2007) and lactose (Holden & Mace, 1997). Note that these cases are examples of stable shifts in the cultural environment—e.g., once milk becomes a stable part of the diet, there is consistent positive selection for genes that allow for the digestion of lactose. These cases are entirely consistent with our position—coevolution can and does occur where culture provides a stable target (see Christiansen, Reali, & Chater, 2011, for supporting simulations). But this could not be the case for the arbitrary regularities purported to be encoded in UG. By contrast, specific cognitive mechanisms may have coevolved with language in this way: they may have been positively selected because of their *functional* role in language acquisition and use (e.g., reasoning about other's communicative intentions; increased memory capacity allowing a large vocabulary; or the structure of the vocal apparatus, and so on).

Problem 2: The dispersion of human populations

The thesis that a genetically encoded UG arose through adaptation not only must deal with the issue of language being a moving target for genetic assimilation, but also faces the problem of the dispersion of human populations.

Consider, once more, Pinker and Bloom's communication protocol analogy. In the absence of sufficient top-down pressures for standardization, such protocols may rapidly diverge. Even relatively minor breakdowns in such standardization can have quite catastrophic consequences, as illustrated by the space-age example of how confusions about different measurement protocols caused the loss of the Mars Climate Orbiter on September 23, 1999 (Stephenson et al., 1999). The problem arose because one piece of software at Lockheed Martin in Colorado calculated thrust adjustments in English units (pounds/second), while the software used to determine flight trajectory based on these calculations at NASA in California expected metric units (Newtons/second). Even though the metric system had been NASA's official unit system since the 1980s, failure to enforce the same measurement protocol caused the spacecraft to get too close to Mars and disintegrate in its atmosphere.

Given that language presumably evolved without top-down pressures for standardization, divergence between languages seems inevitable. To assume that "universal" arbitrary features of language would emerge from adaptation by separate groups of language users would be analogous to assuming that the same set of specific features for computer communication protocols might spontaneously emerge from separate teams of scientists, working in separate laboratories (e.g., that different Internet networking designers independently alight on 128-bit IP addresses, IPv6 packets, Stateless Address Auto Configuration, and so on). Note that this problem would be almost as severe, even if the teams of scientists emerged from a single group. Once cut off from each other, groups would develop in independent ways. Indeed, in biology, adaptations appear to rapidly evolve to deal with a specific local environment, as Darwin noted when interpreting the rich patterns of variations in fauna (e.g., finches) across the Galapagos Islands as adaptation to local island conditions (see box 2.1. for a discussion of how even our immune system shows evidence of geographical adaptations).

Baronchelli, Chater, Pastor-Satorras, and Christiansen (2012) conducted a series of simulations to determine the plausibility of evolving an innate UG in light of the dispersion of human populations. They adapted the Chater et al. (2009) model of the evolution of language and genes, mentioned above, by splitting an initially stable "mother population" into two geographically separated groups. Their results revealed two different patterns: When language changes rapidly, it becomes a moving target, and neutral genes are favored in both populations—that is, no UG emerged. However, when language changes slowly, two isolated subpopulations that originally spoke the same language will diverge, first linguistically, and then biologically through genetic assimilation to the diverging languages. In the latter case, we do see biological adapta-

Box 2.1
Lessons from the Evolution of the Human Immune System

Comparisons with the immune system have been used by proponents of UG to argue in favor of the idea of a "language organ" (e.g., Chomsky, 2000; Jenkins, 2000; Piattelli-Palmarini, 1989; Piattelli-Palmarini & Uriagereka, 2004). Indeed, in his Nobel Prize lecture, Niels Jerne (1985) proposed an analogy between the immune system and language: both allow for the possibility of creating an unbounded number of structures from a limited vocabulary. However, although this analogy may be useful in terms of thinking about the combinatorial power of the immune system, it breaks down when it comes to the evolution of the two systems. In response to a very rapidly changing microbial environment, the immune system can build new antibody proteins (and the genetic mechanisms from which antibody proteins are constructed) without having to eliminate old antibody proteins (Goldsby, Kindt, Osborne, & Kuby, 2003). Therefore, natural selection will operate to *enrich* the coverage of the immune system (though such progress will not always be cumulative, of course); there is no penalty for the immune system following a fast-moving "target" (defined by the microbial environment). But the case of acquiring genes coding for regularities in language is very different because, at any one time, there is just one language (or at most two or three) that must be acquired, and hence a bias that helps learn a language with property *P* will thereby *inhibit* learning languages with *not-P*. The fact that language change is so fast (so that whether the current linguistic environment has property *P* or not will vary rapidly in the time scale of biological evolution) means that such biases will, on balance, be counterproductive.

Given that the immune system does coevolve with the microbial environment, different coevolutionary paths have been followed when human populations have diverged. Therefore populations that have coevolved to their local microbial environment are often poorly adapted to other microbial environments. For example, when Europeans began to explore the New World, they succumbed in large numbers to the diseases they encountered, and conversely, European diseases caused catastrophic collapse in indigenous populations (e.g., Diamond, 1997). If an innate UG had coevolved with the linguistic environment, similar radically divergent coevolutionary paths might be expected—different populations of language learners should have coevolved with their own linguistic environment, just as different populations of immune systems have coevolved with their own microbial environments. Yet, as we have noted, the contrary appears to be the case.

tions for language, but with two different UGs, each specifically adapted to its local linguistic context. Again, there is a tempting counter-argument that, for humans, all of these sublanguages will, nonetheless, obey *universal* grammatical principles, thus providing some constancy in the linguistic environment—but this argument falls into the circularity trap, as we are attempting to explain the *origin* of such presumed universal principles.

Additional simulations showed that incorporating gene-language coevolution simply resulted in more rapid divergence of UGs. Because the linguistic environment in each population is itself shaped by the different language-genes in that subpopulation, increasing differences between linguistic environments become amplified. This, of course, does not fit with the key assumption of the generative linguistics perspective that the language endowment does not vary across human subpopulations but is universal across the species (Chomsky,

1980a, 2011; Pinker, 1994). Instead, Baronchelli et al.'s simulations suggest that the evolution of a genetic predisposition to accommodate rapid cultural evolution of linguistic structure may be the key to reconciling the diversity of human language with a largely uniform biological basis for learning language.

The problem of divergent populations further arises across a range of other possible scenarios concerning the relationship between language evolution and the dispersion of human populations. One scenario is that language evolution pre-dates the dispersion of modern humans. If so, then it is conceivable that prior dispersions of hominid populations, perhaps within Africa, might have led to the emergence of diverse languages and diverse UGs adapted to learning and processing such languages, but that a single local population subsequently proved to be adaptively most successful, and came to displace other hominid populations. On this account, our current UG might conceivably be the only survivor of a larger family of such UGs due to a population "bottleneck." The universality of UG would arise, then, because it was genetically encoded in the sub-population from which modern humans descended[7]. This viewpoint is not without difficulties. Some interpretations of the genetic and archeological evidence suggest that the last bottleneck in human evolution occurred between 500,000 and 2,000,000 years ago (e.g., Hawks, Hunley, Lee, & Wolpoff, 2000); few researchers in language evolution believe that language, in anything like its modern form, is this old. Moreover, even if we assume a more recent bottleneck, any such bottleneck must at least predate the 100,000 years or so since the geographical dispersion of human populations, and 100,000 years still seems to provide sufficient time for substantial linguistic, and presumably consequent genetic, divergence to occur (as this scenario assumes that genes are shaped by language).

But perhaps early language change was slow and only later sped up to reach the fast pace observed today? After all, the archeological record indicates very slow cultural innovation in, for example, tool use until 40,000–50,000 years ago (Mithen, 1996). Perhaps UG coevolved with an initially slow-changing language—a *protolanguage* (Bickerton, 1995)—the genes for which were then conserved through later periods of increased linguistic change? Baronchelli et al. (2012) simulated the effects of initially slow, but accelerating, language change across generations. The results suggested a genetic assimilation of the

7. One prominent view is that language emerged within the last 100,000 to 200,000 years (e.g., Bickerton, 2003). Hominid populations over this period and before appear to have undergone waves of spread; "… modern languages derive mostly or completely from a single language spoken in East Africa around 100 kya … it was the only language then existing that survived and evolved with rapid differentiation and transformation." (Cavalli-Sforza & Feldman, 2003: p. 273).

original protolanguage. However, with increasing speed of linguistic change, the number of neutral alleles also increased. This trend continued after the population split, with the consequence that the genetic make-up of the two subpopulations ended up consisting predominantly of neutral genes, thus undoing the initial genetic adaptation to the early, more stable language. That is, even if a UG could emerge as an adaptation to a supposedly fixed protolanguage, it would subsequently be eliminated in favor of general learning strategies, once languages became more labile. This argues against a "Prometheus" scenario (Chomsky, 2010) in which a single mutation (or very few) gave rise to the language faculty in an early human ancestor, whose descendants then dispersed across the globe. The simulation results further imply that current languages are unlikely to carry within them significant "linguistic fossils" (Jackendoff, 1999) of a purported initial protolanguage.

Thus, the dispersion of human populations poses fundamental problems for adaptationist accounts of UG: because processes of genetic adaptation to language would be likely to continue to operate throughout human history[8], different genetic bases for language would be expected to evolve across geographically separated populations. That is, the evolution of UG by adaptation would appear to require adaptations for language *prior* to the dispersion of human populations, implausibly followed by an abrupt cessation of such adaptation for a long period after dispersion. The contrast between the evolution of the presumed "language organ" and that of biological processes, such as digestion, is striking. The digestive system is evolutionarily very old, and many orders of magnitude older than the recent divergence of human populations. Nonetheless, digestion appears to have adapted in important ways to recent changes in the dietary environment, for example, with apparent coevolution of lactose tolerance and the domestication of milk-producing animals (Beja-Pereira et al., 2003).

Problem 3: What is genetically encoded?

Even if the first two difficulties for adaptationist accounts of UG could be solved, the view still faces a further puzzle: why is it that genetic adaptation occurred only to very abstract properties of language, rather than also occurring to its superficial properties? Given the spectacular variety of surface forms of the world's languages, in both syntax (including every combination of basic orderings of subject, verb, and object, and a wide variety of less constrained

8. Human genome-wide scans have revealed evidence of recent positive selection for more than 250 genes (Voight, Kudaravalli, Wen, & Pritchard, 2006), making it very likely that genetic adaptations for language would have continued in this scenario.

word orders) and phonology (including tone and click languages, for example), why did language genes not adapt to these surface features? Why should genes become adapted to capture the extremely rich and abstract set of possibilities countenanced by the principles of UG, rather than merely encoding the actual linguistic possibilities in the specific language that was being spoken (e.g., the phonological inventory and particular morphosyntactic regularities of the early click-language, from which the Khoisan family originated and which might be the first human language; e.g., Pennisi, 2004)? The unrelenting abstractness of the universal principles makes them difficult to reconcile with an adaptationist account.

One of the general features of biological adaptation is that it is driven by the constraints of the immediate environment. It can have no regard for distant or future environments that might one day be encountered. For example, the visual system is highly adapted to the laws of optics as they hold in normal environments. Human vision mis-estimates the length of a stick in water, because it does not correct for the refraction of light through water (this being not commonly encountered in the human visual world). By contrast, the visual system of the archerfish, which must strike air-born flies with a water jet from below the water surface, does make this correction (Rossel, Corlija, & Schuster, 2002). Biological adaptation produces systems designed to fit the environment to which adaptation occurs; there is, of course, no selectional pressure to fit environments that have not occurred, or might do so at some point in the future. Hence, if a UG did adapt to a past linguistic environment, it seems inevitable that it would adapt to that language environment *as a whole*, thus adapting to its *specific* word order, phonotactic rules, inventory of phonemic distinctions, and so on. In particular, it seems implausible that an emerging UG would be selected primarily for extremely abstract features, which apply equally to all possible human languages (not just the language evident in the linguistic environment in which selection operates). This would be analogous to an animal living in a desert environment somehow developing adaptations that are not specific to desert conditions, but that are equally adaptive in all terrestrial environments.

Statistical analyses of typological and genetic variation across Old World languages pose a dilemma for UG proponents in this regard. Dediu and Ladd (2007) found that two recently evolved alleles of the genes *ASPM* (Mekel-Bobrov et al., 2005) and *Microcephalin* (Evans et al., 2005)—both involved in brain development—were strongly associated with the absence of tone in a language, even when controlling for geographical factors and common linguistic history (see also Dediu, 2011). On the one hand, this could be taken as pointing to biological adaptations for a surface feature of phonology: the adoption of a single-tier phonological system relying only on phoneme-sequence information to differentiate between words instead of a two-tier system incorporating both phonemes and tones (i.e., pitch contours). On the other hand, given that the relevant mutations would have had to occur independently several times across different populations, the causal explanation plausibly goes in the opposite direction, from genes to language. The two alleles may have been selected for other reasons relating to brain development but once in place they made it harder to acquire phonological systems involving tonal contrasts, which, in turn, allowed languages without tonal contrasts to evolve more readily. This *genetic-biasing* perspective (also advocated by Dediu and Ladd) dovetails with our suggestion that language is shaped by the brain, as discussed below. However, neither of these interpretations would be consistent with an adaptationist account of the emergence of a UG encoding for *all* possible languages.

In light of this dilemma, it may be tempting to claim that the principles of UG are just those that are invariant across languages, whereas contingent aspects of word order or phonology will vary across languages. Thus, one might suggest that only the highly abstract, language-universal principles of UG will provide a stable basis upon which natural selection can operate. But this argument is again, of course, a further instance of the circularity trap. We are trying to explain how an assumed UG might become genetically fixed, and hence we cannot assume UG is already in place. Thus, this counterargument is blocked.

We are not, of course, arguing that abstract structures cannot arise by adaptation. Indeed, abstract patterns, such as the body plans of mammals or birds, are conserved across species, and constitute a complex and highly integrated system. Notice, though, that such abstract structures are still tailored to the specific environment of each species. Thus, while bats, whales, and cows have a common abstract body plan, these species embody dramatically different instantiations of this pattern, adapted to their ecological niches in the air, in water, or on land. Substantial modifications of this kind can occur quite rapidly, due to changes in a small number of genes and/or their pattern of expression. For example, the differing beak shape in Darwin's finches, adapted to different habitats in the Galapagos Islands, may be largely determined by as few as two genes: *BMP4*, the expression of which is associated with the width as well as depth of beaks (Abzhanov, Protas, Grant, Grant, & Tabin, 2004), and *CaM*, the expression of which is correlated with beak length (Abzhanov et al., 2006). Again, these adaptations are all related closely to the local environment in which an organism exists. In contrast, adaptations for UG are hypothesized to be for abstract principles that apply across all linguistic environments, with no adaptation to the local environment of specific languages and language users.

In summary, Pinker and Bloom (1990), as we have seen, draw a parallel between the adaptationist account of the development of the visual system and an adaptationist account of a putative language faculty. But the above arguments indicate that the two cases are profoundly different. The principles of optics, and the structure of the visual world, have many invariant features across environments (e.g., Simoncelli & Olshausen, 2001), but the linguistic environment is vastly different from one population to another. Moreover, the linguistic environment, unlike the visual environment, will itself be altered in line with any genetic changes in the propensity to learn and use languages, thus further amplifying differences between linguistic environments. We conclude, then, that linguistically-driven biological adaptation cannot underlie the evolution of language.

It remains possible, though, that the development of language did have a substantial impact on biological evolution. The arguments given here merely preclude the possibility that aspects of linguistic conventions that originally *differ* across linguistic environments could somehow become universal across all linguistic communities by virtue of biological adaptation to the linguistic environment. This is because, in the relevant respects, the linguistic environment for the different populations is highly variable, and hence any biological adaptations would only serve to entrench such differences further. But there might be features that are universal across linguistic environments that might lead to biological adaptation (such as the means of producing speech, Lieberman, 1984; the need for enhanced memory capacity, Wynne & Coolidge, 2008; or complex pragmatic inferences, Givón & Malle, 2002). However, these language features are likely to be functional, i.e., they facilitate language *use*, and thus would typically not be considered part of UG.

It is consistent with our arguments that the emergence of language influenced biological evolution in a more indirect way. As we noted at the beginning of this chapter, the possession of language likely would have fundamentally changed the patterns of collective problem solving and other social behavior in early humans, with a consequent shift in the selectional pressures on humans engaged in these new patterns of behavior (see, Levinson & Dediu, 2013, for discussion). But universal, arbitrary constraints on the structure of language cannot emerge from biological adaptation to a varied pattern of linguistic environments. Thus, the adaptationist account of the biological origins of UG is not viable.

2.3 Evolution of Universal Grammar by Non-adaptationist Means

Some theorists advocating a genetically-based UG might concur with our arguments against adaptationist accounts of language evolution (e.g., Berwick, 2009). Indeed, Chomsky (1972, 1988, 1993) has for more than two decades expressed strong doubts about neo-Darwinian explanations of language evolution, hinting that UG may be a by-product of increased brain size or yet unknown physical or biological evolutionary constraints (see also Chomsky, 2010). Further arguments for a radically non-adaptationist perspective have been advanced by Berwick et al. (2013), Jenkins (2000), Lanyon (2006), Lightfoot (2000), and Piattelli-Palmarini (1989, 1994).

Non-adaptationists typically argue that UG is both highly complex, and radically different from other biological machinery (though see Berwick et al., 2013; Hauser et al., 2002). They suggest, moreover, that UG appears to be so unique in terms of structure and properties, that it is unlikely to be a product of natural selection among random mutations. However, we argue that non-adaptationist attempts to explain a putative language-specific genetic endowment also fail.

To what extent can any non-adaptationist mechanism account for the development of a genetically encoded UG, as traditionally conceived? In particular, can such mechanisms account for the appearance of genetically specified principles that are presumed to be (a) idiosyncratic to language, and (b) of substantial complexity? We argue that the probability that non-adaptationist factors played a substantial role in the evolution of UG is vanishingly small.

The argument involves a straightforward application of information theory. Suppose that the constraints embodied in UG are indeed language-specific, and hence do not emerge as side-effects of existing processing mechanisms. This means that UG would have to be generated *at random* by non-adaptationist processes. Suppose further that the information required to specify a language acquisition device, so that language can be acquired and produced, over and above the pre-linguistic biological endowment, can be represented as a binary string of N bits (this particular coding assumption is purely for convenience, and not essential to the argument). Then the probability of generating this sequence of N bits by chance is 2^{-N}. If the language-specific information could be specified using a binary string that would fit on one page of normal text (which would presumably be a considerable underestimate, from the perspective of most linguistic theory), then N would be over 2500. Hence the probability of generating the grammar by a random process would be less than 2^{-2500}. So to generate this machinery by chance (i.e., without the influence of the forces of adaptation) would be expected to require on the order of 2^{2500} individuals. But the total population of hominins over the last two million or so years, including the present, is measured in billions, and is much smaller than 2^{35}. Hence, the probability of non-adaptationist mechanisms "chancing" upon a specification of a language organ or language instinct through purely non-adaptationist means is astronomically low[9].

It is sometimes suggested, apparently in the face of this type of argument, that the recent evo-devo (evolutionary-developmental) revolution in biology can provide a foundation for a non-adaptationist account of UG (e.g., Chomsky, 2010; Piattelli-Palmarini, Hancock, & Bever, 2008). The growing evo-devo literature has revealed how local genetic changes, e.g., on homeobox genes, can influence the expression of other genes, and through a cascade of developmental influences, can result in extensive phenotypic consequences (e.g., Gerhart & Kirschner, 1997; Laubichler & Maienschein, 2007; for a discussion of evo-devo as applied to language, see Dediu & Christiansen, in press). Yet suppose that UG arises from a small "tweak" to pre-linguistic cognitive machinery. Then the general cognitive machinery will provide the bulk of the explanation of language structure—without this machinery, the impact of the tweak would be impossible to understand. Thus, the vision of universal grammar as a language-specific innate faculty or language organ would have to be retracted.

But the idea that a simple tweak might lead to a novel complex, highly interdependent, and intricately organized system, such as a UG, is extremely implausible. Small genetic changes lead to modifications of existing complex systems (and these modifications can be quite wide-ranging); they do not lead to the construction of entirely new forms of complexity. A mutation might lead to an insect having an extra pair of legs, and a complex set of genetic modifications (almost certainly resulting from strong and continuous selectional pressure) may modify a leg into a flipper; but no single gene creates an entirely new means of locomotion, from scratch. Indeed, proponents of evo-devo emphasize that biological structures built *de novo* appear invariably to be shaped by long periods of adaptation (Finlay, 2007). Thus, while an antenna may be a modification of the insect leg (Carroll, 2005), it is *not* an insect leg, or anything like one. It is exquisitely crafted to play its new role—and such

9. We have presented the argument in informal terms. A more rigorous argument is as follows. We can measure the amount of information embodied in UG, U, over and above the information in pre-existing cognitive processes, C, by the length of the shortest code that will generate U from C. This is the conditional Kolmogorov complexity $K(U|C)$ (Li & Vitányi, 1997). By the coding theorem of Kolmogorov complexity theory (Li & Vitányi, 1997), the probability of randomly generating U from C is approximately $2^{-K(U|C)}$. Thus, if UG has any substantial complexity, then it has an exceedingly low probability of being encountered by a random process, such as a non-adaptational mechanism.

apparent design is universally explained within biology as an outcome of Darwinian selection. The impact of evo-devo is to help us understand the elaborate structure and constraints of the space of organisms over which the processes of variation and natural selection unfold; it is not an *alternative* to the operation of natural selection (Carroll, 2001).

The whole burden of the classic arguments for UG is that UG is both highly organized and elaborate, and utterly distinct from general cognitive principles. The emergence of such a UG requires the construction of a new complex system; and the argument sketched above shows that the probability of even modest new complexity arising by chance is incredibly small. The implication of this argument is that it is extremely unlikely that substantial quantities of linguistically idiosyncratic information have been specified by non-adaptationist means. Indeed, the point applies more generally to the generation of any complex, functional biological structures without adaptation. Thus, it is not clear how any non-adaptationist account can explain the emergence of something as intricate and complex as a fully functioning neural *system* of any kind, let alone one embodying UG.

Some authors who express skepticism concerning the role of adaptation implicitly recognize this kind of theoretical difficulty. One line of attack is to argue that many apparently complex and arbitrary aspects of cognition and language have emerged from the constraints on building any complex information processing system, given perhaps currently unknown physical and biological constraints (e.g., Chomsky, 1993; see Kauffman, 1995, for a related viewpoint on evolutionary processes), or minimum principles that give rise to soap bubbles and snowflakes (Chomsky, 2005). A related perspective is proposed by Gould (1993), who views language as a spandrel—i.e., as emerging as a byproduct of other cognitive processes. Another option would be to appeal to *exaptation* (Gould & Vrba, 1982) whereby a biological structure that was originally adapted to serve one function is put to use to serve a novel function. Yet the non-adaptationist attracted by these or other non-adaptationist mechanisms is faced with a dilemma. If language can emerge from general physical, biological, or cognitive factors, then the complexity and idiosyncrasy of UG is illusory—indeed, the usefulness of the term UG seems to evaporate. The idea that language emerges from non-linguistic factors that pre-date the emergence of language is one that we advocate here, but it is difficult to reconcile this viewpoint with anything remotely resembling the traditional picture of UG. If, by contrast, UG is maintained to be *sui generis* and not readily derivable from general processes, the complexity argument bites: i.e., the probability of a new and highly complex adaptive system emerging by chance is infinitesimally low.

The dilemma is equally stark for the non-adaptationist who attempts to reach for other non-adaptationist mechanisms of evolutionary change, such as epigenesis, which produces heritable cell changes as a result of environmental influences without corresponding changes to the basic DNA (Jablonka & Lamb, 1989). Such mechanisms offer a richer picture of the processes of evolutionary change—but provide no answer to the question of how novel and highly complex adaptive systems, such as UG, might emerge *de novo* (see Dediu & Christiansen, in press, for a discussion). From the point of view of the construction of a highly complex adaptive system, these mechanisms amount, in essence, to random perturbations (Schlosser & Wagner, 2004). However, if language is viewed as embodying substantial and novel complexity, then the emergence of this complexity by non-adaptationist (and hence, from an adaptive point of view, random) mechanisms is astronomically unlikely.

Our discussion so far would appear to present us with a paradox. It seems clear that the mechanisms involved in acquiring and processing language are very complex and moreover intimately connected to the structure of natural languages. The complexity of these mechanisms rules out, as we have seen, a non-adaptationist account of their origin. However, if these mechanisms arose through adaptation, this adaptation cannot, as we argued above, have been adaptation *to language*. But if the mechanisms that currently underpin language acquisition and processing were originally adapted to carry out other functions, then how is their apparently intimate relationship with the structure of natural language to be explained? How, in particular, are we to explain that the language acquisition mechanisms seem particularly well-adapted to learning natural languages, but not to any of a vast range of conceivable non-natural languages (e.g., Chomsky, 1980b)? As we now argue, the paradox can be resolved if we recognize that the "fit" between the mechanisms of language acquisition and processing, on the one hand, and natural language, on the other, has arisen because natural languages themselves have "evolved" to be as easy to learn and process as possible: language has been shaped by the brain, rather than vice versa.

2.4 Language as Shaped by the Brain

We propose, then, to invert the traditional perspective on language evolution, shifting the focus from the evolution of *language users* to the evolution of *languages*. Figure 2.1 provides a conceptual illustration of these two perspectives (see also Andersen, 1973; Hurford, 1990; Kirby & Hurford, 1997). The UG adaptationists (a) suggest that selective pressure toward better language

a)

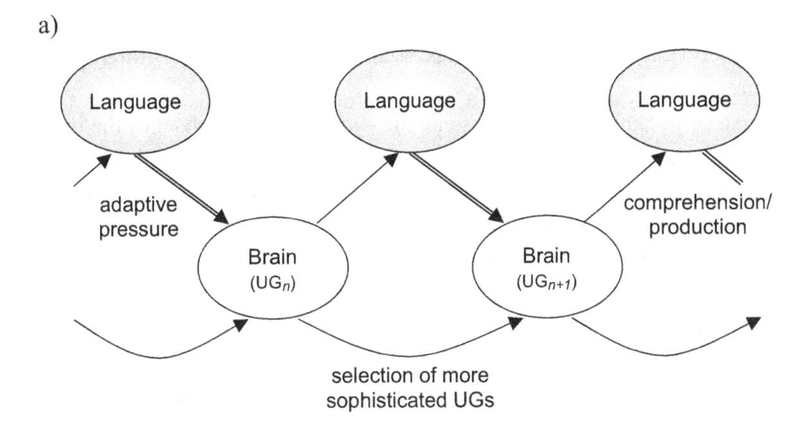

b)

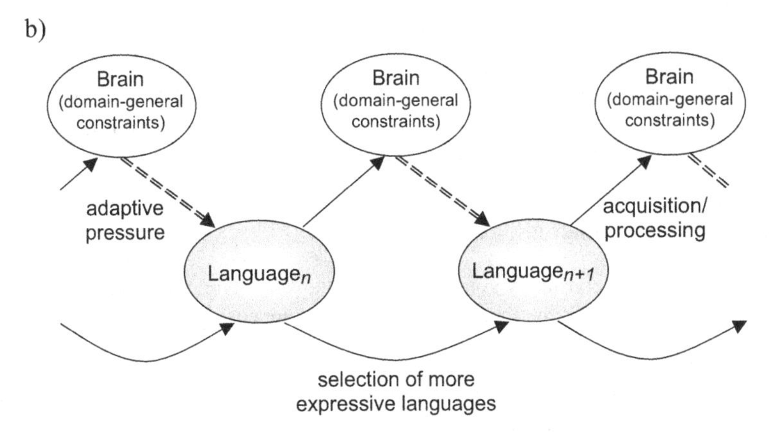

Figure 2.1
Illustration of two different views on the primary direction of causation in language evolution: a) biological adaptations of the brain to language (double arrows), resulting in gradually more complex UGs (curved arrows) to provide the basis for increasingly advanced language production and comprehension (single arrows); b) cultural adaptation of language to the brain (dashed double arrows), resulting in increasingly expressive languages (curved arrows) that are well suited to being acquired and used given largely domain-general constraints deriving from the brain (single arrows).

abilities gradually led to the selection of more sophisticated UGs. In contrast, (b) we propose to view language as an evolutionary system in its own right (see also e.g., Christiansen, 1994; Christiansen & Chater, 2008; Deacon, 1997; Keller, 1994; Kirby, 1999; Ritt, 2004), subject to adaptive pressures from the human brain. As a result, linguistic adaptation allows for the evolution of increasingly expressive languages that can nonetheless still be learned and processed by domain-general mechanisms. From this perspective, we argue that the mystery of the fit between human language acquisition and processing mechanisms and natural language may be unraveled, and we might, furthermore, understand how language has attained its apparently "idiosyncratic" structure.

Instead of puzzling that humans can only learn a small subset of the infinity of mathematically possible languages, we take a different starting point: the observation that natural languages exist only because humans can produce, learn, and process them. In order for languages to be passed on from generation to generation, they must adapt to the properties of the human learning and processing mechanisms. The key to understanding the fit between language and the brain is to understand how language has been shaped by the brain, not the reverse.

The process of cultural evolution by which language has been shaped is, in important ways, akin to Darwinian selection—indeed, we suggest that it is a productive metaphor to view languages as analogous to biological species, adapted through natural selection to fit a particular ecological niche: the human brain. This perspective on language evolution has a long historical pedigree, as summarized in box 2.2. Indeed, as indicated by the quote at the beginning of this chapter, Darwin (1871), too, recognized the similarities between linguistic and biological change. He cited an elegant passage written by the philologist, Max Müller (1870, p. 257):

> A struggle for life is constantly going on among the words and grammatical forms in each language. The better, the shorter, the easier forms are constantly gaining the upper hand. …

Darwin concluded that "The survival and preservation of certain favored words in the struggle for existence is natural selection." (1871, p. 59–60)

This viewpoint does not, of course, rule out the possibility that language may have played a role in the biological evolution of hominins. Good language skills may indeed enhance reproductive success. But the pressures working on language to adapt to humans are significantly stronger than the selection pressures on humans to use language: a language can *only* survive if it is learnable and can be processed by humans. By contrast, adaptation toward language use is merely *one of many* selective pressures working on hominin evolution (including, for example, avoiding predators, finding a mate, and obtaining food). Whereas humans can survive without language, the opposite is not the case; humans are the creators of language across the timescales of evolution, the human lifespan, and in-the-moment conversational exchange. Thus, prima facie, language is more likely to have been shaped to fit the human brain than the other way round. Languages that are too hard for humans to learn and process will most probably never come into existence at all, but if they did,

Box 2.2
Historical Parallels between Linguistic and Biological Change

One of the earliest proponents of the idea that languages evolve diachronically was the eighteenth-century language scholar Sir William Jones, the first Western scholar to study Sanskrit and note its affinity with Greek and Latin (Cannon, 1991). Later, nineteenth-century linguistics was dominated by an "organistic" view of language (McMahon, 1994). Franz Bopp, one of the founders of comparative linguistics, regarded language as an organism that could be dissected and classified (Davies, 1987). Wilhelm von Humboldt—the father of generative grammar (Chomsky, 1965; Pinker, 1994)—argued that "... language, in direct conjunction with mental power, is a fully-fashioned *organism* ..." (von Humboldt, 1836/1999, p. 90; original emphasis). Languages were viewed as having life-cycles that included birth, progressive growth, procreation, and eventually decay and death. However, the notion of evolution underlying this organistic view of language was largely pre-Darwinian. This is perhaps reflected most clearly in the writings of another influential linguist, August Schleicher. Although he explicitly emphasized the relationship between linguistics and Darwinian theory (Schleicher, 1863; quoted in Percival, 1987), Darwin's principles of mutation, variation, and natural selection did not enter into his theorizing about language evolution (Nerlich, 1989).

Darwin (1871), too, recognized the similarities between linguistic and biological change, even noting the role of memory constraints in the quote with which we began this chapter. Thus, natural language can be construed metaphorically as akin to an organism whose evolution has been constrained by the properties of human learning and processing, as well as other communicative and social factors. A similar perspective on language evolution was revived, within a modern evolutionary framework, by Stevick (1963) and later by Nerlich (1989) and Sereno (1991). Christiansen (1994) pushed the analogy a little further, suggesting that language may be viewed as a "beneficial parasite" engaged in a symbiotic relationship with its human hosts, without whom it cannot survive (see also Deacon, 1997). Symbiotic parasites and their hosts tend to become increasingly co-adapted (e.g., Dawkins, 1976). But note that this co-adaptation will be very lopsided, because the rate of linguistic change is far greater than the rate of biological change. This suggestion is further corroborated by work in evolutionary game theory, showing that when two species with markedly different rates of adaptation enter a symbiotic relationship, the rapidly evolving species adapts to the slowly evolving species, but not the reverse (Frean & Abraham, 2004).

they would quickly be eliminated (see also Culicover, 2013a, for a similar perspective).

2.4.1 Cultural Recycling of Neural Substrates during Development

If language is a collective cultural creation, then its properties may be expected to reflect the neural and cognitive machinery of language users. In other cultural domains, this is a familiar observation. Musical patterns appear to be rooted, in part at least, in the machinery of the human auditory and motor systems (Sloboda, 1985); art is partially shaped by the properties of human visual perception (Gombrich, 1960); tools, such as scissors or spades, are built around the constraints of the human body; aspects of religious belief may connect, among other things, with the human propensity for folk-psychological explanation (Boyer, 1994). But how might such effects of cultural evolution be instantiated at the neurobiological level?

Over the past decade, a new perspective on the functional architecture of the brain has emerged (see Anderson, 2010, for a review), which would appear to fit naturally with our perspective on language evolution. Instead of viewing various brain regions as being dedicated to broad cognitive domains such as language, vision, memory, or reasoning, it is proposed that low-level neural circuits that have evolved for one specific purpose are redeployed as part of another neuronal network to accommodate a new function. This general perspective has been developed independently in several different theoretical proposals, including the *neural exploitation* theory (Gallese, 2008), the *shared circuits model* (Hurley, 2008), the *neuronal recycling* hypothesis (Dehaene & Cohen, 2007), and the *massive redeployment hypothesis* (Anderson, 2010). The basic premise is that reusing existing neural circuits to accomplish a new function is more likely from an evolutionary perspective than evolving a completely new circuit from scratch (cf. Jacob, 1977).

Research on brain development and evolution provides initial support for the idea of cultural recycling of neural substrates. Large-scale analyses of allometric data ranging from mammals (Finlay & Darlington, 1995) to sharks (Yopak et al., 2010) indicate that as brains become larger some areas grow proportionally more than others due to the highly conserved order of neurogenesis during development (roughly going from the back of the brain to the front): more anterior structures experience relatively longer periods of neural production as the brain grows larger overall (Finlay, Darlington, & Nicastro, 2001). These analyses provide little evidence that specific brain areas have been selective targets for adaptation to specific cognitive niches (e.g., as suggested by Pinker, 1997; Tooby & Cosmides, 2005); rather, the results appear to reveal only a general pressure for larger brains (though it is in principle possible that selection for a specific brain area could lead to the enlargement of the whole brain).

Increased social complexity in early hominin groups—potentially combined with increased environmental stresses—may have set a coevolutionary process in motion, in which more and more complex cultural evolution of social structures, technology, belief systems, and socio-pragmatic communication (see contributions in Richerson & Christiansen, 2013) provided a continual pressure favoring larger brains. The subsequent differential increase in the size of brain areas would have made it possible for some regions to do more complex computations and potentially take on additional roles (Schoenemann, 2009). This would have allowed more brain circuits to be recruited into new networks to accommodate novel functions that, in turn, could support further cultural evolution, and so on. In this context, development is likely to have played a key role in the emergence of new networks through culturally supported learning of new cognitive functions. As these new culturally mediated functions increased in number and complexity, more time and effort would be needed to master them, which might partially explain the very protracted period of development that modern humans go through (Locke & Bogin, 2006). Thus, a process of developmentally "scaffolded" cultural evolution could have resulted in increased brain size in the hominid lineage without necessarily involving selection for specific traits.

Importantly, if the neural recycling hypothesis is correct, we should expect most brain areas to play a role in a diverse range of behavioral functions. Supporting this prediction, Anderson (2010) reviews results from analyses of 1,469 subtraction-based fMRI experiments involving eleven different task domains, ranging from action execution, vision, and attention to memory, reasoning, and language: a particular cortical region is typically active for most of these task domains. That is, a specific neural circuit that is active in a particular cognitive task, such as language, is generally also active for multiple other tasks.

The cultural recycling hypothesis further predicts that cognitive functions that have emerged more recently in human evolution should be more widespread across the brain than older functions. Older traits have had a longer period over which dedicated, special purpose neural machinery can develop—and we should expect such machinery to be "massed" in particular brain areas, to minimize communication costs in passing information across different elements of the relevant computations. By contrast, more recent traits will recruit from a wider variety of existing neural circuits, capitalizing on their different properties, which may be usefully coopted to deal with different aspects of the novel task. We should expect that the most optimal network for this novel function will draw together a variety of pre-existing neural circuits; and there is no a priori reason why the most useful pre-existing neural circuits for recruitment for a novel task should happen to be spatially adjacent in the brain (Anderson, 2010). Thus, if the neural mechanisms involved in language are primarily the product of recycling of older neural substrates, as proposed by cultural evolution theorists, then we would expect to find the brain areas involved in language to be widely distributed across the brain. Analyzing the co-activation of Brodmann areas for eight different task domains in 472 fMRI experiments, Anderson (2008) found that language was the task domain for which co-activation patterns were the most widely scattered across the brain. Next in terms of the degree of distribution of neural co-activation patterns came reasoning, memory, emotion, mental imagery, visual perception, action and, lastly, attention. Indeed, language was significantly more widely distributed than the latter three task domains: visual perception, action, and attention.

Importantly, as existing neural circuits take on new roles by participating in new networks to accommodate novel functions, they typically retain their original function (though the latter may in some cases be affected by properties of the new function through developmental processes). The limitations and computational constraints of the original workings of those circuits will therefore be inherited by the new function, creating a *neuronal niche* (Dehaene & Cohen, 2007) for cultural evolution. In other words, the emerging new function will be shaped by constraints deriving from the recycled neural circuits as it evolves culturally. Thus, as in the case of reading (see Box 2.3), we argue that language has been shaped by the brain through the cultural recycling of preexisting neural substrates.

Box 2.3
Reading as a Product of Cultural Recycling

Writing systems are only about 7,000 years old, and for most of this time the ability to read and write was confined to a small group of individuals. Thus, reading is a culturally evolved ability for which humans would be unlikely to have any specialized biological adaptations. This marks reading as a prime candidate for a cognitive skill that is the product of cultural recycling of prior neural substrates.

Dehaene and Cohen (2007) argue that skills resulting from culturally mediated neuronal recycling, such as reading, should have certain characteristics. First, variability in the neural representations of the skill should be limited across individuals and cultures. With regard to reading, the so-called visual word form area, which is located in the left occipitotemporal sulcus, has been consistently associated with word processing across different individuals and writing systems (see Bolger, Perfetti & Schneider, 2005, for a review). Second, there should be considerable similarity across cultures in the manifestation of the skill itself. Consistent with this prediction, Dehaene and Cohen (2007) note that individual characters in alphabetic writing systems across the world consist of an average of three strokes, and the intersection contours of the parts of these characters follow the same frequency distribution (e.g., T, Y, Z, Δ). Third, there should be some continuity in terms of both neural biases and abilities for learning in non-human primates. That reading might build (at least in part) on the recruitment of evolutionary older mechanisms for object recognition is supported by recent results from a study of "orthographic processing" in baboons (Grainger et al., 2012) indicating that they could be trained to distinguish English words from nonsense words above chance, although of course without any understanding of the meaning of these words.

The available data regarding the neural representation of reading, combined with analyses of writing systems and experiments with non-human primates, suggest that writing systems have been shaped by a neuronal niche that includes the left ventral occipitotemporal cortex. In line with Dehaene and Cohen (2007), we argue that language similarly might be the product of cultural recycling of neural substrates.

2.4.2 Language as a System

We have proposed that the evolution of language is best understood in terms of the cultural recycling of neural substrates, suggesting that language has recruited pre-existing networks to support the evolution of various language functions. But why should language be viewed as akin to an integrated *system*, rather than as a collection of separate traits, evolving relatively independently? The reason is that language is highly *systematic*—so much so, indeed, that much of linguistic theory is concerned with tracking the systematic relationships between different aspects of linguistic structure. Yet although language is an integrated system, it can, nonetheless, be viewed as comprising a complex set of "features" or "traits," which may or may not be passed on from one generation to the next (concerning lexical items, idioms, aspects of phonology, syntax, and so on; Beckner, Blythe, Bybee, Christiansen, Croft, et al., 2009). To a first approximation, traits that are easy for learners to acquire and use will become more prevalent; traits that are more difficult to acquire and use will disappear. Selectional pressure from language learners and users will shape the way in which language evolves. The systematic character of linguistic traits means that, to some degree at least, the fates of different traits in a language are intertwined. That is, the degree to which any particular trait is easy to learn or process will, to some extent, depend on the other features of the language because language users will tend to learn and process each aspect of the language in light of their experience with the rest. This picture is familiar in biology—the selectional impact of any gene depends crucially on the rest of the genome; the selectional forces on each gene, for good or ill, are tied to the development and functioning of the entire organism.

Construing language as an evolutionary system has implications for explanations of *what* is being selected in language evolution. From the viewpoint of generative grammar, the unit of selection would seem to be either specific UG principles (in PPT; Newmeyer, 1991), particular parts of the UG toolkit (in SS; Culicover & Jackendoff, 2005), or recursion in the form of Merge (in MP; Hauser et al., 2002). In all cases, selection would seem to take place at a high level of abstraction that cuts across a multitude of specific linguistic constructions. But a different perspective is suggested by the "lexical turn" in linguistics (e.g., Combinatory Categorial Grammar, Steedman, 2000; Head-driven Phrase Structure Grammar, Sag, & Pollard, 1987; Lexical-Functionalist Grammar, Bresnan, 1982), focusing on specific lexical items with their associated syntactic and semantic information. Moreover, from a Construction Grammar perspective on language (e.g., Croft, 2000, 2001; Goldberg, 2006; O'Grady, 2005), it is natural to propose that individual constructions consisting of words or combinations of words are among the basic units of selection.

According to a Construction Grammar viewpoint, the long-term survival of any given construction is affected both by its individual properties (e.g., frequency of usage) and how well it fits into the overall linguistic system (e.g., syntactic, semantic, or pragmatic overlap with other constructions). In a series of linguistic and corpus-based analyses, Bybee (2007) has shown how frequency of occurrence plays an important role in shaping language from phonology to morphology to morphosyntax, due to the effects of repeated processing experiences with specific examples (either types or tokens). Additionally, groups of constructions overlapping in terms of syntactic, semantic, and/or pragmatic properties emerge to form the basis of usage-based generalizations (e.g., Goldberg, 2006; Tomasello, 2003). Fundamentally, though, these groupings lead to a distributed system of *local* generalizations across partially overlapping constructions, rather than the abstract, mostly global, generalizations of current generative grammar. This idea will be further developed in chapter 4.

In psycholinguistics, the effects of frequency and pattern overlap have also been observed in so-called Frequency × Regularity interactions (discussed in chapter 6). As an example, consider the acquisition of the English past tense. Frequently occurring mappings, such as *go* → *went*, are learned more easily than more infrequent mappings, such as *lie* → *lay*. However, low-frequency patterns may be more easily learned if they partially overlap with other patterns. The overlap in the mappings from stem to past tense in *sleep* → *slept, weep* → *wept, keep* → *kept* (i.e., *-eep* → *-ept*) makes learning these mappings relatively easy, even though none of the words individually has a particularly high frequency. Importantly, the two factors—frequency and regularity (i.e., degree of partial overlap)—interact with each other. High-frequency patterns are easily learned independent of whether they are regular or not, whereas the learning of low-frequency patterns suffers if they are not regular (i.e., if they do not have partial overlap with other patterns). Results from psycholinguistic experimentation and computational modeling have observed such Frequency × Regularity interactions across many aspects of language, including auditory word recognition (Lively, Pisoni, & Goldinger, 1994), visual word recognition (Seidenberg, 1985), English past tense acquisition (Hare & Elman, 1995), and sentence processing (Juliano & Tanenhaus, 1994; MacDonald & Christiansen, 2002; Pearlmutter & MacDonald, 1995; Wells, Christiansen, Race, Acheson, & MacDonald, 2009; see also chapter 6).

We suggest that similar interactions between frequency and pattern overlap are likely to play an important role in language evolution more broadly. Individual constructions may survive through frequent usage or because they participate in usage-based generalizations through syntactic, semantic, or pragmatic overlap with other similar constructions. Further support for this suggestion comes from artificial language learning studies with human subjects, which demonstrate that certain combinations of artificial-language structures are more easily learned than others given human sequence learning biases (e.g., Christiansen, 2000; Reeder, 2004; Saffran, 2001; see also chapter 7). For example, Ellefson and Christiansen (2000) compared human learning across two artificial languages that only differed in the order of words in two out of six sentence types. They found that not only was the more "natural" language learned better overall, but also that the four sentence types common to both languages were also learned more easily. This suggests that the artificial languages were learned as integrated systems, rather than as collections of independent items. Further corroboration comes from a study by Kaschak and Glenberg (2004) who had adult participants learn the *needs* construction (e.g., *The meal needs cooked*), a feature of the American English dialect spoken in the northern midlands region of the US from western Pennsylvania across Ohio, Indiana, and Illinois to Iowa. The training on the *needs* construction facilitated the processing of related modifier constructions (e.g., *The meal needs cooked vegetables*), again suggesting that constructions form an integrated system that can be affected by learning new constructions. Thus, although constructions are selected independently, they also provide an environment for each other within which selection takes place, just as the selection of individual genes are tied to the survival of the other genes that make up an organism.

2.4.3 Natural Selection for Functional Aspects of Language?

Before going on to discuss the specific constraints that shape the cultural evolution of language, we want to pause and emphasize what our arguments are *not* intended to show. In particular, we are not suggesting that biological adaptation is irrelevant to the emergence of language. Indeed, it seems likely that a number of adaptations occurred that might have been crucial to the emergence of language (see Hurford, 2003, for a review), such as the ability to represent discrete symbols (Deacon, 1997; Tomasello, 2003), to reason about other minds (Malle, 2002), to understand and share intentions (Tomasello, 2003; Tomasello, Carpenter, Call, Behne, & Moll, 2005), and to perform pragmatic reasoning (Levinson, 2000); there may also be a connection with the emergence of an exceptionally prolonged childhood and the possibility of transmitting language across generations (Locke & Bogin, 2006). Similarly, biological adaptations might have led to improvements to the cognitive systems that support language, including increased working memory capacity (Gruber, 2002), domain-general capacities for word learning (Bloom, 2001), and complex sequence-learning abilities (e.g., Calvin, 1994; Conway & Christiansen, 2001; Greenfield, 1991), though these adaptations are likely to have been for improved cognitive skills in general rather than for language in particular.

Some language-specific adaptations may, moreover, also have occurred, but given our arguments above, these would only be for functional features of language, and not the arbitrary features of UG. For example, changes to the human vocal tract may have resulted in more intelligible speech (Lieberman, 1984, 1991, 2003—though see also Hauser & Fitch, 2003); selectional pressure for this functional adaptation might apply relatively independently from the particular language being spoken by a population. Similarly, it remains possible that the Baldwin effect may be invoked to explain cognitive adaptations to language, provided that these adaptations are to functional aspects of language, rather than arbitrary linguistic structures. For example, it has been suggested that there might be a specialized perceptual apparatus for speech (e.g., Vouloumanos & Werker, 2007), or enhancement of the motor control system for articulation (e.g., Studdert-Kennedy & Goldstein, 2003). But explaining innate adaptations even in these domains is likely to be difficult—because, if adaptation to language occurs at all, it is likely to occur not merely to functionally universal features (e.g., the fact that linguistic signals can be segmented into words), but to specific cues for those features (e.g., for segmenting those words in the current linguistic environment, which differ dramatically across languages; Cutler, Mehler, Norris, & Segui, 1986; Otake, Hatano, Cutler, & Mehler, 1993). Hence, adaptationist explanations, even for functional aspects of language and language processing, should be treated with considerable caution.

2.5 Constraints on Language Structure

We have proposed that language has adapted to non-linguistic constraints from language learners and users, and that this helps explain the observed patterns of similarity across languages. But how far can these constraints be identified? To what extent can linguistic structure previously ascribed to an innate UG be identified as having a non-linguistic basis? Clearly, establishing a complete answer to this question would require a vast program of research. In this section, we illustrate how research from different areas of the language sciences can be brought together to capture aspects of language previously explained by UG. For the purpose of exposition, we divide the constraints guiding language evolution into four groups relating to thought, perceptuo-motor factors, cognition, and pragmatics. These constraints derive from the limitations and idiosyncratic properties of the human brain and other parts of our body involved in language (e.g., the vocal tract for spoken language, our hands for sign language). However, as we note below, any given linguistic phenomenon is likely to arise from a combination of multiple constraints that cut across these groupings, and thus across different kinds of mechanism.

2.5.1 Constraints from Thought

The relationship between language and thought is potentially abundantly rich, but also extremely controversial. The analytic tradition in philosophy can be viewed as attempting to understand thought through a careful analysis of language (e.g., Blackburn, 1984); and it has been widely assumed that the structure of sentences, and the inferential relations over them, provide an analysis of thought. An influential assumption is that thought is largely prior to, and independent of, linguistic communication. Accordingly, fundamental properties of language such as compositionality, function-argument structure, quantification, aspect, and modality, are proposed to arise from the structure of the thoughts language is required to express (e.g., Schoenemann, 2009). Moreover, presumably language provides a reasonably efficient mapping of the mental representation of thoughts with these properties into phonology. This perspective can be instantiated in a variety of ways. For example, the emphasis on incremental interpretation (e.g., that successive partial semantic representations are constructed as the sentence unfolds—the thought that a sentence expresses is built up piecemeal) is one motivation for Combinatory Categorial Grammar (e.g., Steedman, 2000). From a very different standpoint, the aim of finding a "perfect" relationship between thought and phonology is closely related to the goals of MP (Chomsky, 1995). Indeed, Chomsky (e.g., 2005) has suggested that language may have originated as a vehicle for thought, and only later become exapted to serve as a system of communication. This viewpoint would not, of course, explain the content of a presumed UG, which concerns principles for mapping mental representations of thought into phonology; and this mapping surely *is* specific to communication: inferences are, after all, presumably defined over semantic representations of thoughts (i.e., representations of meaning), rather than phonological representations or, for that matter, syntactic trees.

Vocabulary is presumably also strongly constrained by processes of perception and categorization—the meanings of words must be both learnable and cognitively useful (e.g., Murphy, 2002); indeed, the philosophical literature on lexical meaning, from a range of theoretical perspectives, sees cognitive constraints as fundamental to understanding word meaning, whether these constraints are given by innate systems of internal representation (Fodor, 1975), or primitive mechanisms of generalization (Quine, 1960). Cognitive linguists (e.g., Croft & Cruise, 2004) have argued for a far more intimate relation between thought and language: for example, basic conceptual machinery (e.g., concerning spatial structure) and the mapping of such structure into more abstract domains (e.g., via metaphor) is, according to some accounts, evident in language (e.g., Lakoff & Johnson, 1980). And from a related perspective (e.g., Croft, 2001), some linguists have argued that semantic categories of thought (e.g., of objects and relations) may be shared between languages, whereas syntactic categories and constructions are defined by language-internal properties, such as distributional relations, so that the attempt to find cross-linguistic syntactic universals is doomed to failure.

2.5.2 Perceptuo-motor Constraints

The motor and perceptual machinery underpinning language seems inevitably to have some influence on language structure. The seriality of vocal output, most obviously, forces a sequential construction of messages. Perceptual and memory systems are likely to engage in "greedy" processing, and have a very limited capacity for storing "raw" sensory input of any kind (e.g., Haber, 1983). This may force a code that can be interpreted incrementally (rather than the many practical codes in communication engineering, in which information is stored in large blocks, e.g., MacKay, 2003). The noisiness and variability (both with context and speaker) of vocal (or, indeed, signed) signals may, moreover, force a "digital" communication system with a small number of basic messages: e.g., one that uses discrete units (phonetic features, phonemes, or syllables). The basic phonetic inventory is transparently related to deployment of the vocal apparatus, and it is also possible that it is tuned, to some degree, to respect "natural" perceptual boundaries (Kuhl, 1987). Some theorists have argued for more far-reaching connections. For example, MacNeilage (1998) argues that aspects of syllable structure emerge as a variation on the jaw movements involved in eating, and for some cognitive linguists, the perceptual-motor system is a critical part of the machinery on which the linguistic system is built (e.g., Hampe, 2006). The depth of the influence of perceptual and motor control on more abstract aspects of language is controversial—but it seems plausible that such influence may be substantial, as we discuss further in chapter 4.

2.5.3 Cognitive Constraints on Learning, Memory, and Processing

Cognitive constraints clearly affect how we use language. One suggestion from O'Grady (2005) is that the language processing system seeks to resolve linguistic dependencies (e.g., between verbs and their arguments) at the first opportunity—a tendency that might not be syntax-specific, but instead an instance of a general cognitive tendency to attempt to resolve ambiguities rapidly whether for linguistic (Clark 1975) or perceptual input (Pomerantz & Kubovy, 1986). In a similar vein, Hawkins (1994, 2004) and Culicover (1999) propose specific measures of processing complexity (roughly, the number of linguistic constituents required to link syntactic and conceptual structure), which they assume underpin judgments concerning linguistic acceptability. The collection of studies in Bybee (2007) further underscores the importance of frequency of use in shaping language. Importantly, this line of work has begun to detail learning and processing constraints that can help explain specific linguistic patterns, such as the aforementioned examples of pronoun binding (examples 1–4 above; see O'Grady, 2005) and heavy NP-shift (examples 5–6; see Hawkins, 1994, 2004), and illustrates an increasing emphasis on performance constraints within linguistics (e.g., Culicover, 2013a; Kempson, Meyer-Viol, & Gabbay, 2001; O'Grady, 2013).

In turn, a growing body of empirical research in computational linguistics, cognitive science, and psycholinguistics has begun to explore how these theoretical constraints may be instantiated in terms of computational and psychological mechanisms. For instance, basic word-order patterns may derive from memory constraints related to sequence learning and processing, as indicated by computational simulations (e.g., Christiansen & Devlin, 1997; Kirby, 1999; Lupyan & Christiansen, 2002; Reali & Christiansen, 2009; Van Everbroeck, 2003), human experimentation involving artificial languages (e.g., Christiansen, 2000; Reeder, 2004), and cross-linguistic corpus analyses (e.g., Bybee, 2002; Hawkins, 1994, 2004). Similarly, behavioral experiments and computational modeling have provided evidence for general processing constraints (instead of innate subjacency constraints) on complex question formation (Berwick & Weinberg, 1984; Ellefson & Christiansen, 2000).

Importantly, though, in our framework, we do not see learning and processing as being separate from one another: language acquisition is viewed not as learning a distant and abstract grammar, but as *learning how to process* language. As we noted in chapter 1, constraints on learning and processing are often treated as largely or even entirely distinct (e.g., Bybee, 2007; Hawkins, 2004; Tomasello, 2003). By contrast, we suggest that they are fundamentally intertwined, subserved by the very same underlying mechanisms. We will return to this issue repeatedly in the chapters to come, and will highlight the importance of sequence learning for language acquisition and processing in our usage-based account of recursion in chapter 7.

2.5.4 Pragmatic Constraints

Language is likely, moreover, to be substantially shaped by the pragmatic constraints involved in linguistic communication. The program of developing and extending Gricean implicatures (Grice, 1967; Levinson, 2000; Sperber & Wilson, 1986) has revealed enormous complexity in the relationship between the literal meaning of an utterance and the message that the speaker intends to convey. Pragmatic processes may, indeed, be crucial to understanding many aspects of linguistic structure, as well as the processes of language change. For example, Levinson (2000) notes that discourse and syntactic anaphora have interesting parallels, which can provide the starting point for a detailed theory of anaphora and binding. Levinson argues that initially pragmatic constraints may, over time, become "fossilized" in syntax, leading to some of the complex syntactic patterns described by binding theory. This kind of shift, over time, from default constraint to rigid rule is widespread in language change and much studied in the sub-field of grammaticalization (as we shall discuss below). In sum, pragmatic principles may at least partly explain both the structure and origin of linguistic patterns that are often viewed as solely formal, and hence arbitrary (in the next chapter, we apply this perspective to explaining aspects of the binding constraints as exemplified by the examples 1–4 above).

2.5.5 The Impact of Multiple Constraints

In this section, we have discussed four types of constraints that have shaped the evolution of language. Importantly, we see specific linguistic phenomena as resulting from the interaction of these constraints. For example, as we discuss in chapter 3, the patterns of binding phenomena are likely to require explanations that cut across the four types of constraints, including constraints on cognitive processing (O'Grady, 2005) and pragmatics (Levinson, 1987a; Reinhart, 1983). That is, the explanation of any given aspect of language is likely to require the inclusion of multiple overlapping constraints deriving from thought, perceptual-motor factors, cognition, and pragmatics.

The idea of explaining language structure and use through the integration of multiple constraints goes back at least to early functionalist approaches to the psychology of language (e.g., Bates & MacWhinney, 1979; Bever, 1970; Slobin, 1973). It plays an important role in current constraint-based theories of sentence comprehension (e.g., MacDonald, Pearlmutter, & Seidenberg, 1994; Tanenhaus & Trueswell, 1995). Experiments have demonstrated how adults' interpretations of sentences are sensitive to a variety of constraints, including specific world knowledge relating to the content of an utterance (e.g., Kamide, Altmann, & Haywood, 2003), the visual context in which the utterance is produced (e.g., Tanenhaus, Spivey-Knowlton, Eberhard, & Sedivy, 1995), the sound properties of individual words (Farmer, Christiansen, & Monaghan, 2006), prior experience with particular constructions (e.g., Wells et al., 2009), and various pragmatic factors (e.g., Fitneva & Spivey, 2004). Similarly, the integration of multiple constraints—or "cues"—also figures prominently in contemporary theories of language acquisition (see e.g., contributions in Golinkoff et al., 2000; Morgan & Demuth, 1996; Weissenborn & Höhle, 2001; for a review, see Monaghan & Christiansen, 2008). We discuss such multiple-cue integration in language acquisition further in chapter 5.

But if the mechanisms shaping language evolution are to a large extent not specific to language, then what differentiates human cognition and communication from that of other primates? That is, why do humans have languages whereas other primates do not? The multiple-constraint satisfaction perspective on language evolution offers an answer to this question: as a cultural product, language has been shaped by constraints from many mechanisms, some of which have properties unique to humans. Specifically, we suggest that language may not necessarily be underpinned in humans by qualitatively different mechanisms compared to extant apes, but instead by a number of quantitative evolutionary refinements of older primate systems (e.g., for intention sharing and understanding, Tomasello et al., 2005; or complex sequence learning and processing, Conway & Christiansen, 2001—see chapter 7, for further discussion). These changes could be viewed as providing necessary non-linguistic adaptations that, once in place, allowed language to emerge through cultural transmission (e.g., Elman, 2005). As indicated by figure 2.2, only once a critical mass of language-enabling traits was in place, was the cultural evolution of language able to "take off" (albeit initially in a very modest form).

It is also conceivable that initial changes, if functional, could have been subject to further amplification through the Baldwin effect, perhaps resulting in multiple quantitative shifts in human evolution. The key point is that none of these changes would result in the evolution of UG. The species-specificity of a given trait, such as language, does not necessitate postulating specific biological adaptations for that trait. For example, even though playing the children's game of tag may be species-specific and perhaps even close to universal, few people, if any, would argue that humans have evolved adaptations specifically for playing this game. Thus, the uniqueness of language is better viewed as part of the larger question: why are humans different from other primates? It seems clear that considering language in isolation is not going to give us the answer to this question.

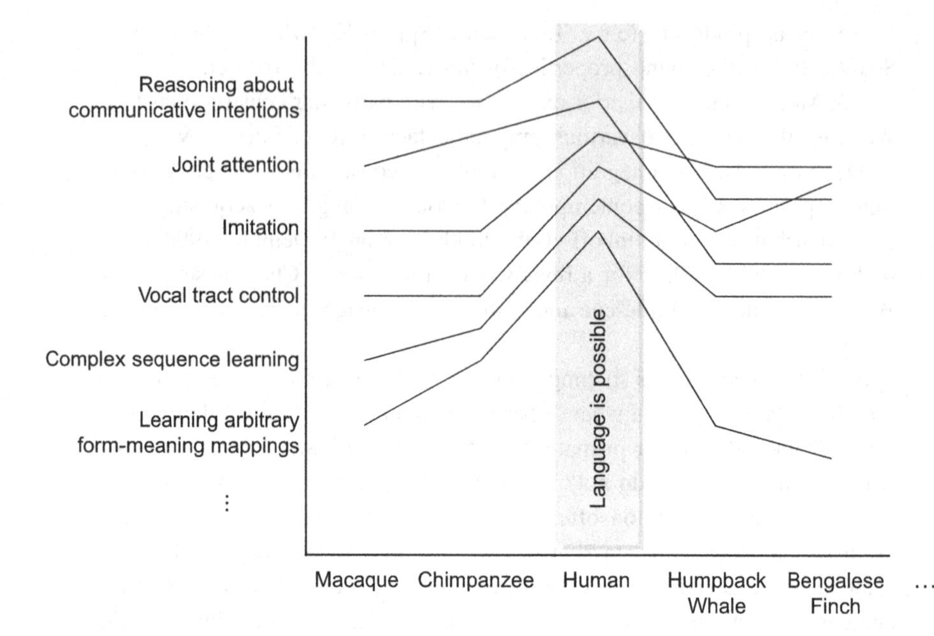

Figure 2.2
Language depends on many mechanisms that we share with other species (note that the relative level of skill across species is purely hypothetical). Natural selection has likely improved some of these skills in humans, allowing language to emerge through their interaction. This figure is inspired by Elman (2005).

2.6 How Constraints Shape Language over Time

According to the view that language evolution involves the emergence of UG (e.g., Berwick et al., 2013), there is a sharp divide between questions of language evolution (how the genetic endowment could arise through natural selection), and historical language change (which is viewed as variation within the genetically determined limits of possible human languages). By contrast, if language has evolved to fit prior cognitive and communicative constraints, then it is plausible that historical processes of language change provide a *model* of language evolution; indeed, historical language change may be language evolution in microcosm. This perspective is consistent with much work in functional and typological linguistics (e.g., Bever & Langendoen, 1971; Croft, 2000; Givón, 1998; Hawkins, 2004; Heine & Kuteva, 2002a).

At the outset, it is natural to expect that language will be the outcome of competing selectional forces. On the one hand, as we shall note, there will be a variety of selectional forces that make the language "easier" for speakers/

hearers; on the other, it is likely that expressivity is a powerful selectional constraint, tending to increase linguistic complexity over evolutionary time. For instance, it has been suggested that the use of hierarchical structure and limited recursion to express more complex meanings may have arisen in later stages of language evolution (Jackendoff, 2002; Johansson, 2006). Indeed, the modern Amazonian language, Pirahã, lacks recursion and has one of the world's smallest phoneme inventories (though its morphology is complex), limiting its expressivity (Everett, 2005; but see also the critique by Nevins, Pesetsky, & Rodrigues, 2007, and Everett's, 2007, response).

While expressivity is one selectional force that may tend to increase linguistic complexity, it will typically stand in opposition to others: specifically, ease of learning and processing will tend to favor linguistic simplicity. And the picture may be more complex: in some cases, ease of learning and ease of processing may stand in opposition. For example, regularity makes items easier to *learn*; the shortening of frequent items, and consequent irregularity, may make aspects of language easier to *say*. There are similar tensions between ease of production (which favors simplifying the speech signal), and ease of comprehension (which favors a richer, and hence more informative, signal). Moreover, whereas constraints deriving from the brain provide pressures toward simplification of language, processes of grammaticalization can add complexity to language (e.g., by the emergence of morphological markers). Thus, part of the complexity of language, just as in biology, may arise from the complex interaction of competing constraints.

The tension between multiple constraints can in some cases allow for other processes to affect how specific aspects of language become conventionalized. Consider, for example, color names, which, to some extent at least, appear to be driven by perceptual considerations (Regier, Kay, & Khetarpal, 2007). Nonetheless, color names show considerable variation across languages, perhaps because the drive to align with other group members may outweigh the drive to fit perceptual biases (Wallentin & Frith, 2008). Thus, the goodness of a classification may partially be defined by agreement with other group members, which may potentially lead to rapid and unpredictable runaway processes. Indeed, simulations by Baronchelli, Gong, Puglisi, & Loreto (2010) demonstrate that cross-linguistic patterns of color naming evidenced by World Color Survey data (Cook, Kay, & Regier, 2005) can be accounted for by cultural evolutionary processes, weakly constrained by a single well-established perceptual constraint: Just Noticeable Differences (Long, Yang, & Purves, 2006). We suggest that such processes may be particularly likely where there is a large range of alternative solutions, which are roughly equally "good" from the point of view of the individual agent, and particularly when it is difficult to shift from one type of solution to another. Many aspects of the world's languages, ranging from the inventory of phonemes and the variety of syntactic categories to the functioning of pronouns, seem to exhibit considerable variation. These variants are, perhaps, roughly equally good solutions, and moving between solutions is slow and difficult (although historical linguistics does sometimes indicate that change does occur between such forms, McMahon, 1994). In such cases, the selection pressure on language from the brain may impose only a relatively weak constraint on the solution that is reached. Conversely, the functional pressure for the emergence of other aspects of language, such as arbitrariness in form-meaning mappings (Monaghan, Shillcock, Christiansen, & Kirby, 2014) or word order regularities (de Vries, Christiansen, & Petersson, 2011), may be so strong that these factors are not disturbed by social forces (as discussed in chapters 5 and 7, respectively).

2.6.1 Language Evolution as Linguistic Change

Recent theory in diachronic linguistics has focused on grammaticalization (e.g., Bybee, Perkins, & Pagliuca, 1994; Heine, 1991; Hopper & Traugott, 2003): the process by which functional items, including closed class words and morphology, develop from what are initially open-class items. This transitional process involves a "bleaching" of meaning, phonological reduction, and increasingly rigid dependencies with other items. Thus, the English number *one* is likely to be the root of the indefinite article *a*(*n*). And, to pick a morphological example, the Latin *cantare habeo* (*I have* (something) *to sing*) mutated into *chanterais, cantaré, cantarò* (*I will sing* in French, Spanish, Italian). The suffix corresponds phonologically to *I have* in each language (respectively, *ai*, *he*, *ho*—the *have* element has collapsed into inflectional morphology, Fleischman, 1982). The same processes of grammaticalization can also cause certain content words over time to be bleached of their meaning and become grammatical particles. For example, the use of *go* and *have* as auxiliary verbs (as in *I am going to sing* or *I have forgotten my hat*) have been bleached of their original meanings concerning physical movement and possession (Bybee et al., 1994). The processes of grammaticalization appear gradual, and follow consistent historical patterns, suggesting that there are systematic selectional pressures operating in language change. Moreover, these processes provide a possible origin of grammatical structure from a protolanguage initially involving perhaps unordered and uninflected strings of content words (Heine & Kuteva, 2012).

Understanding the cognitive and communicative basis for the direction of grammaticalization and related processes is an important challenge. But equally, the suggestion that this type of observable historical change may be continuous with language evolution opens up the possibility that research on the origin of language may not be a theoretically isolated island of speculation, but may connect directly with one of the most central topics in linguistics: the nature of language change (e.g., Zeevat, 2006). Indeed, grammaticalization has become the center of many recent perspectives on the evolution of language as mediated by cultural transmission across hundreds (perhaps thousands) of generations of learners (e.g., Bybee et al., 1994; Givón, 1998; Heine & Kuteva, 2002a; Schoenemann, 1999; Tomasello, 2003). It is important, though, not to see processes of cultural transmission as an *alternative to* theories of brain and cognition, but as *grounded in* such theories. For example, Bybee (2009) contends that many properties of language arise from purely historical processes of language change (such as grammaticalization), hypothesized to be fairly independent of underlying brain mechanisms. But we argue that processes of historical language change depend critically on the cognitive and neural machinery of the speakers involved. Even if language is a cultural product, created by processes of cultural transmission, it is nonetheless shaped by the brains that create and transmit linguistic structure. The brain, and the cognitive and learning constraints that it embodies, is centrally important after all, shaping the selectional pressures operating on language change. Thus, as discussed further in chapter 4, we see grammaticalization and other processes of linguistic change as heavily influenced by the nature of learning and processing.

2.6.2 Language Evolution through Cultural Transmission

How far can language evolution and historical processes of language change be explained by general mechanisms of cultural transmission? And how might language be selectively distorted by such processes? Central to any such model, whether concerning language or not, are assumptions about the channel over which cultural information is transmitted; the structure of the network of social interactions over which transmission occurs; and the learning and processing mechanisms that support the acquisition and use of the transmitted information (Boyd & Richerson 2005).

A wide range of recent computational models of the cultural transmission of language has been developed, with different points of emphasis (see Smith, 2014, for a review). Some of these models have considered how language is shaped by the process of transmission over successive generations, by the nature of the communication problem to be solved and/or by the nature of the learners (e.g., Batali, 1998; Kirby, 1999). For example, Kirby, Dowman, and Griffiths (2007) show that, if information is transmitted directly between individual learners, and learners sample grammars from the Bayes posterior distribution of grammars given that information, then language asymptotically converges to match the priors initially encoded by the learners. In contrast, Smith, Brighton, and Kirby (2003), using a different model of how information is learned, indicate how compositional structure in language might have resulted from the complex interaction of learning constraints and cultural transmission, resulting in a "learning bottleneck." Moreover, a growing number of studies have started to investigate the potentially important interactions between biological and linguistic adaptation in language evolution (e.g., Baronchelli, Chater, Christiansen, & Pastor-Satorras, 2013; Christiansen et al., 2011; Hurford, 1990; Hurford & Kirby, 1999; Kvasnicka & Pospichal, 1999; Livingstone & Fyfe, 2000; Munroe & Cangelosi, 2002; Niyogi, 2006; Reali & Christiansen, 2009; Smith, 2002, 2004; Yamauchi, 2001).

Of particular interest are simulations indicating that apparently arbitrary aspects of linguistic structure may arise from constraints on learning and processing (e.g., Kirby, 1998; Van Everbroeck, 2003). For example, it has been suggested that subjacency constraints may arise from cognitive limitations on sequence learning (Ellefson & Christiansen, 2000). Moreover, using rule-based language induction, Kirby (1999) accounted for the emergence of typological universals as a result of domain-general learning and processing constraints. Finally, note that in line with the arguments described earlier in this chapter, a range of recent studies has challenged the plausibility of biological adaptation to arbitrary features of the linguistic environment (e.g., Baronchelli et al., 2012; Chater et al., 2009; Kirby et al., 2007; Kirby & Hurford, 1997; Munroe & Cangelosi, 2002; Yamauchi, 2001).

The range of factors known to be important in cultural transmission (e.g., group size, networks of transmission between group members, and fidelity of transmission) has been explored relatively little in simulation work (though see Culicover & Nowak, 2003; Nettle, 1999; Reali, Chater, & Christiansen, 2014). Furthermore, to the extent that language is shaped by the brain, enriching models of cultural transmission of language to take account of learning and processing constraints will be an important direction for the study both of historical language change and language evolution. However, viewing language as shaped by cultural transmission (Arbib, 2005; Bybee, 2002; Donald, 1998) provides no more than a starting point for an explanation of linguistic regularities. The real challenge, we suggest, is to delineate the wide range of constraints, from perceptuo-motor to pragmatic (as sketched above), that operate on language evolution. Detailing these constraints is likely to be crucial to explaining the origin of complex linguistic regularities and how they can readily be learned and processed. In chapter 4, we discuss how the fact that both speakers and listeners have to deal with language in the "here-and-now" has profound implications for language processing, acquisition, and evolution, and even for the structure of language.

We note here that this perspective on the adaptation of language differs from the processes of cultural change that operate through deliberate and conscious innovation and/or evaluation of cultural variants. On the present account, the processes of language change operate to make languages easier to learn and process, and more communicatively effective. But these changes do not operate through processes either of "design" or deliberate adoption by language users. Thus, following Darwin, we view the origin of the adaptive complexity in language as analogous to the origin of adaptive complexity in biology. Specifically, the adaptive complexity of biological organisms is presumed to arise from random genetic variation, winnowed by natural selection (a "blind watchmaker"; Dawkins, 1986); we argue that the adaptive complexity of language arises, similarly, from random linguistic variation winnowed by selectional pressures, though here arising from learning, processing, and communicative use (again, we have a *blind* watchmaker).

By contrast, for aspects of cultural changes for which variants are either invented or selected by deliberate choice, the picture is different. Such cultural products can be viewed instead as arising from the incremental action of processes of intelligent design, and more or less explicit evaluations, and decisions to adopt (see Chater, 2005). Many phenomena discussed by evolutionary theorists concerning culture (e.g., Campbell, 1965; Richerson & Boyd, 2005; Richerson & Christiansen, 2013)—including those described by meme-theorists (e.g., Blackmore, 1999; Dawkins, 1976; Dennett, 1995)—fall into this latter category: explanations of fashions (e.g., wearing baseball caps backward), catch-phrases, memorable tunes, engineering methods, cultural conventions and institutions (e.g., marriage, revenge killings), scientific and artistic ideas, religious views, and so on, seem patently to be products of *sighted* watchmakers; i.e., they are products, in part at least, of many generations of intelligent designers, imitators, and critics.

Our focus here concerns, instead, the specific and interdependent constraints operating on particular linguistic structures and of which people have no conscious awareness. Presumably, speakers do not deliberately contemplate syntactic reanalyses of existing structures, bleach the meaning of common verbs so that they play an increasingly syntactic role, or collapse discourse structure into syntax or syntactic structure into morphology. Of course, there is some deliberate innovation in language (e.g., people do occasionally consciously invent new words and phrases). But such deliberate innovations are very different from the unconscious operation of the basic learning and processing biases that have shaped the phonological, syntactic, and semantic regularities of language.

2.6.3 Language Change "in Vivo"

We have argued that language has evolved over time to be compatible with the human brain. However, it might be objected that it is not clear that languages become better adapted over time given that they all seem capable of expressing a similar range of meanings (Sereno, 1991). In fact, the idea that all languages are fundamentally equal and independent of their users—uniformitarianism—is widely presupposed in linguistics, preventing many linguists from thinking about language evolution (Newmeyer, 2003). Yet much variation exists in how easy it is to use a given language to express a particular meaning, given the limitations of human learning and processing mechanisms.

The recent work on creolization in sign language provides a window onto how pressures toward increased expressivity interact with constraints on learning and processing "in vivo." In less than three decades, a sign language has emerged in Nicaragua, created by deaf children with little exposure to established languages. Senghas, Kita, and Özyürek (2004) compared signed expressions for complex motions produced by deaf signers of Nicaraguan Sign Language (NSL) with the gestures of hearing Spanish speakers. The results showed that the hearing individuals used a single simultaneous movement combining both manner and path of motion, whereas the deaf NSL signers tended to break the event into two consecutive signs: one for the path of motion and another for the manner. Moreover, this tendency was strongest for the signers who had learned NSL more recently, indicating that NSL has changed from denoting motion events holistically to a more sequential, compositional format. Although such creolization may be considered as evidence of UG (e.g., Bickerton, 1984; Pinker, 1994), these findings may be better understood in terms of cognitive constraints on cultural transmission. Indeed, as we have noted, computational simulations have demonstrated how iterated learning in cultural transmission can gradually change a language starting as a collection of holistic form-meaning pairings into a more compositional format, in which sequences of forms are combined to produce meanings previously expressed holistically (see Kirby & Hurford, 2002, for a review). Similarly, "cross-generational" iterated learning experiments with human participants—in which the output of one artificial-language learner is used as the input for subsequent "generations" of language learners have shown that over many generations the selection pressure caused by such learning biases can change the structure of artificial languages from holistic mappings to a compositional format (Kirby, Cornish & Smith, 2008). This allows language to become more expressive, while still being learnable from exposure to a finite set of form-meaning pairings. Thus, the shift toward using sequential compositional forms to describe motion events in NSL can be viewed as a reflection of similar processes of learning and cultural transmission.

In a similar vein, the emergence of a regular SOV (subject-object-verb) word order in the Al-Sayyid Bedouin Sign Language (ABSL; Sandler, Meir, Padden, & Aronoff, 2005) can be interpreted as arising from constraints on learning and processing. ABSL has a longer history than NSL, going back some 70 years (Sandler, 2012). The Al-Sayyid Bedouin group forms an isolated community with a high incidence of congenital deafness, located in the Negev desert region of southern Israel. In contrast to NSL, which developed within a school environment, ABSL has evolved in a more natural setting and is recognized as the second language of the Al-Sayyid village. Two key features of ABSL are that it has developed a basic SOV word order within sentences (e.g., *boy apple eat*), and that modifiers follow heads (e.g., *apple red*). Notice first that the fact that these abstract regularities form at all is interesting—for example, modifier/head order could be unconstrained, or could be entirely dependent on the modifiers concerned. Second, note that SOV languages typically also have modifiers following heads, rather than the reverse. ABSL fits this pattern. But although this type of word order is very common across the world (Dryer, 1992), it is found neither in the local spoken Arabic dialect nor in Israeli Sign Language, suggesting that ABSL has developed these grammatical regularities *de novo*. In a series of computational simulations, Christiansen and Devlin (1997) found that languages with consistent word order were easier to learn by a sequence learning device compared to languages with inconsistent word orders (see also Van Everbroeck, 2003). Thus, a language with a grammatical structure such as ABSL should be easier to learn than one, say, in which an SOV word order is combined with a modifier-head (rather than head-modifier) order within phrases. Similar results were obtained when people were trained on artificial languages with either consistent or inconsistent word orders (Christiansen, 2000; Reeder, 2004). Further simulations have demonstrated how biases on sequence learning can lead to the emergence of languages with regular word orders through cultural transmission—even when starting from a language with a completely random word order (Christiansen & Dale, 2004; Reali & Christiansen, 2009; see also chapters 5 and 7).

Differences in learnability are not confined to newly emerged languages but can also be observed in well-established languages. For example, Slobin and Bever (1982) found that when children learning English, Italian, Turkish, or Serbo-Croatian were asked to act out reversible transitive sentences, such as *the horse kicked the cow*, using familiar toy animals, language-specific differences in performance emerged. Turkish-speaking children performed very well already at two years of age, most likely due to the regular case markings in this language, indicating who is doing what to whom. Young English and Italian-speaking children initially performed slightly worse than the Turkish children but caught up by around three years of age, relying on the relatively consistent word-order information available in these languages, with subjects preceding objects. The children acquiring Serbo-Croatian, on the other hand, had problems determining the meaning of the simple sentences, most likely because this language uses a combination of case markings and word order to indicate agent and patient roles in a sentence. However, only masculine and feminine nouns take on accusative or nominative markings and can occur in any order with respect to one another. Sentences with one or more unmarked neuter nouns are typically ordered as SVO. Of course, Serbo-Croatian children eventually catch up with the Turkish, English, and Italian-speaking children. But these results show that some meanings are harder to learn and process in some languages compared to others, indicating differential fitness across languages (see Lupyan & Christiansen, 2002, for corroborating computational simulations). Moreover, as we discuss in chapter 6, substantial differences even exist between individual idiolects within a specific language.

In sum, we have argued that human language has been shaped by selectional pressure over thousands of generations of language learners and users. Linguistic variants that are easier to learn to understand and produce will be favored, and so will linguistic variants that are more economical, expressive, and generally effective in communication, persuasion, and perhaps signaling status within a social group and identity within that group. Just as with the multiple selectional pressures operative in biological evolution, the matrix of factors at work in driving the evolution of language is complex. Nonetheless, as we have seen, candidate pressures can be proposed (e.g., the pressure for incrementality, minimizing memory load, regularity, brevity, and so on), and regular patterns of language change that may be responses to those pressures can be identified (e.g., the processes of successive entrenchment, generalization, and erosion of structure evident in grammaticalization); we expand on this viewpoint in chapter 4. In particular, note that the logical problem of language evolution that appears to confront attempts to explain how a genetically specified linguistic endowment could become encoded, does not even arise if we view language evolution as primarily cultural, rather than biological, in nature. It is not the brain that has somehow evolved to language, but the reverse.

2.7 Summary

In this chapter, we have presented a theory of language evolution as shaped by the brain. From this perspective, the close fit between language learners and the structure of natural language that motivates many theorists to posit a language-specific biological endowment may instead arise from processes of adaptation operating on language itself. Moreover, we have argued that there are fundamental difficulties with postulating a language-specific biological endowment. It is implausible that such an endowment could evolve through adaptation because the prior linguistic environments would be too diverse to give rise to universal principles. It is also unlikely that a language-specific endowment of any substantial complexity arose through non-adaptational genetic mechanisms because the probability of a functional language system arising essentially by chance is vanishingly small.

Instead, we have suggested that seemingly arbitrary aspects of language structure may arise from the interaction of a range of factors, from general constraints on learning, to impacts of semantic and pragmatic factors, and concomitant processes of grammaticalization and other aspects of language change. But, intriguingly, it is also possible that many apparently arbitrary aspects of language can be explained by relatively natural cognitive constraints, and hence that language may be less arbitrary than first supposed (e.g., Bates & MacWhinney, 1979, 1987; Bybee, 2007; Elman, 1999; Kirby, 1999; Levinson, 2000; O'Grady, 2005; Tomasello, 2003). And perhaps even more compelling, as we discuss in the next chapter, this perspective further promises to simplify the problem of language acquisition. When children acquire their native language(s), their biases will be the right biases because language has been optimized by past generations of learners to fit those very biases (Chater & Christiansen, 2010; Zuidema, 2003). This does not, however, trivialize the problem of language acquisition but instead suggests that children tend to make the right guesses about how their language works—not because of an innate UG—but because language has been shaped by cultural evolution to fit the non-linguistic constraints that they bring to bear on language acquisition. The brain has not been shaped to process language but, rather, language is shaped by the brain.

Summary

[illegible]

3 Language Acquisition Meets Language Evolution

… the debate … is not about Nature vs. Nurture, but about the "nature of Nature"
—Liz Bates, 1999, *On the Nature and Nurture of Language*

At first blush, it may seem odd to think that language evolution would be relevant to language acquisition. After all, even though language change is relatively fast, as we have noted in the previous chapter, it is generally on the scale of hundreds or thousands of years, and thus very much slower than human development. Of course, some aspects of language actually change considerably faster. Words, for example, can enter into the language and then disappear almost as quickly. This is especially true of words related to technology, such as the word *fax*, which refers to a device that scans and transmits documents over telecommunication lines, to the output of such a device, or to the act of sending a document. According to Google Books Ngram Viewer[1] (Michel et al., 2011), the word *fax* rarely occurred before the early 1980s but then became relatively popular in the late 1980s and early 1990s—peaking in 1999 (0.00118% of word use)—after which its use was halved in less than a decade. If the current downward trend holds up, then the word *fax* will likely disappear from common use altogether in a few years, along with the fax machines themselves. But besides our ever-changing vocabulary, even the way we speak can change over our lifetime. This is illustrated by the measurable shifts in the vowels produced by the Queen of England in her yearly Christmas messages recorded by the BBC from the 1950s through the 1980s (Harrington, Palethorpe, & Watson, 2000). Apparently, the British Queen no longer speaks the Queen's English of the 1950s.

Nonetheless, vocabulary and minor pronunciation shifts aside, the linguistic environment is typically fairly stable during the period of primary linguistic

1. Google Books Ngram Viewer can be found at: https://books.google.com/ngrams. The results for *fax* were accessed on May 26, 2015.

development. Consequently, researchers have treated language as, in essence, fixed, for the purposes of understanding language acquisition. In contrast, we argue that new light can be thrown on the problem of language acquisition by taking an evolutionary perspective: if language is largely the product of cultural evolution, as we argued in chapter 2, then language will have been shaped to fit whatever biases children bring to bear on acquisition, by previous learners with the very same biases (see also Wexler & Culicover, 1980, for a precursor to this perspective). In this sense, language is similar to other human inventions (see contributions in Richerson & Christiansen, 2013): it is closely adapted to being used by us.

Consider, for example, how modern scissors are exquisitely adapted to the shape of our hands, so much so that the standard right-handed scissors are hard to use for left-handers and vice versa (and even most ambidextrous scissors are actually easier for right-handers to use because the orientation of the blades will block a left-hander's view of the cutting line). The close adaptation of scissors to the human hand by way of cultural evolution makes them easy for us to use, and allows children to pick up the skill with little or no instruction (though being able to cut things neatly, of course, does require some practice). Thus, the ease with which we are able to use scissors could seem quite puzzling indeed, if we ignored the fact that this tool has been adapted by previous generations of scissors users (and makers) to function optimally given the shape of our hands. Similarly, we argue that understanding how language changes over time provides important constraints on theories of language acquisition and recasts, and substantially simplifies, the problem of induction in the context of language acquisition. We find scissors easy to use, and to learn to use, because they have been shaped by generations of past users. So, too, for language.

In this chapter, then, we aim to throw light on the problem of language acquisition by taking an evolutionary perspective, concerning both the biological evolution of putative innate domain-specific constraints and, more importantly, the cultural evolution of human linguistic communication. An evolutionary perspective casts many apparently intractable problems of induction in a new light. Where the child aims to learn an aspect of human culture (rather than an aspect of the natural world), the learning problem is dramatically simplified, because culture (including language) is the product of past learning from previous generations. In learning about the cultural world, we are learning to "follow in each other's footsteps," so that our wild "guesses" are likely to be the right ones, because the right guess is the most popular guess made by previous generations of learners. This is how considerations from language *evolution* dramatically shift our understanding of the problem of language *acquisition*. We suggest, too, that an evolutionary perspective may also require rethinking theories of the acquisition of other aspects of culture. In particular, in the context of learning about culture, rather than constraints from the natural world, we suggest that the conventional nativist picture, stressing domain-specific, innately-specified modules, cannot be sustained. Adapting Dobzhansky's (1964) famous phrase, we are thus suggesting that *nothing in development makes sense except in the light of evolution*.

3.1 C-induction and N-induction

Human development involves overcoming two inter-related challenges: acquiring the ability to understand and manipulate the natural world (N-induction), and acquiring the ability to co-ordinate with each other (C-induction). Pure cases of these two types of problem are very different. In *N-induction*, the world imposes an external standard, against which performance is assessed. In *C-induction*, the standard is not externally imposed, but socially created: the key is that we do the *same* thing, not that we all do an objectively "right" thing. In reality, most challenges facing the child involve a complex mixture of N- and C-induction—and teasing apart the elements of the problem that involve understanding the world, versus coordinating with others, may be very difficult. Nonetheless, we suggest that the distinction is crucially important, both in understanding human development in general, and in understanding the acquisition of language in particular.

To see why the distinction between N- and C-induction is important, consider the difference between learning the physical properties of the everyday world, and learning how to indicate agreement or disagreement using head movements. In order to interact effectively with the everyday world, the child needs to develop an understanding of persistent objects, exhibiting constancies of color and size, which move coherently, have weight and momentum, and have specific patterns of causal influences on other objects. The child's perceptuo-motor interactions with the everyday world (e.g., catching a ball, Dienes & McLeod, 1993) depend crucially on such understanding, and do so individualistically—in the sense that success or failure is, to a first approximation, independent of how other children, or adults, understand the everyday world. The child is, in this regard at least, a lone scientist (A. Gopnik, Meltzoff, & Kuhl, 1999; Karmiloff-Smith & Inhelder, 1975).

By contrast, in C-induction, the aim is to do as others do. Thus, when learning how to deploy a repertoire of head movements or facial expressions to indicate agreement, disagreement, puzzlement, surprise, and so on, the child must learn to follow in the footsteps of others. Whereas there are objective constraints, derived from physics, on catching a ball, the problem of communication via head movements is much less constrained—from an abstract point of view, vast numbers of mappings between overt communicative acts and underlying mental states may be equivalent. For example, in Northern Europe nodding one's head indicates "yes," but in Greece nodding signals "no." Similarly, in many places across the world, shaking one's head is used for "no," but in Sri Lanka it indicates general agreement (Wang & Li, 2007). What is crucial for the child is that it comes to adopt the *same* pattern of head movement to indicate agreement as those around it. The child is here not a lone scientist, but a musician whose objective is not to attain any absolute pitch, but to be "in tune" with the rest of the orchestra.

Before we turn to the question of why C-induction is dramatically easier than N-induction, note that the distinction between N- and C-induction is conceptually distinct from the debate between nativist and empiricist accounts of development (although it has striking implications for these accounts, as we shall see). Table 3.1 illustrates this point with a range of examples from animal behavior. In many species, innate constraints appear fundamental to solving N- and C-induction problems. Innate solutions concerning problems of N-induction include basic processes of flying, swimming, and catching prey, as well as highly elaborate and specific behaviors such as nest building. And such innate constraints are equally dominant in determining coordination between animals. For example, from a functional point of view, patterns of movement might translate into information about food sources in a range of ways, but genetic constraints specify that honeybees employ a *particular* dance (Dyer, 2002). This amounts to solving a problem of C-induction, i.e., of coordination (although solving it over phylogenetic time, via natural selection,

Table 3.1
A tentative classification of a sample of problems of understanding and manipulating the world (N-induction) vs. coordinating with others (C-induction) in non-human animals

	Innate constraints dominant	*Learning dominant*
N-induction	Locomotion and perceptual-motor control[a], hunting, foraging and feeding[b], nest building[c]	Learning own environment[d], identifying kin[e], learned food preferences and aversion[f]
C-induction	Insect social behavior[g], fixed animal communication systems[h], including the bee waggle dance[i], many aspects of play[j] and mate choice[k]	Social learning[l], including imitative songbirds[m]

Note: a. Alexander (2003); b. Stephens, Brown, & Ydenberg (2007); c. Healy, Walsh & Hansell (2008); d. Healy & Hurly (2004); e. Holmes & Sherman (1982); f. Garcia, Kimeldorf & Koelling (1955); g. Wilson (1971); h. Searcy & Nowicki (2001); i. Dyer (2002); j. Bekoff & Byers (1998); k. Andersson (1994); l. Galef & Laland (2005); m. Marler & Slabbekoorn (2004).

rather than solving it over ontogenetic time, via learning): the bees must adopt the *same* dance with the same interpretation (and indeed dances do differ slightly between bee species). Courtship, rutting, and play behaviors may often have the same status—the "rules" of social interactions in a species are shaped by genetic biases—but they are also somewhat arbitrary. The key is that these rules are coordinated across individuals—that a male courtship display is recognizable by relevant females, for example.

In other cases, problems of both N- and C-induction may be solved not by natural selection of relevant genes across generations, but by mechanisms of learning within the lifespan of an organism. Animals learn about their immediate environment, where food is located, what is edible, and, in some cases, how to identify their conspecifics—this is N-induction, concerning objective aspects of the world. Indeed, some learned behaviors (such as milk-bottle pecking in blue tits or food preparation techniques in chimpanzees or gorillas) may be learned from conspecifics, although whether by processes of emulation, imitation, or simpler mechanisms, is not clear (Hurley & Chater, 2005). To a modest degree, some non-human animals also learn to coordinate their behavior. For example, some songbirds and whales learn their songs from others. Reproductive success depends on producing a "good" song defined in terms of the current dialect (Marler & Slabbekoorn, 2004), rather than targeting any "objective" standard of singing.

The distinction between problems of C- and N-induction is, then, conceptually separate from the question of whether an induction problem is solved over phylogenetic time by natural selection (and specifically, by the adaptation of genetically encoded constraints), or over ontogenetic time by learning. Nonetheless, the distinction has two striking implications for the theory of development, and, in particular, for language acquisition. First, as we shall argue, C-induction is dramatically easier than N-induction, and many aspects of language acquisition seem paradoxically difficult because a problem of C-induction is mischaracterized as a problem of N-induction. Second, the child's ability to solve C-induction problems, including language acquisition, must primarily be based on cognitive and neural mechanisms *that predate the emergence of the cultural form to be learned.* That is, natural selection cannot lead to the creation of dedicated, domain-specific learning mechanisms for solving C-induction problems (e.g., innate modules for language acquisition). By contrast, such mechanisms may be extremely important for solving N-induction problems. Table 3.2, somewhat speculatively, considers examples from human cognition. Rather than focusing in detail on each of these cases, we focus here on the general distinction between N-induction and C-induction, before turning to a brief illustrative example: the binding constraints, which we mentioned in the previous chapter.

Table 3.2
A tentative classification of sample problems of understanding and manipulating the world (N-induction) vs. coordinating with others (C-induction) in human development

	Innate constraints dominant	*Learning dominant*
N-induction	Low-level perception, motor control[a], perhaps core naïve physics[b]	Perceptual, motor and spatial learning[c], science and technology[d]
C-induction	Understanding other minds[e], pragmatic interpretation[f], social aspects of the emotions[g], basic principles of cooperation, reciprocation and punishment[h]	Language, including syntax, phonology, word learning and semantics[i], linguistic categorization[j]; other aspects of culture[k], including music, art, social conventions, ritual, religion, and moral codes

Note: a. Crowley & Katz (1999); b. Carey & Spelke (1996); c. Shadmehr & Wise (2005), Johnson (2010), Twyman & Newcombe (2010); d. Cartwright (1999); e. Tomasello, Carpenter, Call, Behne & Moll (2005); f. de Ruiter & Levinson (2008); g. Frank (1988); h. Olson & Spelke (2008), Fehr & Gächter, (2002); i. Smith, Colunga & Yoshida (2010); j. Sloutzky (2010); k. Geertz (1973).

3.2 Implications for Learning and Adaptation

Suppose that some natural process yields the sequence 1, 2, 3… How does it continue? Of course, we have far too little data to know. It might oscillate (1, 2, 3, 2, 1, 2, 3, 2, 1…), become "stuck" (1, 2, 3, 3, 3, 3…), exhibit a Fibonacci structure (1, 2, 3, 5, 8…), or follow an infinity of more or less plausible alternatives. This indeterminacy makes the problem of N-induction of structure from the natural world difficult, although not necessarily hopelessly so, in the light of recent developments in statistics and machine learning (Chater & Vitányi, 2007; Harman & Kulkarni, 2007; Li & Vitányi, 1997; Tenenbaum, Kemp, & Shafto, 2007).

But consider the parallel problem of C-learning—we need not guess the "true" continuation of the sequence. We only have to *coordinate* our predictions with those of other people in the community. This problem is very much easier. From a psychological point of view, the overwhelmingly natural continuation of the sequence is "…4, 5, 6…." That is, most people are likely to predict this. Thus, coordination emerges easily and unambiguously on a specific infinite sequence, even given a tiny amount of data.

Rapid convergence of human judgments from small samples of data, is observed across many areas of cognition. For example, Feldman (1997) and Tenenbaum (1999) show that people converge on the same categories incredibly rapidly, given a very small number of perceptual examples; and rapid convergence from extremely limited data is presupposed in intelligence testing, where the majority of problems are highly indeterminate, but responses nonetheless converge on a single answer (e.g., Barlow, 1983). Moreover, when people are allowed to interact, they rapidly align their choice of lexical items and frames of reference, even when dealing with novel and highly ambiguous perceptual input (e.g., Clark & Wilkes-Gibbs, 1986; Pickering & Garrod, 2004).

Finally, consider a striking and important class of examples from game theory in economics. In a typical coordination game, two players simultaneously choose a response; if it is the same, they both receive a reward, otherwise, they do not. Even when given very large sets of options, people often converge in "one shot." Thus, if asked to select a time and a place to meet in New York City, Schelling (1960) found that people generated several highly frequent responses (so-called *focal points*) such as "twelve noon at Grand Central Station," so that players might potentially meet successfully, despite choosing from an almost infinite set of options. The corresponding problem of N-induction (i.e., of guessing the time and place of an arbitrarily chosen event in New York City) is clearly hopelessly indeterminate; but as a problem of C-induction, where each player aims to coordinate with the other, it is nonetheless readily solved.

C-induction is, then, vastly easier than N-induction, essentially because, in C-induction, human cognitive biases inevitably work in the learner's favor, since those biases are shared with other people with whom coordination is to be achieved. In N-induction, the aim is to predict Nature—and here, our cognitive biases will often be an unreliable guide.

Language acquisition is a paradigm example of C-induction. There is no human-independent "true" language to which learners aspire. Rather, today's language is the product of yesterday's learners; and hence language acquisition requires *coordinating* with those learners, present and past. What is crucial is not *which* phonological, syntactic, or semantic regularities children prefer when confronted with linguistic data, it is that they prefer the *same* linguistic regularities as each other—each generation of learners needs only to follow in the footsteps of the last. And in doing so, the members of each generation will follow the *same* path, and will be able to coordinate successfully not only with previous generations, but with each other.

Note that the existence of very strong cognitive biases is evident across a wide range of learning problems—from categorization, to series completion, to coordinating a meeting in New York City (or anywhere else). Thus, the mere existence of strong biases in no way provides evidence for a dedicated innate "module" embodying such biases. From this point of view, a key research question concerns the nature of the biases that influence language acquisition—these biases will help explain the structures that are, or are not, observed in the world's languages. Moreover, the *stronger* these biases (e.g., flowing from the interaction of perceptuo-motor factors, cognitive limitations on learning and processing, and constraints from thought and pragmatics, as described in chapter 2), the *greater* the constraints on the space of possible languages, and hence the *easier* the problem of language acquisition.

Language and other cultural phenomena are evolving systems, and one of the most powerful determinants of which linguistic or cultural patterns are invented, propagated or stamped out, is how readily those patterns are learned and processed. Hence, the learnability of language, or other cultural structures, is not a puzzle demanding the presence of innate information, but rather an inevitable consequence of the process of the incremental creation of language, and culture more generally, by successive generations (Christiansen, 1994; Deacon, 1997; Kirby & Hurford, 2002; Zuidema, 2003).

The first implication we have drawn from the distinction between C- and N-induction is that C-induction is dramatically easier than N-induction. But there is a second important implication, concerning the feasibility of the biological adaptation of specific inductive biases, i.e., whether genetically-encoded domain-specific modules could have arisen through Darwinian selection. This possibility looks much more plausible for problems of N-induction than for C-induction. We have considered this issue in some detail in the previous chapter, but in the context of understanding language acquisition it deserves a second look.

Many aspects of the natural world are fairly stable. Thus, across long periods of evolutionary time, there is little change in the low-level statistical regularities in visual images (Field, 1987), in the geometric properties of optic flow, stereo, or structure-from-motion (Ullman, 1979), or in the coherence of external visual and auditory "objects" (e.g., Bregman, 1990). These aspects of the environment therefore provide a stable selectional pressure over which natural selection can operate—often over times scales of tens or hundreds of millions of years. Just as the sensory and motor apparatus is exquisitely adapted to deal with the challenges of the natural environment, so it is entirely plausible that the neural and cognitive machinery required to operate this apparatus is equally under genetic control, at least to a substantial degree (e.g., Crowley & Katz, 1999). Indeed, in many organisms, including mammals, much complex perceptual-motor behavior is functioning within hours of birth—a newborn giraffe, for example, takes only hours before it first staggers unsteadily to its feet, and it is soon walking and feeding. By contrast, perceptuo-motor function appears to be considerably delayed in human infancy, but it is nonetheless entirely plausible that some innate neural structures are conserved, or perhaps even elaborated, in humans. More broadly, it is at least prima facie plausible that biases regarding many problems of N-induction might be established by natural selection.

Consider, by contrast, the case of C-induction. While the natural world is stable, the behaviors on which people coordinate are typically *un*stable. The choice of meeting place in New York City will, clearly, depend on contingent historical and cultural factors; but, more importantly, cultural and linguistic conventions are in general highly labile—recall that the entire Indo-European language group has diverged in about 9,000 years (Gray & Atkinson, 2003). Moreover, focal points on which people converge can emerge very rapidly during an experiment, e.g., different pairs of participants rapidly develop one of a wide range of classifications in a task involving novel tangrams (Clark & Wilkes-Gibbs, 1986); and complex patterns of conventions can arise very rapidly in the emergence of languages. For example, as we noted in the last chapter, the rich conventions of an entire language, Nicaraguan sign language, was spontaneously created in less than three decades by deaf children with little exposure to established languages (Senghas, Kita, & Özyürek, 2004). This reinforces our point from chapter 2 that the often-invoked analogy between the evolution of vision and language (e.g., Pinker & Bloom, 1990; Pinker & Jackendoff, 2009) breaks down. Specifically, one crucial disanalogy is that our understanding of the perceptual world is primarily a problem of N-induction, whereas learning a language is a problem of C-induction (as discussed in box 3.1, the same issue arises for the limited notion of UG implied by the Minimalist Program).

To summarize, C-induction involves learning what others will do; but what others will do is highly variable and, crucially, changes far more rapidly than genetic change. Suppose that a particular set of cultural conventions is in play (a specific language, or a particular religious or moral code). Learners with an inductive bias that, by chance, makes these conventions particularly easy to acquire will be favored. But as discussed in chapter 2, there is no opportunity for those innate biases to spread throughout the population, because long before substantial natural selection can occur, those conventions will no longer apply, and a bias to adopt them will, if anything, be likely to be a disadvantage (Baronchelli, Chater, Christiansen & Pastor-Satorras, 2013; Chater, Reali, & Christiansen, 2009). Hence, Darwinian selection will favor agents that are generalists—i.e., who can adapt to the changing cultural environment (Baronchelli, Chater, Pastor-Satorras, & Christiansen, 2012). Rapid cultural evolution (e.g., fast-changing linguistic, moral, or social systems) will automatically lead to a fit between culture and learners—because cultural patterns can only be created and propagated if they are easy to learn and use. But cultural evolution can work *against* biological (co)evolution in the case of malleable aspects of culture—rapid cultural change leads to a fast-changing cultural environment, which serves as a "moving target" to which biological adaptation cannot occur (cf. Ancel, 1999).

Box 3.1
On the Possibility of a Minimal Universal Grammar

As we noted in chapter 1, the generativist concept of Universal Grammar (UG) as a genetic endowment of language-specific principles or constraints has recently been in a state of flux, ranging from Principles and Parameters Theory (e.g., Crain, Goro, & Thornton, 2006; Crain & Pietroski, 2006), to Simpler Syntax (Culicover & Jackendoff, 2005; Pinker & Jackendoff, 2009), to the Minimalist Program (e.g., Boeckx, 2006; Chomsky, 1995, 2011). It might be objected, in light of the minimalist program in linguistics, that only a very modest biological adaptation specific to language—recursion—may be required (Hauser, Chomsky & Fitch, 2002).

This response appears to fall on the horns of a dilemma. On the one hand, if UG consists only of the operation of recursion (Merge), then traditional generativist arguments concerning the poverty of the stimulus, and the existence of language universals, have been greatly exaggerated—and indeed, an alternative, non-UG-based explanation of the possibility of language acquisition and the existence of putative language universals, is required. This position, if adopted, seems to amount to a complete retraction of the traditional generativist position (as argued in Pinker & Jackendoff, 2005). On the other hand, if the argument from the poverty of the stimulus is still presumed to hold good, with its implication that highly-specific regularities such as the binding constraints (see section 3.3) must be part of an innate UG, then there is the problem that the probability of such complex, arbitrary systems of constraints arising by chance is vanishingly small. To be sure, the minimalist explanation of many linguistic regularities is based on the recursive operation Merge—but, in reality, explanations of specific linguistic data require drawing on extensive and highly abstract linguistic machinery, which goes far beyond simple recursion (Adger, 2003; Boeckx, 2006). This is also the case for the account of binding offered in Reuland (2011) to which Chomsky (2011) referred in response to this dilemma (first posed in Chater & Christiansen, 2010). However, the postulation of both a derivational level (*Narrow Syntax*) and a structural relation level (*Logical Syntax*) as well as the use of a so-called *Theta System* would seem to involve highly complex, language-specific mechanisms that are unlikely to be the result of general laws of nature or external data (cf. Chomsky, 2011), leaving UG on the second horn of the dilemma.

There has, indeed, been extensive computational and mathematical analysis of the process of cultural evolution, including some models of language change (e.g., Batali, 1998; Richerson & Boyd, 2005; Kirby, Dowman, & Griffiths, 2007; Hare & Elman, 1995; Nettle, 1999; Niyogi, 2006; Nowak, Komarova, & Niyogi, 2001—for reviews, see Smith, 2014; chapter 2). Learning or processing constraints on learners provide one source of constraint on how such cultural evolution proceeds. Under some restricted conditions, learning biases specify a "fixed" probability distribution of linguistic/cultural forms, from which cultural evolution can be viewed as sampling (Griffiths & Kalish, 2005). In the general case, though, historical factors can also be crucially important—once a culture/language has evolved in a particular direction, there may be no way to reverse the process. This observation seems reasonable in light of numerous unidirectional clines observed in empirical studies of language change (Comrie, 1989; Hopper & Traugott, 2003). For example, lexical items often follow the direction:

Content word → grammatical word → clitic → derivational affix

This has been observed for English, where 'will' has shifted from being a content word expressing intention, to a grammatical auxiliary verb indicating future tense; in Spanish, the parts of a grammatical auxiliary verb 'haber' indicating the future have turned into future tense suffixes. Languages rarely, if ever, exhibit changes in the opposite direction.

While arbitrary conventions, in language or other aspects of culture, typically change rapidly, and hence do not provide a stable target upon which biological evolution can operate, there may be important aspects of language and culture that are *not* arbitrary, i.e., for which certain properties have functional advantages. For example, the functional pressure for communicative efficiency might explain why frequent words tend to be short (Zipf, 1949), and the functional pressure to successfully engage in repeated social interactions with the same people may explain the tendency to show reciprocal altruism (Trivers, 1971). Such aspects of culture could potentially provide a stable environment against which biological selection might take place (Christiansen et al., 2011). Moreover, "generalist" genes for dealing with a fast-changing cultural environment may also be selected for (Baronchelli et al., 2012). Thus, it is in principle possible that the human vocal apparatus, memory capacity, and perhaps the human auditory system, might have developed specific adaptations in response to the challenges of producing and understanding speech, although the evidence that this actually occurred is controversial (e.g., Lieberman, 1984; but see also Hauser & Fitch, 2003). But genes encoding aspects of culture that were initially freely varying, and not held constant by functional pressure, could not have arisen through biological evolution, as we have stressed in chapter 2.

While the issues discussed above apply across cognitive domains, we illustrate the pay-off of this standpoint by considering a particularly central aspect of language—binding constraints—which previously has been viewed as especially problematic for non-nativist approaches to language acquisition. We want to stress, though, that we are not proposing a complete theory of binding; rather, drawing together prior work by others, we aim to outline a framework within which such a theory might be developed.

3.3 The Emergence of Binding Constraints

The problem of binding, especially between reflexive and non-reflexive pronouns and noun phrases, has long been a theoretically central topic in generative linguistics (Chomsky, 1981); and the principles of binding appear both complex and arbitrary, as noted in chapter 2. Binding theory is thus a paradigm case of the type of information that has been proposed to be part of an innate

UG (e.g., Crain and Lillo-Martin, 1999; Reuland, 2008), and provides a challenge for theorists who do not assume the existence of UG. As we illustrate, however, a range of alternative approaches provides promising starting points for understanding binding as arising from domain-general factors. If such approaches can make substantial in-roads into the explanation of key binding phenomena, then the assumption that binding constraints are arbitrary language universals, and must arise from an innate UG, is undermined. Indeed, according to the latter explanation, apparent links between syntactic binding principles and pragmatic factors must presumably be viewed as mere coincidences (and coincidences that would seem, prima facie, to be astonishingly unlikely to arise by chance), rather than as originating from the "fossilization" of pragmatic principles into syntactic patterns by processes such as grammaticalization (Hopper & Traugott, 2003).

The principles of binding capture patterns of use of, among other things, reflexive pronouns (e.g., *himself*, *themselves*) and accusative pronouns (e.g., *him*, *them*). Consider the following examples, where subscripts indicate coreference and asterisks indicate ungrammaticality.

(1) That John_i enjoyed himself_i /*him_i amazed him_i/*himself_i.
(2) John_i saw himself_i/*him_i/*John_i
(3) *He_i/he_j said John_i won

Why is it possible for the first, but not the second, pronoun to be reflexive, in (1)? According to generative grammar, the key concept here is *binding*. Roughly, a noun phrase *binds* a pronoun if it *c-commands* that pronoun and they are co-referring. Employing the familiar analogy between linguistic and family trees, we can define c-command as follows: an element c-commands its siblings and all their descendants. A noun phrase, NP, *A-binds* a pronoun if it binds it and, roughly, if the NP is in either subject or object position. Now we can state simplified versions of Chomsky's (1981) three binding principles:

Principle A. Reflexives must be A-bound by an NP
Principle B. Pronouns must not be A-bound by an NP
Principle C. Full NPs must not be A-bound

Informally, Principle A says that a reflexive pronoun (e.g., *herself*) must be used, if co-referring to a "structurally nearby" item (defined by c-command), in subject or object position. Principle B says that a non-reflexive pronoun (e.g., *her*) must be used otherwise. These principles explain the pattern in (1) and (2). Principle C rules out coreference such as (3). *John* cannot be bound to *he*. For the same reason, *John likes John*, or *the man likes John* do not allow coreference between subject and object.

Need the apparently complex and arbitrary principles of binding theory be part of the child's innate UG? Or can these constraints be explained as a product of more basic perceptual, cognitive, or communicative constraints? One suggestion, due to O'Grady (2005), considers the possibility that binding constraints may in part emerge from processing constraints (we will discuss language processing in much more detail in chapter 4). Specifically, he suggests that the language processing system seeks to resolve linguistic dependencies (e.g., between verbs and their arguments) at the first opportunity. This tendency might not be specific to syntax but, rather, might be an instance of a general cognitive tendency to resolve ambiguities rapidly in linguistic (Clark, 1975) and perceptual input (Pomerantz & Kubovy, 1986). The use of a reflexive is assumed to signal that the pronoun corefers with an available NP, given a local dependency structure. Thus, in processing (1), the language system reaches *That John enjoyed himself…* and creates the first available dependency relationship between *enjoyed*, *John*, and *himself*. The use of the reflexive *himself* signals that coreference with the available NP *John* is intended (cf. Principle A). With the dependencies now resolved, the internal structure of the resulting clause is closed off and the language system moves on: *[That [John enjoyed himself]] surprised him/*himself.* The latter *himself* is not possible because there is no appropriate NP available to connect with—the only NP is [*that John enjoyed himself*], which is used as an argument of *surprised*, but which clearly cannot corefer with the later use of *himself*. But in *John enjoyed himself*, *John* is available as an NP when *himself* is encountered.

O'Grady (2015) stresses that his processing account of coreference does not imply that the nearest referent, in terms of the linear order of the sentence, is the preferred coreferent for a reflexive pronoun. He argues that "interpretation of reflexive pronouns is engineered by the processing operations that create the mapping between form and meaning in the real-time course of language use. This is very different from a system in which coreference is determined by scanning a string of words and selecting the linearly closest antecedent" (O'Grady 2015, p. 107). So, for example, in *the sister of the bride thought herself beautiful*, the reflexive *herself* does not refer to the most recent referent in the sentence (*the bride*). Instead, as *the sister of the bride* is incrementally processed, the referent of this noun phrase is immediately computed as part of the semantic representation, and is taken as the argument of the main verb (*thought*); it is the referent of *the sister of the bride*, not the referent of *the bride*, that is therefore available for coreference from the reflexive, when it is encountered. O'Grady (2015) shows how this approach deals with apparent counterexamples to processing-based accounts of binding (e.g., Chomsky, 2011). Specifically, these critiques incorrectly presuppose that processing accounts merely focus on linear order with no access to semantic structure.

We have so far considered reflexive pronouns. By contrast, non-reflexive pronouns, such as *him*, are used in roughly complementary distribution to reflexive pronouns (cf. Principle B). It has been argued that this complementarity may originate from pragmatic principles (Levinson, 1987; Reinhart, 1983). Specifically, given that the use of reflexives is highly restrictive, they are, where appropriate, highly informative. Hence, by not using them, the speaker signals that coreference is not appropriate. The phenomenon by which a more informative form pre-empts another through pragmatic forces occurs widely in language. So for example, given that a person could have said *Everyone at the party ate cake*, the fact that they actually said *Some people at the party ate cake* will typically be taken by listeners to indicate that some people did not eat cake (or at least that the speaker cannot rule out this possibility), otherwise the speaker would have used the more specific *everyone*. This is an example of inference according to Grice's (1975) *maxim of quality*: speakers aim to be as informative as possible, and listeners assume that this is so. Similarly, then, if the highly specific reflexive is not used, then listeners will assume that it could not be used, so that the non-reflexive pronoun must be a complementary distribution of possible referents.

So there may be powerful *pragmatic* constraints on anaphor resolution. And such pragmatic constraints, while perhaps originally optional, may lead to frozen patterns of discourse; and what begin as discourse constraints may generalize to apply within the sentence and, ultimately, may become obligatory syntactic constraints. Pursuing this line of thought, Levinson (2000) notes that the patterns of discourse anaphora (4) and syntactic anaphora (5) have interesting parallels.

(4) a. John arrived. He began to sing.
b. John arrived. The man began to sing.
(5) a. John arrived and he began to sing.
b. John arrived and the man began to sing.

In both (4) and (5), the first form (a) indicates preferred coreference of *he* and *John*; the second form (b) prefers non-coreference. The general pattern is that brief expressions encourage coreference with a previously introduced item. Drawing on pragmatic theory, we note that Grice's (1975) *maxim of quantity*—that, by default, an overly verbose expression will not be used where a brief one would suffice—implies that more prolix expressions are typically taken to imply non-coreference with previously introduced entities. Where the referring expression is absent, then coreference may be required as in (6), in which the singer can only be John:

(6) John arrived and began to sing.

It is natural to assume that syntactic structures emerge diachronically from reduction of discourse structures—reflecting Givón's (1979) hypothesis that today's syntax is yesterday's discourse.[2] The shift, over time, from default constraint to rigid rule is wide-spread in language change and much studied in the sub-field of grammaticalization (Hopper & Traugott, 2003; see also chapters 2 and 4). Levinson (2000), building on related work by Reinhart (1983), provides a comprehensive account of the binding constraints, and putative exceptions to them, purely on pragmatic principles (see also Huang, 2000, for a cross-linguistic perspective).

Finally, simple cases of Principle C can be explained by similar pragmatic arguments. To use the sentence *John sees John* (see (2) above), where the object can, in principle, refer to any individual named John, would be pragmatically infelicitous if coreference were intended, because the speaker should instead have chosen the more informative *himself* in object position. O'Grady (2005) and Reinhart (1983) consider more complex cases related to Principle C in terms of a processing bias toward so-called *upward feature-passing*, though we do not consider this further in this book.

As should be clear, the linguistic phenomena involved in binding are extremely complex, and not fully captured by *any* theoretical account.[3] Yet it seems quite possible that the complexity of the binding constraints arises from subtle interactions of *multiple* constraints, including from processing, pragmatics, and semantics. For example, Culicover and Jackendoff (2005) have recently argued that many aspects of binding may be semantic in origin (see also Culicover, 2013b). Thus, *John found himself in the school photograph* may be an acceptable sentence in English due to semantic principles concerning representation (the photograph is a representation of John, and many other people), rather than any syntactic factors.

Note, too, that we can say *Looking up, Tiger was delighted to see himself at the top of the leaderboard,* where the reflexive refers to the name "Tiger," not Tiger himself. And violations appear to go beyond mere representation—e.g., *After a wild tee-shot, Rory found himself in a deep bunker*, where the

2. Although Givón (1979) discussed how syntactic constructions might derive from previous pragmatic discourse structure, he did not coin the phrase "today's syntax is yesterday's discourse." Instead, it has been ascribed to him via paraphrasings of his maxim that "today's morphology is yesterday's syntax" from Givón (1971), an idea he attributes to the Chinese philosopher Lao Tse.

3. Indeed, the minimalist program (Chomsky, 1995) has no direct account of binding, but relies on the hope that the principles and parameters framework, in which binding phenomena have been described, can eventually be reconstructed from a minimalist point of view. Reuland (2011) may be seen as an important step in this direction (e.g., Chomsky, 2011), seeking to replace binding theory by more general principles—though see box 3.1 on reservations about whether this solution can be seen as relying only on machinery that is not specific to language.

reflexive here refers to his golf ball. More complex cases involving pronouns and reflexives are also natural in this type of context, e.g., *Despite his*$_{\text{(i Tiger's)}}$ *mis-cued drive, Angel*$_{\text{j}}$ *still found himself*$_{\text{(j's golf ball)}}$ *ten yards behind him*$_{\text{(i's golf ball)}}$. There can, of course, be no purely syntactic rules connecting golfers and their golf balls, and presumably no general semantic rules either, unless such rules are presumed to be sensitive to the rules of golf (among other things, that each player has exactly one ball). Rather, the reference of reflexives appears to be determined by pragmatics and general knowledge—e.g., we know from context that a golf ball is being referred to, that golf balls and players stand in one-to-one correspondence, and hence that picking out an individual could be used to signal the corresponding golf ball.

The very multiplicity of constraints involved in the shaping of language structure, which arises naturally from the present account, may be one reason why binding is so difficult to characterize in traditional linguistic theory. But the sheer number of constraints does not pose any challenges for the child because these constraints are the very constraints with which the child is equipped. If learning the binding constraints were a problem of N-induction (e.g., if the linguistic patterns were drawn from the language of intelligent aliens, or deliberately created as a challenging abstract puzzle), then learning would be extraordinarily hard. But it is not: it is a problem of C-induction. To the extent that binding can be understood as emerging from a combination of processing, pragmatic, or other constraints, operating on past generations of learners, then binding will be readily learned by the new generations of learners, who will necessarily embody those same constraints.

If binding constraints arise from the interaction of a multiplicity of constraints, one might expect that binding principles across historically unrelated languages would show strong family resemblances (they might, in essence, be products of cultural *coevolution*), rather than being strictly identical, as is implicit in the claim that binding principles are universal across human languages. And indeed, it turns out that the binding constraints, like other apparently "strict" language universals, may not be universal at all, when a suitably broad range of languages is considered (e.g., Evans & Levinson, 2009). Thus, Levinson (2000) notes that, even in Old English, the equivalent of *He saw him* can optionally allow coreference (apparently violating Principle B) (see also Reuland, 2005, for an analysis of such cross-linguistic cases from a viewpoint closer to a classical generative framework). Counterexamples to binding constraints, including the semantic/pragmatic cases outlined above, can potentially be fended off, by introducing further theoretical distinctions, but such moves risk stripping the claim of universality of any empirical bite (Evans and

Levinson, 2009). If we take cross-linguistic data at face value, the pattern of data seems, if anything, more compatible with the present account—that binding phenomena result from the operation of multiple constraints during the cultural evolution of language—than with the classical assumption that binding constraints are a rigid part of a fixed UG, ultimately rooted in biology.

To sum up: binding has been seen as paradigmatically arbitrary and specific to language, and the learnability of binding constraints has been viewed as requiring a language-specific UG. Any apparent relations to processing, semantics, and pragmatics as discussed above are presumably viewed as nothing more than a misleading, if astonishingly improbable, coincidence. If the problem of language learning were a matter of N-induction—i.e., if the binding constraints were merely a human-independent aspect of the natural world—then this viewpoint might be persuasive. But language learning is a problem of C-induction—people have to learn the *same* linguistic system as each other. Hence, the patterns of linguistic structure will themselves have adapted, through processes of cultural evolution, to be easy to learn and process, or, more broadly, to fit with the multiple perceptual, cognitive, and communicative constraints governing the adaptation of language. From this perspective, binding is, in part, determined by innate constraints, but those constraints pre-date the emergence of language (de Ruiter & Levinson, 2008).

In the domain of binding, as elsewhere in linguistics, this type of cultural evolutionary story is, of course, incomplete—though to no greater degree, arguably, than is typical in evolutionary explanations in the biological sciences. We suggest, however, that viewing language as a cultural adaptation provides a powerful and fruitful framework within which to explore the evolution of linguistic structure and its consequences for language acquisition.

3.4 Broader Implications for Language Acquisition and Development

So far in this chapter, we have stepped back from language acquisition to argue that only by viewing language acquisition through the lens of language evolution can we fully understand the nature of the task that children are facing during acquisition. Language acquisition fundamentally involves C-induction, and not N-induction as assumed by proponents of UG. We also briefly sketched how this perspective might be applied to re-evaluate a key puzzle for language acquisition: the emergence of the binding constraints, which has traditionally been interpreted as providing strong support for the existence of an innate UG. We now turn our attention to some of the broader implications of our approach for language acquisition and human development.

3.4.1 The Logical Problem of Language Acquisition Reconsidered

We have argued that viewing the evolution of language as the outcome of cultural, rather than biological, evolution (and as a problem of C-induction, rather than N-induction) leads us to consider language acquisition in an entirely new light. The ability to develop complex language from what appears to be such poor input has traditionally led many to speak of the "logical" problem of language acquisition (e.g., Baker & McCarthy, 1981; Hornstein & Lightfoot, 1981). One solution to the problem is to assume that learners have some sort of biological "head-start" in language acquisition—that their learning apparatus is precisely meshed with the structure of natural language. This viewpoint is, of course, consistent with theories according to which there is a genetically specified language module, language organ, or language instinct (e.g., Chomsky, 1986; Crain, 1991; Piattelli-Palmarini, 1989; Pinker, 1994; Pinker & Bloom, 1990). But if we view language acquisition as a problem of C-induction, then the learner's objective is merely to follow prior learners, and hence the patterns in language will inevitably be those that are most readily learnable. It is not that people have evolved to learn language; rather, language has evolved to fit the multiple constraints of our learning and processing abilities.

Whatever learning biases people have, so long as these biases are *shared* across individuals, learning should proceed successfully. Moreover, the viewpoint that children learn language using general-purpose cognitive mechanisms, rather than language-specific mechanisms, has also been advocated on independent grounds (e.g., Bates & MacWhinney, 1979, 1987; Chater, Alexander Clark, Goldsmith, & Perfors, 2015; Deacon, 1997; Elman et al., 1996; Monaghan & Christiansen, 2008; Seidenberg & MacDonald, 2001; Tomasello, 2000, 2003).

This alternative characterization of language acquisition additionally offers a different perspective on linguistic phenomena that have typically been seen as requiring a UG account for their explanation, such as specific language impairment (SLI) and creolization. These phenomena are beyond the scope of this chapter, so we can only sketch how they may be approached. For example, the acquisition problems in SLI may, on our account, be largely due to deficits in underlying sequence learning mechanisms that support language (see Ullman & Pierpont, 2005, for a similar perspective, and box 6.3 in chapter 6), rather than impaired language-specific modules (e.g., M. Gopnik & Crago, 1991; Pinker, 1994; Van der Lely & Battell, 2003). Consistent with this perspective, recent studies have shown that children and adults with SLI have impaired sequence learning abilities (e.g., Evans, Saffran, & Robe-Torres, 2009; Hsu, Tomblin, & Christiansen, 2014; Tomblin, Mainela-Arnold, & Zhang, 2007).

Although processes of creolization, in which children acquire consistent linguistic structure from noisy and inconsistent input, have been seen as evidence of UG (e.g., Bickerton, 1984), we suggest that creolization may be better construed as arising from cognitive constraints on learning and processing. As discussed in chapter 2, the rapid emergence of a consistent SOV word order in the Al-Sayyid Bedouin Sign Language (Sandler, Meir, Padden, & Aronoff, 2005) is consistent with this suggestion. Additional research is required to flesh out these accounts in detail, but a growing bulk of work indicates that such general learning-based accounts are indeed viable (e.g., Chater & Vitányi, 2007; Goldberg, 2006; Hudson Kam & Newport, 2005; O'Grady, 2005; Reali & Christiansen, 2005; Tomasello, 2003).

3.4.2 On the Nature of Language Universals

It has been widely argued that an innate UG must be postulated to explain two key observations: first, that children acquire language so readily from an apparently impoverished linguistic input (the poverty of the stimulus argument); and second, that languages share supposedly universal patterns, which appear arbitrary from a functional, communicative point of view. In this chapter, we have provided an alternative perspective on the first observation: language is relatively easy for children to acquire not because of an innate UG, but rather because language itself has been adapted by previous generations of learners to fit exactly the (largely non-linguistic) biases that children bring to the task of language acquisition. But what about the second observation? That is, how can we explain the existence of apparent language universals, that is, regularities in language structure and use?

The very starting point for the discussion is not, of course, straightforward; it is by no means clear exactly what counts as a language universal. Rather, the notion of language universals differs considerably across language researchers (e.g., as exemplified by the variety in perspectives among contributions in Christiansen, Collins, & Edelman, 2009). Many linguists working within the generative grammar framework see universals as primarily, and some times exclusively, deriving from UG (e.g., Hornstein & Boeckx, 2009; Pinker & Jackendoff, 2009). Functional linguists, on the other hand, view universals as arising from patterns of language usage due to pragmatic, processing, and other constraints, and amplified in diachronic language change (e.g., Bybee, 2009). Moreover, even within the same theoretical linguistic framework, there is often little agreement about what the exact universals are. For example, when surveying specific universals proposed by different proponents of UG, Tomasello (2004) found little overlap between proposed universals.

Although there may be little agreement about specific universals, some consensus can nonetheless be found with respect to their general nature. Thus, within mainstream generative grammar approaches (including MP and PPT), language universals are seen as arising from the inner workings of UG. Hornstein and Boeckx (2009) refer to such UG-based universals as internalist or I-Universals. They note that:

> on this conception I-Universals are likely to be (and have been found to be) quite abstract. They need not be observable. Thus, even were one to survey thousands of languages looking for commonalities, they could easily escape detection. In this they contrast with Greenbergian Universals, which we would call E(xternalist)-Universals. In fact, on this conception, the mere fact that every language displayed some property P does not imply that P is a universal in the I-sense. Put more paradoxically, the fact that P holds universally does not imply that P is a universal. Conversely, some property can be an I-Universal even if only manifested in a single natural language. The only thing that makes something an I-Universal on this view is that it is a property of our innate ability to grow a language (p. 81).

Thus, from the perspective of MP and PPT, language universals are by definition properties of UG; that is, they are what have been termed *formal* universals (Chomsky, 1965; Katz & Postal, 1964). A similar view of universals also figures within the SS framework (Culicover & Jackendoff, 2005), defined in terms of the universal toolkit encoded in UG. Because different languages are hypothesized to use different subsets of tools, the SS approach—like MP and PPT—suggests that some universals may not show up in all languages (Pinker & Jackendoff, 2009). However, both notions of universals face the logical problem of language evolution discussed in chapter 2: how could the full set of UG constraints have evolved if any single linguistic environment only ever supported a subset of them?

The solution to this problem, we suggest, is to adopt a non-formal conception of universals in which they emerge from processes of repeated language acquisition and use. We see universals as products of the interaction between constraints deriving from the four sources we described in chapter 2: the nature of thought, perceptuo-motor factors, cognitive limitations on learning and processing, and pragmatic considerations. This view implies that most universals are unlikely to be found across all languages; rather, "universals" are more akin to statistical trends tied to patterns of language use. Consequently, specific candidate universals fall on a continuum, ranging from being attested only in some languages to being found across most languages. An example of the former is the universal that verb-final languages *tend* to have postpositions (Dryer, 1992), whereas the presence of nouns and verbs (minimally as typological prototypes; Croft, 2001) in most, though perhaps not all (Evans & Levinson, 2009), languages is an example of the latter.

Individual languages, on our account, are seen as evolving under the pressures from multiple constraints deriving from the brain, as well as cultural-historical factors (including language contact and sociolinguistic influences), resulting over time in the breathtaking linguistic diversity that characterizes the roughly 7,000 currently existing languages (see also Dediu et al., 2013). Languages variously employ tones, clicks, or manual signs to signal differences in meaning; some languages appear to lack the noun-verb distinction (e.g., Straits Salish), whereas others have a proliferation of fine-grained syntactic categories (e.g., Tzeltal); some languages do without morphology (e.g., Mandarin), while others pack a whole sentence into a single word (e.g., Cayuga). Cross-linguistically recurring patterns do emerge due to similarity in constraints and culture/history, but such patterns should be expected to be probabilistic tendencies, not the rigid properties of UG (Christiansen & Chater, 2008). From this perspective it seems unlikely that the world's languages will fit within a single parameterized framework (e.g., Baker 2001), and more likely that languages will provide a diverse, and somewhat unruly, set of solutions to a hugely complex problem of multiple constraint satisfaction, as appears consistent with research on language typology (Comrie, 1989; Evans, 2013; Evans & Levinson 2009). Thus, we construe recurring patterns of language along the lines of Wittgenstein's (1953) notion of "family resemblance": although there may be similarities between pairs of individual languages, there is no single set of features common to them all (see box 3.2).

3.4.3 Cultural Evolution Meets Evolutionary Psychology

How far do these arguments generalize from language acquisition to the development of the child's knowledge of culture more broadly? How far might this lead to a new perspective in evolutionary psychology, in which the fit between the brain and cultural forms is not explained in terms of domain-specific modules, but by the shaping of cultural forms to pre-existing biological machinery?

Human development involves the transmission of an incredibly elaborate culture from one generation to the next. Children acquire not just language, of course, but also lay theories and concepts about the natural, artificial, and psychological worlds, social and moral norms, a panoply of practical lore, skills, and modes of expression, including music, art, and dance. The absorption of this information is all the more remarkable given that so much of it appears to be acquired incidentally, rather than being a topic of direct instruction.

As with language, it might initially seem unclear how this astonishing feat of learning is accomplished. One natural line of explanation is to assume that

Box 3.2
Family Resemblance: Chairs and Language Universals

In contrast to animal communication systems, diversity is characteristic of almost every aspect of human language (Dediu et al., 2013; Evans & Levinson, 2009). Yet there are, of course, also commonalities among languages, often referred to as language universals. We argue that cross-linguistic regularities among languages are best construed as "family resemblances" (Wittgenstein, 1953), rather than as rigid formal universals (Chomsky, 1965).

Chairs provide a good illustration of what we mean by family resemblance. They come in all shapes and sizes, often "adapted" to local conditions and particular functions (dentist's chair, tennis referee chair, wheelchair, etc.). Intuitively, it might seem straightforward to come up with a set of defining features for a chair, but on reflection it is far from easy. Although many chairs have legs, this is not true of all chairs: beanbag chairs don't have legs, and neither does the hanging egg chair, the dentist's chair or the wheelchair. Similarly, many chairs have backs but not all (indeed so-called "backless" chairs have become increasingly popular as a way to improve posture while sitting). And, of course, not all chairs are comfortable to sit on as comically illustrated in the waiting room scene from Jacques Tati's 1967 movie *Playtime*, in which Monsieur Hulot tries in vain to make himself comfortable in the awkwardly low, minimalist chairs.

If we, following Wittgenstein, were to line all types of chairs up in a row according to similarity, each pair of chairs would overlap in multiple features (such as having four legs or a back) but the chairs at the two endpoints would have no single feature in common. Likewise, if we were to write a list of all languages, in which languages were placed next to one another according to how similar they were in terms of their phonological, morphological, and syntactic properties, we would get the same result as with our chair line-up: no single feature might apply to all languages even though specific subsets of languages may have many overlapping properties. But of course most, or perhaps even all, chairs and languages do have *some* features in common—chairs have a mass, are not made of gas, exist in space and time (unlike, say, numbers). And natural languages allow an infinite number of messages to be conveyed by a finite number of components, have a number of levels of organization, and so on. Yet these features are very different from the abstract and arbitrary features postulated in UG.

there is a close fit between the cultural information to be transmitted, and the prior assumptions of the child, whether implicit or explicit. The strongest form of this position is that some, and perhaps the most central, elements of this information is actually innately "built-in" to each learner—and hence that cultural transmission is constructed over a skeleton of genetically fixed constraints (e.g., Hauser, 2006). Generalizing from the case of UG, some evolutionary psychologists have likened the mind to a Swiss Army Knife, consisting of a variety of special-purpose tools (Barkow, Cosmides, & Tooby, 1992). The design of each of these special-purpose tools is presumed to have arisen through biological selection. More broadly, the key suggestion is that there is a close mesh between genes and culture, and that this mesh helps explain how cultural complexity can successfully be transmitted from generation to generation.

The processes by which any assumed connection between genes and culture might arise are central to the study of human development, and understanding such processes is part of the wider project of elucidating the relationship between biological and cultural explanation in psychology, anthropology, and throughout the neural and social sciences. But here we wish to take a wider view of these familiar issues from the point of view of historical *origins*: how did the mesh between genes and culture arise?

The origin of a close mutual relationship between any two systems raises the question: which came first? A natural line, in considering this type of problem, is to consider the possibility of coevolution—and hence that the claim that one, or the other, must come first is misleading. As we have argued in chapter 2, in the case of genes and language, the conditions under which such coevolution can occur are surprisingly limited; but the same issues arise in relation to the supposed coevolution of genes and any cultural form. Let us now broaden the argument, and consider the two clear-cut options: that culture comes first, and biological adaptation brings about the fit with cultural structure; or that biological structures come first, and cultural adaptation brings about the fit with these biological structures. As a short hand, let us call these the *biological evolution* and *cultural evolution* perspectives.

How might the biological story work? If cultural conventions have a particular form, then people within that culture will, we may reasonably assume, have a selective advantage if they are able to acquire those conventions rapidly and easily. So, for example, suppose that human cultures typically (or even always) fit some specific moral, social, or communicative pattern. Hence, children who are able to rapidly learn these constraints will presumably have a selective advantage. Thus, it is possible that, after a sufficiently long period of biological adaptation to an environment containing such constraints, learners who are genetically biased to conform to those constraints would be favored by natural selection. These learners might then be able to acquire these constraints from very little cultural input; and, at the extreme, learners might be so strongly biased that they require no cultural input at all (this is, of course, the Baldwin [1896] effect that we discussed in chapter 2; see also, Weber & Depew, 2003, for discussion).

If, though, we assume that genetic (or more generally biological) structure is *developmentally* prior (i.e., that learners acquire their culture via domain-specific genetic constraints, adapted to cultural patterns), then it appears that culture must be *historically* prior. After all, the cultural structure (e.g., the pattern of specific regularities) provides the key aspect of the environment to which genes have adapted. Thus, if a genetically-specified and domain-specific system containing specific cultural knowledge has arisen through Darwinian processes of selection, then such selection appears to require a pre-existing cultural environment, to which biological adaptation occurs. However, this conclusion is in direct contradiction to the key assumption of the biological approach, because it presupposes that the cultural forms do *not* arise from biological constraints, but pre-date them. If cultural patterns must pre-date biological constraints in order to provide the putative selectional pressures that are then supposed to entrench such biological constraints, then the reason to postulate such biological constraints almost entirely evaporates.[4]

But it is clear, in light of the arguments above, that there is an alternative evolutionary approach: that *biological* structure is taken to be prior, and that it is cultural forms that adapt, through processes of cultural transmission and variation (e.g., Boyd & Richerson, 2005; Mesoudi, Whiten, & Laland, 2006) to fit biological structure as well as possible. Specifically, the culture is viewed as shaped by endless variation and winnowing, in which forms and patterns that are readily learned and processed are adopted and propagated, whereas forms that are difficult to learn or process are eliminated. Not merely language, but culture in general, is shaped by the brain, rather than the reverse.

Cultural forms will, of course, also be shaped by functional considerations; just as language has been shaped to support flexible and expressive communication, tool use may have been shaped by efficacy in hunting, flaying, and food preparation. But according to this viewpoint, the fit between learners and culture is underpinned by prior biological "machinery" *that pre-dates that culture, and hence is not itself shaped to deal with cultural problems*. This biological machinery may very well be the product of Darwinian selection, but in relation to pre-existing goals. For example, the perceptuo-motor and planning systems may be highly adapted for processing complex structured sequences (e.g., Byrne & Byrne, 1993); and such abilities may then be coopted as a partial basis for producing and understanding language (Conway & Christiansen, 2001; see also chapter 7). Similarly, the ability to "read" other minds may have developed to deal with elaborate social challenges in societies with relatively little cultural innovation (as in non-human primates), but such mind reading might be an essential underpinning for language, and the development of social and moral rules (Tomasello, Carpenter, Call, Behne, & Moll, 2005).

3.5 Summary

In this chapter, we have argued that one cannot think about language acquisition without placing it in its proper evolutionary context, which allows the

4. Although possible coevolutionary processes between genes and culture may complicate the argument, they do not change the conclusion, as discussed in chapter 2.

operation of cultural as well as biological evolution. That is, our argument involved stepping back from questions concerning the acquisition of language, to take an evolutionary perspective concerning both the biological evolution of putative innate constraints and the cultural evolution of human linguistic communication. Based on an evolutionary analysis, we proposed reconsidering development in terms of two types of inductive problems: N-induction, where the problem involves learning some aspect of the natural world, and C-induction, where the key to solving the learning problem is to coordinate with others.

A key challenge for future research will be to identify the specific biological, cognitive, and social constraints that have shaped the structure of language through cultural transmission, to show how the selection pressures imposed by these constraints lead to specific patterns in the world's languages, and to demonstrate how these constraints can explain particular patterns of language acquisition and processing. If we generalize our evolutionary approach to other aspects of cultural evolution and human development, then similar challenges will also lie ahead in identifying specific constraints, and explaining how these capture cross-cultural patterns in development. Importantly, this perspective on human evolution and development does not view the mind as a blank slate (Pinker, 2003a)—far from it: we need genetic constraints to explain the various patterns observed across phylogenetic and ontogenetic time. Instead, we have argued that there are many biological constraints that shape language and other culturally-based human skills, but that these are unlikely to be domain-specific. Thus, as the late Liz Bates (1999) put it so elegantly in the quote with which we began this chapter, "… the debate … is not about Nature vs. Nurture, but about the 'nature of Nature.'"

4 The Now-or-Never Processing Bottleneck

The baby, assailed by eyes, ears, nose, skin, and entrails at once, feels it all as one great blooming, buzzing confusion …
—William James, 1890, *Principles of Psychology*

Language happens in the here-and-now. Although, as we shall see, this provides one of the most important constraints on language, we rarely notice it during normal language use. Nonetheless, we have all felt the effects of this constraint at one point or another. This is perhaps most obvious when listening to someone speaking an unfamiliar language, which often forces us to re-experience some of the "great blooming, buzzing confusion" that William James (1890) so famously noted is his classic work on psychology. In many cases, it seems that the speaker of the foreign language is talking very fast, making it close to impossible to segment the speech stream into recognizable sounds, let alone figure out where one word ends and another one begins. Some of the difficulty we face can be illustrated intuitively by the following Hindi version[1] of the above William James quote, written in the Devanagari script (but with the white spaces between words removed to better approximate the continuous nature of fluent speech):

जबबच्चादेखनेसुननेसूंघनेऔरस्परशकीसंवेदनाओंकाअनुभवएकसाथकरताहैतोउसकादि
मागबौखलाजाताहै

Of course, to more accurately "simulate" actual spoken Hindi, each character should appear one at a time, at a rate of about ten to fifteen characters per second.

1. We thank Professor Rajeshwari Pandharipande for help with the Hindi translation of the William James quote.

Providing a Romanized version of the Hindi sentence does not improve things much:

Jababaccādēkhanēsunanēsūṅghanēaurasparśakīsanvēdanā’ōṅkāanubhavaēkas āthakaratāhaitōusakādimāgabaukhalājātāhai

Even with mostly familiar letters, segmenting the string into words is still very challenging—especially, when considering that in normal speech the whole string would be produced in about six to eight seconds. But understanding spoken language requires much more than discovering the words. To understand the full meaning of the string, further processing is needed to determine the morphological and grammatical relations that allow for a semantic interpretation of the sentence as well as relevant pragmatic inferences. And this, again, must be done very rapidly before the next sentence begins (or the listener is expected to produce a response).

The fundamental real-time character of language processing challenges us not only when we listen to foreign languages but also when we process our native language(s). For example, if someone was reading this book aloud to you, then you would likely not retain much from passages during which you were somehow distracted (e.g., because someone else was talking to you at the same time). These examples highlight the fleeting nature of language. As we hear a sentence unfold, our memory for preceding material is rapidly lost. Speakers, too, soon lose track of the details of what they have just said. Language processing is therefore *Now-or-Never:* if linguistic information is not processed rapidly, that information is lost for good. Importantly, though, while fundamentally shaping language, the Now-or-Never bottleneck is not language-specific but arises from general principles of perceptuo-motor processing and memory (Christiansen & Chater, in press).

The existence of a Now-or-Never bottleneck is relatively uncontroversial, although its precise character may be debated. However, in this chapter, we argue that the *consequences* of this constraint for language are remarkably far-reaching, touching on issues that include the multi-level organization of language, the prevalence of *local* linguistic dependencies, the incrementality of language processing, the use of prediction in language comprehension, the nature of what is learned during language acquisition, and even the structure of language itself. Thus, we argue that the Now-or-Never bottleneck has fundamental implications for key questions in the language sciences. The consequences of this constraint are, moreover, incompatible with many theoretical positions in linguistic, psycholinguistic, and language acquisition research, as we shall see.

Note, though, that arguing that a phenomenon follows from the Now-or-Never bottleneck does not necessarily undermine alternative explanations of that phenomenon (although it may make some accounts less important). Many phenomena in language may simply be over-determined. For example, we argue that incrementality follows from the Now-or-Never bottleneck. But it is also possible that, irrespective of memory constraints, language understanding would still be incremental on functional grounds, to extract the linguistic message as rapidly as possible. Such counterfactuals are, of course, difficult to evaluate. By contrast, the properties of the Now-or-Never bottleneck arise from basic information processing limitations that are directly testable by experiment. Moreover, the Now-or-Never bottleneck should, we suggest, have methodological priority to the extent that it provides an *integrated* framework for explaining many aspects of language structure, acquisition, processing, and evolution that have previously been treated separately.

4.1 The Now-or-Never Bottleneck

Language input is highly transient. Speech sounds, like other auditory signals, are short-lived. Classic speech perception studies have shown that very little of the auditory trace remains after 100 msec (Elliott, 1962), with more recent studies indicating that much of the information is already lost after just 50 msec (Remez et al., 2010). Similarly, of relevance for the perception of sign language, studies of visual change detection suggest that the ability to maintain visual information beyond 60–70 msec is highly reduced (Pashler, 1988). Thus, sensory memory for language input is quickly overwritten by new incoming information and lost forever *unless* the perceiver is able to process what has been heard or seen right away.

The problem of the rapid loss of the speech or sign signal is further exacerbated by the sheer speed of the incoming linguistic input. At a normal speech rate, speakers produce about ten to fifteen phonemes per second, corresponding to roughly five to six syllables every second or 150 words per minute (Studdert-Kennedy, 1986). However, the resolution of the human auditory system for discrete auditory events is only about ten sounds per second, beyond which the sounds fuse into a continuous buzz (Miller & Taylor, 1948). Even at normal rates of speech, the language system needs to work beyond the limits of the auditory temporal resolution for non-speech stimuli. Remarkably, listeners can learn to process speech in their native language at up to twice the normal rate without much decrease in comprehension (Orr, Friedman, & Williams, 1965). Although the production of signs appears to be slower than the production of speech (at least when comparing the production of ASL signs and spoken English; Bellugi & Fischer, 1972), signed words are still very brief visual events, with the duration of an ASL syllable being about a quarter of a second (Wilbur & Nolkn, 1986).

Making matters even worse, our memory for sequences of auditory input is also very limited. For example, it has been known for more than four decades that naïve listeners are unable correctly to recall the temporal order of just four distinct arbitrary sounds—e.g., hisses, buzzes, and tones—even when they are perfectly able to recognize and label each individual sound in isolation (Warren, Obusek, Farmer, & Warren, 1969). Our ability to recall well-known auditory stimuli is not substantially better, with estimates ranging from 4±1 (Cowan, 2000) to 7±2 (Miller, 1956). A similar limitation applies to visual memory for sign language (Wilson & Emmorey, 2006). We see interference as the primary cause of these memory limitations rather than fixed capacity buffers, because the same patterns of forgetting and memory errors are observed over many timescales (e.g., Brown, Neath, & Chater, 2007). From this perspective, apparent capacity limitations are a side-effect of interference, rather than stemming from, for example, a fixed number of "slots" in memory (see also Van Dyke & Johns, 2012). The resulting poor memory for auditory and visual information, combined with the fast and fleeting nature of linguistic input, is the source of the Now-or-Never bottleneck. If the input is not processed immediately, it will quickly be subject to interference from new information.

Importantly, the Now-or-Never bottleneck is not unique to language but applies to other aspects of perception and action as well. Sensory memory is rich in detail but decays rapidly unless it is further processed (e.g., Cherry, 1953; Coltheart, 1980; Sperling, 1960). Likewise, short-term memory for auditory, visual and haptic information is also limited and subject to interference from new input (e.g., Gallace, Tan, & Spence, 2006; Haber, 1983; Pavani & Turatto, 2008). Moreover, our cognitive ability to respond to sensory input is further constrained in a serial (Sigman & Dehaene, 2005) or near-serial (Navon & Miller, 2002) manner, severely restricting our capacity for processing multiple inputs arriving in quick succession. Similar limitations apply to the production of behavior; the cognitive system cannot plan detailed sequences of movements—a long sequence of commands planned far in advance would lead to severe interference, and would be forgotten before it could be carried out (Cooper & Shallice, 2006; Miller, Galanter, & Pribram, 1960). However, the cognitive system adopts several processing strategies to ameliorate the effects of the Now-or-Never bottleneck on perception and action.

First, the cognitive system engages in *eager processing:* the rich perceptual input is recoded as it arrives to capture the key elements of the sensory information as *economically* and as *distinctively* as possible (e.g., Brown, Neath, & Chater, 2007; Crowder & Neath, 1991); and this must be done rapidly, before the sensory information is overwritten by new input. This is a traditional notion, dating back to early work on attention and sensory memory (e.g., Broadbent, 1958; Coltheart, 1980; Haber, 1983; Sperling, 1960; Treisman, 1964). The resulting compressed—or "chunked"—representations are limited relative to the richness of the sensory input (e.g., Pani, 2000). Evidence from the phenomenon of change blindness suggests that these compressed representations can be very selective (see Jensen, Yao, Street & Simons, 2011, for a review), as exemplified by a study in which half of the participants failed to notice that someone to whom they were giving directions, face-to-face, was surreptitiously exchanged with a completely different person (Simons & Levin, 1998; see box 4.1 for further examples and discussion). Information

Box 4.1
When We Don't See What's Right in Front of Us: Change and Inattentional Blindness

Intuition suggests that we create rich and detailed representations to negotiate our physical environment. However, one of the most astonishing discoveries of modern psychology is that the information that we pick up from our perceptual input may be rather limited and short-lived. For example, the literature on *change blindness* has shown that we often miss even large changes in visual input (see Jensen, Yao, Street, & Simons, 2011, for a recent review). When inspecting alternating versions of an image in which a substantial part may flicker in and out of existence (such as a jet engine on the wing of an aircraft), interleaved by a brief mask pattern (simulating an eye blink), people find it very hard to detect the difference (Rensink, O'Regan, & Clark, 1997). Similarly, we also tend to overlook so-called continuity errors in movies, where objects or people may change visibly from one scene to another. Thus, few detect that the epaulettes on Major Strasser's shoulders repeatedly go in and out of existence in the final scenes of the classic movie *Casablanca*.

Similar limitations in visual perception have also been documented in studies of *inattentional blindness*, where subjects fail to notice unexpected stimuli even when they are looking directly at them. Thus, when following people throwing a ball to each other, participants fail to spot otherwise highly salient stimuli, such as a man in a gorilla suit, walking across the middle of the scene and beating his chest (Simons & Chabris, 1999). Importantly, this effect is not merely a curious laboratory phenomenon but has also been observed in more realistic situations. This is exemplified by a flight simulator study, where several experienced pilots failed to notice a clearly visible plane blocking the runway on which they were about to land (Haines, 1991). Clearly such perceptual failures are not only of academic interest, given their potentially catastrophic consequences.

The exact interpretation of change and inattentional blindness effects varies (see Simons & Rensink, 2005; Jensen et al., 2011, for reviews), ranging from suggestions of sparse (Rensink, 2005) or minimal representations (O'Regan & Noë, 2001) to more detailed representations that either are rapidly overwritten (Rensink et al., 1997) or in other ways fail to serve as the basis on which responses are made (Scott-Brown, Baker, & Orbach, 2000). Nonetheless, there is general agreement that our ability to perceive dynamic changes in visual scenes is dramatically more limited than our conscious experience would lead us to believe.

not encoded in the short amount of time during which the sensory information is available will be lost.

Second, because memory limitations also apply to recoded representations, the cognitive system further chunks the compressed encodings into *multiple levels of representation* of increasing abstraction in perception, and decreasing levels of abstraction in action. Consider, for example, memory for serially ordered symbolic information, such as sequences of digits. Typically, people are quickly overloaded and can recall accurately only the last three or four items (e.g., Murdock, 1968). But it is possible to learn to rapidly encode, and recall, long random sequences of digits by successively chunking such sequences into larger units, chunking those chunks into still larger units, and so on. Indeed, an extended study of a single individual, SF (Ericsson, Chase, & Faloon, 1980), found that this strategy allows sequences containing as many as 79 digits to be recalled with high accuracy. But crucially, this requires learning to encode the input into multiple, successive levels of representation; each sequence of chunks at one level must be shifted as a single chunk to a higher level, before the initial chunks are interfered with or overwritten by further chunks. Thus, SF chunked sequences of three or four digits (the natural chunk size in human memory [Cowan, 2000]) into a single unit, corresponding to running times, dates, or human ages, and then grouped sequences of three to four of these chunks into larger chunks. Interestingly, SF also verbally produced items in overtly discernible chunks, interleaved with pauses, indicating how action also follows the reverse process (e.g., Lashley, 1951; Miller, 1956). The case of SF further demonstrates that low-level information is far better recalled when organized into higher level structures than if coded as an unorganized stream. Note, though, that lower-level information is typically forgotten; it seems unlikely that even SF could recall the specific visual details of the digits with which he was presented.

The notion that perception and action involve representational recoding at a succession of distinct representational levels also fits with a long tradition of theoretical and computational models in cognitive science and computer vision (e.g., Bregman, 1990; Marr, 1982; Miller et al., 1960; Zhu, Chen, Torrable, Freeman, & Yuille, 2010; see Gobet et al., 2001, for a review). Our perspective on repeated multi-level compression is also consistent with data from fMRI and intra-cranial recordings, which suggest that cortical hierarchies across vision and audition—from low-level sensory to high-level perceptual and cognitive areas—integrate information at a succession of progressively longer temporal windows (Hasson, Yang, Vallines, Heeger, & Rubin, 2008; Honey et al., 2012; Lerner, Honey, Silbert, & Hasson, 2011).

Third, to facilitate speedy chunking and hierarchical compression, the cognitive system employs *anticipation*, using prior information to constrain the recoding of current perceptual input (see Bar, 2007; Andy Clark, 2013, for reviews). For example, people see the exact same collection of pixels either as a hairdryer when viewed as part of a bathroom scene or as a drill when embedded in a picture of a workbench (Bar, 2004). Using prior information to *predict* future input is likely to be essential *in order to encode that future input sufficiently rapidly* (as well as helping us to react faster to such input). Anticipation allows faster, and hence more effective, recoding, where oncoming information creates considerable time-urgency. Such predictive processing will be most effective to the extent that the greatest possible amount of available information (across different types and levels of abstraction) is integrated as fast as possible. Similarly, anticipation is important for action as well. For example, manipulating an object requires anticipating the grip force needed to deal with the loads generated by the accelerations of the object. Grip force is adjusted too rapidly during the manipulation of an object to rely on sensory feedback (Flanagan & Wing, 1997). Indeed, the rapid prediction of the sensory consequences of actions (e.g., Poulet & Hedwig, 2006) suggests the existence of so-called "forward models," which allow the brain to predict the consequence of its actions in real time. Forward models have been widely argued to be a ubiquitous feature of the computational machinery of motor control and cognition more broadly (e.g., Wolpert, Diedrichsen, & Flanagan, 2011; see also Clark, 2013; Pickering & Garrod, 2013).

The three processing strategies mentioned here—eager processing, computing multiple representational levels, and anticipation—provide the cognitive system with important means to cope with the Now-or-Never bottleneck. Next, we argue that the language system implements similar strategies for dealing with the here-and-now nature of linguistic input and output, with wide-reaching and fundamental implications for language processing, acquisition, and change as well as for how we should view the structure of language itself. Specifically, we propose that our ability to deal with sequences of linguistic information is the result of what we call "Chunk-and-Pass" processing (Christiansen & Chater, in press), by which the language system can ameliorate the effects of the Now-or-Never bottleneck. More generally, our perspective offers a framework within which to integrate language comprehension and production (see also Chater, McCauley, & Christiansen, in press).

4.2 Chunk-and-Pass Language Processing

The fleeting nature of the auditory and visual linguistic input in combination with the impressive speed with which words and signs are produced imposes a severe constraint on the language system as in perception and action (see box 4.2 on rate of information transfer across languages). Each new incoming word or sign will quickly overwrite previously heard and seen input, providing a naturalistic version of the masking used in psychophysical experiments. How, then, is language comprehension possible? Why doesn't interference between successive sounds (or signs) obliterate linguistic input before it can be understood? The answer, we suggest, is that our language system rapidly recodes this input into "chunks," which are immediately "passed" to a higher level of linguistic representation. The chunks at this higher level are then themselves subject to the same Chunk-and-Pass procedure, resulting in progressively larger chunks of increasing linguistic abstraction. Given that the chunks recode increasingly long stretches of input from lower levels of representation, the chunking process enables input to be maintained over larger and larger temporal windows. Thus, it is this repeated chunking of lower level

Box 4.2
Similar Rates of Information Transfer across Languages

When listening to speakers from different languages, we often have the impression that speakers of some languages appear to talk at a particularly rapid pace. For example, Iberian Spanish seems to rattle past far more quickly than spoken German. Recent cross-linguistic analyses provide support for this intuition. Pellegrino, Coupé, and Marsico (2011) measured the number of syllables per second for seven different languages, listed here from fast (7.84 syllables/sec) to slow (5.18 syllables/sec): Japanese, Spanish, French, Italian, English, German, and Mandarin. So, Spaniards really do talk faster than Germans.

However, just because someone speaks faster, it does not necessarily mean that they get more information across. Because Pellegrino et al. used matched texts for each of the seven languages (i.e., texts that expressed the same semantic information), they could calculate the amount of information conveyed by each syllable in a given language using an eighth language, Vietnamese, as a reference for normalization. Strikingly, using this measure of information density, they obtained almost the opposite pattern to that of syllable rate, here listed from high (0.94) to low (0.49): Mandarin, English, German, French, Italian, Spanish, and Japanese. This negative correlation between syllable rate and information density was subsequently replicated with a larger, more typologically diverse set of languages, including Basque, Catalan, Turkish, and Wolof (Oh, Pellegrino, Marsico, & Coupé, 2013). So, although Spaniards talk fast, they convey less information with each syllable they produce.

Indeed, when determining how much semantic information is transferred per second in each language—computed using the mean duration of the texts, normalized with Vietnamese as the reference (i.e., with an information rate of 1)—Pellegrino et al. found that the information rate for six of the languages were very similar, ranging from 0.90 for German to 1.08 for English. Only Japanese with an information rate of 0.74 was different than the reference language. These results suggest a trade-off between the amount of information carried by individual syllables and the speed with which they are produced, while maintaining more or less the same information rate. Importantly, the rate of information transfer per unit of time also appears to be quite similar across both spoken and signed languages (Bellugi & Fischer, 1972). We see the similar information rates across languages, both spoken and signed, as reflecting a balance between speakers and hearers in terms of the perceptual and cognitive constraints imposed by communicative interactions, fine-tuned by cultural evolution.

information that makes it possible for the language system to deal with the continuous deluge of input that, if not recoded, is rapidly lost. This chunking process is also what allows us to perceive speech at a much faster rate than non-speech sounds (Warren et al., 1969): we have learned to chunk the speech stream through long years, and thousands of hours, of language acquisition. Indeed, we can easily understand (and sometimes even repeat back) sentences consisting of many tens of phonemes, despite our severe memory limitations for sequences of non-speech sounds.

What we are proposing is that during comprehension, the language system—similar to SF—must keep on chunking the incoming information into increasingly abstract levels of representation to avoid being overwhelmed by the input. That is, the language system engages in *eager* processing when creating chunks: they must be built right away, or memory for the input will be obliterated by interference from subsequent material. If a phoneme or syllable is recognized, then it is recoded as a chunk, and passed to a higher level of linguistic abstraction. Once recoded, the information is no longer subject to interference from further auditory input. A general principle of perception and memory is that interference arises primarily between overlapping representations (Crowder & Neath, 1991; Treisman & Schmidt, 1982); recoding avoids such overlap. For example, phonemes will interfere with each other; but phonemes will interfere very little with words. At each level of chunking, information from the previous level(s) is compressed and passed up as chunks to the next level of linguistic representation, from sound-based chunks up to complex discourse elements.[2] As a consequence, the rich detail of the original input can no longer be recovered from the chunks, though some key information remains (e.g., certain speaker characteristics; Nygaard, Sommers, & Pisoni, 1994; Remez, Fellowes, & Rubin, 1997).

In production, the process is reversed: discourse-level chunks are recursively broken down into sub-chunks of decreasing linguistic abstraction until the system arrives at chunks with sufficient information to drive the articulators (either the vocal apparatus or the hands). As in comprehension, memory is limited within a given level of representation, resulting in potential interfer-

2. Note that the Chunk-and-Pass framework does not take a stand on whether "coded" (or literal) meaning is necessarily computed before (pragmatically) "enriched" meaning (for discussion, see, Noveck & Reboul, 2008). Indeed, to the extent that familiar "chunks" can be "gestalts" associated with standardized enriched meanings, then the coded meaning could, in principle, be by-passed. So, for example, *could you pass the salt* might be directly interpreted as a request, sidestepping any putative initial representation as a yes/no question. Similarly, an idiom such as *kick the bucket* may directly be associated with the meaning DIE. The same appears to be true for non-idiomatic compositional "chunks" such as *to the edge* (Jolsvai, McCauley, & Christiansen, 2013). This viewpoint is compatible with a variety of perspectives in linguistics that treat multiword chunks in the same way as traditional lexical items (e.g., Croft, 2001; Goldberg, 2006).

ence between the items to be produced (e.g., Dell, Burger, & Svec, 1997). Thus, there is a tendency for higher-level chunks to be passed down immediately to the level below as soon as they are "ready," leading to a bias toward producing easy-to-retrieve utterance components before harder-to-retrieve ones (e.g., Bock, 1982; MacDonald, 2013). For example, if there is a competition between two possible words to describe an object, the word that is retrieved more fluently will immediately be passed on to lower-level articulatory processes. To further facilitate production, speakers will often reuse chunks from the ongoing conversation, which will be particularly rapidly available from memory. This is reflected by the evidence for lexical (e.g., Meyer & Schvaneveldt, 1971) and structural priming (e.g., Bock, 1986; Bock & Loebell, 1990; Pickering & Branigan, 1998; Potter & Lombardi, 1998) within individuals as well as alignment across conversational partners (Branigan, Pickering & Cleland, 2000; Pickering & Garrod, 2004); priming is also extensively observed in text corpora (Hoey, 2005). As noted by MacDonald (2013), these memory-related factors provide key constraints on the production of language and contribute to cross-linguistic patterns of language use.[3]

A useful analogy for the impact of memory constraints on language production is the notion of "just-in-time"[4] stock control, in which stock inventories are kept to a bare minimum during the manufacturing process. Similarly, the Now-or-Never bottleneck requires that only a few chunks are kept in memory at any given level of linguistic representation. For example, low-level phonetic or articulatory decisions cannot be made and stored far in advance and then "reeled off" during speech production, because any buffer in which such decisions can safely be stored would quickly be subject to interference from subsequent material. So the Now-or-Never bottleneck requires that once detailed production information has been assembled, it must be executed straight away, before it is obliterated by the on-coming stream of later low-level decisions, similar to what has been suggested for motor planning (Norman & Shallice, 1986; see also MacDonald, 2013). We call this proposal Just-in-Time language production.[5]

3. Although our account is neutral about how precisely production and comprehension processes are entwined (e.g., Cann, Kempson, & Wedgwood, 2012; Dell & Chang, 2014; Pickering & Garrod, 2013), we see the two as having a very close connection (see Chater et al., in press, for details).

4. The phrase "just-in-time" has been used in the engineering field of speech synthesis in a similar way (Baumann & Schlangen, 2012).

5. We want to stress that we are not advocating for so-called radical incrementality (Wheeldon & Lahiri, 1997; see Ferreira & Swets, 2002, for discussion) in production, in which words are articulated immediately without any planning ahead. Rather, as noted below, we see production as involving planning a few chunks ahead at every level of linguistic abstraction (see Chater & Christiansen, in press, for further discussion).

4.2.1 Incremental Processing

Chunk-and-Pass processing has important implications for comprehension and production: it requires that both take place *incrementally*. Incremental processing is the thesis that representations are built up as rapidly as possible as the input is encountered. By contrast, one might, for example, imagine a parser that waits until the end of a sentence before beginning syntactic analysis, or that meaning is computed only once syntax has been established. However, such processing would require storing a stream of information at a single level of representation, and processing it later; but given the Now-or-Never bottleneck, this is not possible, because of severe interference between such representations. Incremental interpretation and production therefore follow directly from the Now-or-Never constraints on language.

To get a sense of the implications of Chunk-and-Pass processing, it is interesting to relate this perspective to specific computational principles and models. How, for example, do classic models of parsing fit with this framework? A wide range of psychologically-inspired models involve some degree of incrementality in their syntactic analysis, which can potentially support incremental interpretation (e.g., C. Phillips, 1996, 2003; Winograd, 1972). For example, the Sausage Machine (Frazier & Fodor, 1978) proposes that a preliminary syntactic analysis is carried out phrase-by-phrase, but in complete isolation from semantic or pragmatic factors. But for a right-branching language such as English, chunks cannot be built left-to-right, because the leftmost chunks are incomplete until later material has been encountered. Frameworks from Kimball (1973) onwards imply "stacking up" incomplete constituents that may then all be resolved at the end of the clause. This approach is incompatible with the Now-or-Never bottleneck. Reconciling right-branching with incremental chunking and processing is one motivation for the flexible constituency of Combinatory Categorial Grammar (e.g., Steedman, 1987; 2000; see also Johnson-Laird, 1983).

Focusing on comprehension, there is considerable evidence in favor of incremental interpretation, going back more than four decades (e.g., Bever, 1970; Marslen-Wilson, 1975). The language system utilizes all available information to integrate incoming information as quickly as possible to update the current interpretation of what has been said so far. This process not only includes sentence-internal information about lexical and structural biases (e.g., Farmer, Christiansen, & Monaghan, 2006; MacDonald, 1994; Trueswell, Tanenhaus, & Kello, 1993) but also extra-sentential cues from the referential and pragmatic context (e.g., Altmann & Steedman, 1988; Thornton, MacDonald, & Gil, 1999) as well as the visual environment and world knowledge (e.g., Altmann & Kamide, 1999; Tanenhaus, Spivey-Knowlton, Eberhard, &

Sedivy, 1995). As the incoming acoustic information is chunked, it is rapidly integrated with contextual information to recognize words, consistent with a variety of data on spoken word recognition (e.g., Marslen-Wilson, 1975; van den Brink, Brown, & Hagoort, 2001). These words are then, in turn, chunked into larger multi-word units, as evidenced by recent studies showing sensitivity to multi-word sequences in online processing (e.g., Arnon & Snider 2010; Reali & Christiansen, 2007b; Siyanova-Chanturia, Conklin, & Van Heuven, 2011; Tremblay & Baayen, 2010; Tremblay, Derwing, Libben, & Westbury, 2011), and subsequently further integrated with pragmatic context into discourse level structures.

Turning to production, we start by noting the powerful intuition that we speak "into the void"; that is, that we plan only a short distance ahead without knowing quite how our sentences will end. Indeed, experimental studies suggest, for example, that when producing an utterance involving several noun phrases, people plan just one (Smith & Wheeldon, 1999), or perhaps two, noun phrases ahead (Konopka, 2012), and can modify a message during production in the light of new perceptual input (Brown-Schmidt & Konopka, 2015). Moreover, speech-error data (e.g., Cutler, 1982) reveal that, across representational levels, errors tend to be highly local. Phonological, morphemic, and syntactic errors apply to neighboring chunks within each level (where material may be moved, swapped or deleted). Speech planning appears to involve just a small number of chunks, the number of which may be similar across linguistic levels, but which cover different amounts of time depending on the linguistic level in question. For example, planning involving chunks at the level of intonational bursts stretches over considerably longer periods of time than do chunks at the syllabic level. Similarly, processes of reduction to facilitate production (e.g., modifying the speech signal to make it easier to produce, such as reducing a vowel to a schwa, or shortening or eliminating phonemes) can be observed across different levels of linguistic representation, from individual words (e.g., Gahl & Garnsey, 2004; Jurafsky, Bell, Gregory, & Raymond, 2001) to frequent multi-word sequences (e.g., Arnon & Cohen Priva, 2013; Bybee & Scheibman, 1999).

It may be objected that the Chunk-and-Pass perspective's strict notion of incremental interpretation and production leaves the language system vulnerable to the rather substantial ambiguity that exists across many levels of linguistic representation (e.g., lexical, syntactic, pragmatic). So-called garden-path sentences such as the famous *The horse raced past the barn fell* (Bever, 1970) show that people are vulnerable to at least some local ambiguities: they invite comprehenders to take the wrong interpretive path by treating *raced* as the main verb, which leads them to a dead end. Only when the final word *fell* is encountered does it become clear that something is wrong: *raced* should be interpreted as a past participle that begins a reduced relative clause (i.e., *The horse [that was] raced past the barn fell*). The difficulty of recovery in garden path sentences indicates how strongly the language system is geared toward incremental interpretation.

Viewed as a processing problem, garden paths occur when the language system resolves an ambiguity incorrectly. But in many cases, it is possible for an underspecified representation to be constructed online, and for the ambiguity to be resolved later when further linguistic input arrives. This is consistent with Marr's (1976) proposal of the *Principle of Least Commitment*, that the perceptual system only resolves ambiguous perceptual input when it has sufficient data to make it unlikely that such decisions will subsequently have to be reversed. Given the ubiquity of local ambiguity in language, such underspecification may be used very widely in language processing. Note, however, that because of the severe constraints imposed by the Now-or-Never bottleneck, the language system cannot adopt broad parallelism to further minimize the effect of ambiguity (as in many current probabilistic theories of parsing, e.g., Hale, 2006; Jurafsky, 1996; Levy, 2008). Rather, within the Chunk-and-Pass framework, the sole role for parallelism in the processing system is in deciding how the input should be chunked; only when conflicts concerning chunking are resolved can the input be passed on to a higher-level representation. In particular, we suggest that competing higher level codes cannot be activated in parallel. This picture is analogous to Marr's Principle of Least Commitment: while there might be temporary parallelism to resolve conflicts about, say, correspondence between dots in a random dot stereogram, it is not possible to create two conflicting 3D surfaces in parallel; and while there may be parallelism over the interpretation of lines and dots in an image, it is not possible simultaneously to see something as both a duck and a rabbit. More broadly, higher-level representations are only constructed when sufficient evidence has accrued that they are unlikely later to need to be replaced (for stimuli outside the psychological laboratory, at least).

Maintaining, and later resolving, an underspecified representation will create local memory and processing demands that may slow down processing, as is observed, for example, by increased reading times (e.g., Trueswell, Tanenhaus, & Garnsey, 1994) and distinctive patterns of brain activity (as measured by event-related potential [ERP] responses, Swaab, Brown, & Hagoort, 2003). When the input is ambiguous, the language system may require later input to recognize previous elements of the speech stream successfully. The Now-or-Never bottleneck requires that such online *right-context effects* will be highly local, because "raw" perceptual input will be lost if it is not rapidly identified (e.g., Dahan 2010). Right-context effects may arise where the language system can delay resolution of ambiguity or use underspecified representations that do not require resolving the ambiguity right away. Similarly, cataphora, in which, for example, a referential pronoun occurs before its referent (e.g., *He is a nice guy, that John*) require the creation of an underspecified entity (male, animate) when *he* is encountered, which is only resolved to be coreferential with John later in the sentence (e.g., van Gompel & Liversedge, 2003). Overall, the Now-or-Never bottleneck implies that the processing system will build the most abstract and complete representation that is justified, given the linguistic input—a "good-enough" representation (Ferreira, Bailey, & Ferraro, 2002; Ferreira & Patson, 2007)—but, we suggest, it will not go beyond what is justified, because it is so costly and difficult to recover if an analysis later proves to be incorrect.

Of course, outside experimental studies, background knowledge, visual context, and prior discourse will provide powerful cues to help resolve ambiguities in the signal, allowing the system to rapidly resolve many apparent ambiguities without incurring a substantial danger of "garden-pathing." Indeed, although syntactic and lexical ambiguities have been much studied in psycholinguistics, there is increasing evidence that garden paths are not a major source of processing difficulty in practice (e.g., Ferreira, 2008; Jaeger, 2010; Wasow & Arnold, 2003).[6] For example, Roland, Elman, and Ferreira (2006) report corpus analyses showing that in naturally occurring language there is generally sufficient information in the sentential context before the occurrence of an ambiguous verb to specify the correct interpretation of that verb. Moreover, eye-tracking studies have demonstrated that dialogue partners exploit both conversational context and task demands to constrain interpretations to the appropriate referents, thus side-stepping effects of phonological and referential competitors (Brown-Schmidt & Konopka, 2011) that have otherwise been shown to impede language processing (e.g., Allopenna, Magnuson, & Tanenhaus, 1998). These dialogue-based constraints also mitigate syntactic ambiguities that might otherwise disrupt processing (Brown-Schmidt & Tanenhaus, 2008). This information may be further combined with other probabilistic sources of information such as prosody (e.g., Kraljic & Brennan, 2005; Snedeker & Trueswell, 2003) to resolve potential ambiguities within a

6. It is conceivable that the presence of ambiguities may be a necessary component of an efficient communication system in which easy-to-produce chunks are reused—thus becoming ambiguous—in a trade-off between ease of production and difficulty of comprehension (Piantadosi, Tily, & Gibson, 2012).

minimal temporal window. Finally, it is not clear that undetected garden path errors are costly in normal language use because if communication appears to break down, the listener can repair the communication by requesting clarification from the dialogue partner.

4.2.2 Multiple Levels of Linguistic Structure

The Now-or-Never processing bottleneck forces the language system to compress the input into increasingly abstract chunks, covering progressively longer temporal intervals. As an example, consider the chunking of the input illustrated in figure 4.1. The acoustic signal is first chunked into higher-level sound units at the phonological level. To avoid interference between local sound-based units, such as phonemes or syllables, these are recoded as rapidly as possible into higher-level units such as morphemes or words. The same phe-

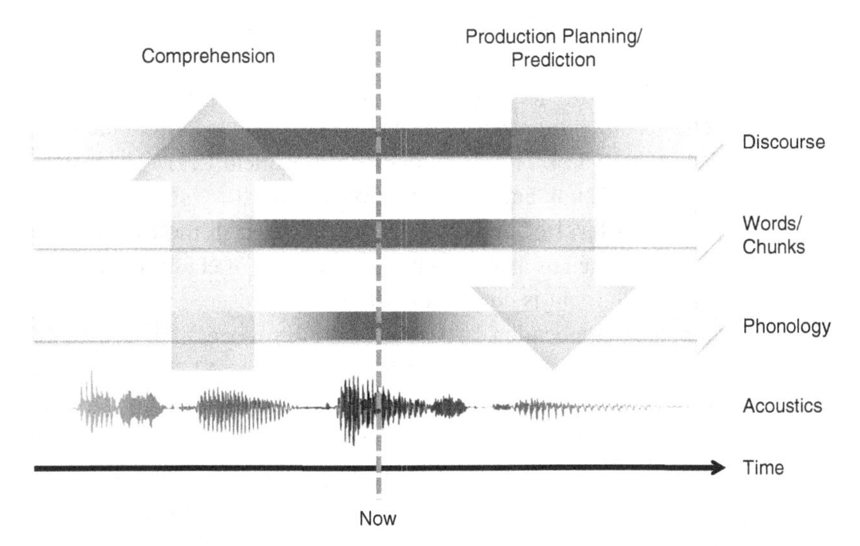

Figure 4.1
Chunk-and-Pass processing across a variety of linguistic levels in spoken language. As input is chunked and passed up to increasingly abstract levels of linguistic representations in comprehension, from acoustics to discourse, the temporal window over which information can be maintained increases, as indicated by the shaded portion of the bars associated with each linguistic level. This process is reversed in production planning, in which chunks are broken down into sequences of increasingly short and concrete units, from a discourse level message to the motor commands for producing a specific articulatory output. More abstract representations correspond to longer chunks of linguistic material, with greater look-ahead in production at higher levels of abstraction. Production processes may further serve as the basis for predictions to facilitate comprehension and thus provide top-down information in comprehension. (Note that the names and number of levels are for illustrative purposes only.)

nomenon occurs at the next level up where local groups of words must be chunked into larger units, possibly phrases or other types of multiword sequences. Subsequent chunking then recodes these representations into higher-level discourse structures (that may themselves be chunked further into even more abstract representational structures). Production requires running the same process in reverse, starting with the intended message and gradually decoding it into increasingly specific chunks, eventually resulting in the motor programs necessary for producing the relevant speech or sign output. As discussed below, the production process may further serve as the basis for prediction during comprehension, allowing higher-level information to influence processing of current input.

Our account is agnostic with respect to the specific characterization of the various levels of linguistic representation, e.g., whether sound-based chunks take the form of phonemes, syllables, etc. Nonetheless, although our account is consistent with the standard linguistic levels, from phonology through syntax to pragmatics, we envisage that a complete model may include finer grained levels, distinguishing, for example, multiple levels of discourse representation. One interesting proposal along these lines, developed from the work of Austin (1962) and Clark (1996), is outlined in Enfield (2013). However, what is central to the Chunk-and-Pass framework is that there should be some form of sound-based level of chunking (or visual-based in the case of sign language), and a sequence of increasingly abstract levels of chunked representations into which the input is continually recoded.

A key theoretical implication of Chunk-and-Pass processing is that the multiple levels of linguistic representations, typically assumed in the language sciences, are a necessary by-product of the Now-or-Never bottleneck. Only by compressing the input into chunks and passing these to increasingly abstract levels of linguistic representation can the language system deal with the rapid onslaught of incoming information. Crucially, though, our perspective also suggests that the different levels of linguistic representations do not have a true part-whole relationship with one another. Unlike participant SF, who was able to unpack chunks from within chunks to reproduce the original string of digits, the language system employs lossy compression to chunk the input. This means that higher-level chunks do not recursively contain complete copies of lower-level chunks. Instead, there is an increasing representational underspecification with higher levels of representation because of the repeated process of lossy compression. Thus, we would expect a growing involvement of extra-linguistic information, such as perceptual input and world knowledge, in processing higher levels of linguistic representation (see, e.g., Altmann & Kamide, 2009).

Whereas our account proposes a lossy hierarchy across levels of linguistic representations, only a very small number of chunks are represented *within* a level, otherwise information is rapidly lost due to interference. This has the crucial implication that chunks within a given level can only interact *locally*. For example, acoustic information must rapidly be coded in a non-acoustic form, say, in terms of phonemes; but this is only possible if phonemes correspond to local chunks of acoustic input. The processing bottleneck therefore enforces a strong pressure toward local dependencies within a given linguistic level. Such representational locality is exemplified across different linguistic levels by the local nature of phonological processes from reduction, assimilation, and fronting, including more elaborate phenomena such as vowel harmony (e.g., Nevins, 2010), speech errors (e.g., Cutler, 1982), the immediate proximity of inflectional morphemes and the verbs to which they apply, and the vast literature on processing difficulty associated with non-local dependencies in sentence comprehension (e.g., Gibson 1998; Hawkins, 2004). As noted above, the higher the level of linguistic representation, the longer the time window within which information can be chunked.

For example, consider the difficulty in keeping track of the dependencies in the following doubly center-embedded sentence (where subscripts indicate subject-verb dependency relations):

(1) The mouse$_1$ the cat$_2$ the dog$_3$ bit$_3$ chased$_2$ ate$_1$ the cheese

Understanding that the sentence means something like *the dog bit the cat that chased the mouse that ate the cheese* is extraordinarily difficult—especially if one had to arrive at an interpretation from spoken input (see also chapter 7). Compare the difficulty of keeping track of dependencies at the syntactic level, with the following example of discourse-level interaction involving a customer (C) and a sales person (S) (where subscripts indicate dependencies between discourse elements, in the form of so-called adjacency pairs):

C$_1$: I would like to purchase some cheese, please.
S$_2$: What kind of cheese would you like?
C$_3$: Do you have any French cheeses?
S$_3$: Yes, we have Camembert, Port Salut, and Rocquefort
C$_2$: I'll take some Camembert, then.
S$_1$: *(wraps up some Camembert for the customer)*

These examples demonstrate that whereas dealing with just two center-embeddings at the sentential level is prohibitively difficult (e.g., de Vries, Christiansen, & Petersson, 2011; Karlsson, 2007; see also chapter 7), we are able to deal with up to four or six embeddings at the multi-utterance discourse level (Levinson 2013). This is because chunking takes place at a much longer time course at the discourse level compared to the sentence level, providing more time to resolve the relevant dependency relations before they are subject to interference.

Finally, as indicated by figure 4.1, processing within each level of linguistic representation takes place in parallel—but with a clear temporal component—as chunks are passed between levels. Note that in the Chunk-and-Pass framework, it is entirely possible that linguistic input can simultaneously, and perhaps somewhat redundantly, be chunked in more than one way. For example, syntactic chunks and intonational contours may be somewhat independent (Jackendoff, 2007). Moreover, we should expect further chunking across different "channels" of communication, including non-verbal input such as gesture and facial expressions.

The Chunk-and-Pass perspective is compatible with a number of recent theoretical models of sentence comprehension, including constraint-based approaches (e.g., MacDonald, Pearlmutter, & Seidenberg, 1994; Trueswell & Tanenhaus, 1994) and certain generative accounts (e.g., Jackendoff's, 2007, parallel architecture). Intriguingly, fMRI data from adults (Dehaene-Lambertz et al., 2006a) and infants (Dehaene-Lambertz et al., 2006b) indicate that activation responses to a single sentence systematically slows down when moving away from the primary auditory cortex either back toward Wernicke's area or forward toward Broca's area, consistent with increasingly long temporal windows for chunking when moving from phonemes to words to phrases. Indeed, the cortical circuits processing auditory input, from lower (sensory) to higher (cognitive) areas, follow different temporal windows, sensitive to increasingly abstract levels of linguistic information from phonemes and words to sentences and discourse (Lerner et al., 2011; Stephens, Honey, & Hasson, 2013). Similarly, the reverse process, going from a discourse-level representation of the intended message to the production of speech (or sign) across parallel linguistic levels, is compatible with several current models of language production (e.g., Chang, Dell, & Bock, 2006; Dell, Burger, & Svec, 1997; Levelt, 2001). Data from intracranial recordings during language production is consistent with different temporal windows for chunk decoding at the word, morphemic, and phonological levels, separated by just over a tenth of a second (Sahin, Pinker, Cash, Schomer, & Halgren, 2009). These results are compatible with our proposal that incremental processing in comprehension and production takes place in parallel across multiple levels of linguistic representation, each involving chunking within a particular temporal window.

4.2.3 Predictive Language Processing

We have already noted that to be able to chunk incoming information as fast and as accurately as possible, the language system exploits multiple constraints in parallel across the different levels of linguistic representation (see chapter 5 for further discussion). Such cues may be used not only to help disambiguate previous input but also to generate expectations for what may come next, potentially further speeding up Chunk-and-Pass processing. Computational considerations indicate that simple statistical information gleaned from sentences provides powerful predictive constraints on language comprehension and can explain many human processing results (e.g., Christiansen & Chater, 1999; Christiansen & MacDonald, 2009; Elman, 1990; Hale, 2006; Jurafsky, 1996; Levy, 2008). Similarly, eye-tracking data suggest that comprehenders routinely use a variety of probabilistic cues—from phonological cues to syntactic context and real-world knowledge—to anticipate the processing of upcoming words (e.g., Altmann & Kamide, 1999; Farmer, Monaghan, Misyak, & Christiansen, 2011; Staub & Clifton, 2006). Results from event-related potential experiments indicate that rather specific predictions are made for upcoming input, including its lexical category (Hinojosa, Moreno, Casado, Muñoz, & Pozo, 2005), grammatical gender (Van Berkum, Brown, Zwitserlood, Kooijman, & Hagoort, 2005; Wicha, Moreno, & Kutas, 2004), and even its onset phoneme (DeLong, Urbach, & Kutas, 2005) and visual form (Dikker, Rabagliati, Farmer, & Pylkkänen, 2010). Thus, there is a growing body of evidence for a substantial role of prediction in language processing (for reviews, see e.g., Federmeier, 2007; Hagoort, 2009; Kamide, 2008; Kutas, Federmeier, & Urbach, 2014; Pickering & Garrod, 2007) and evidence that such language prediction occurs in children as young as two years of age (Mani & Huettig, 2012). Importantly, as well as exploiting statistical relations within a representational level, predictive processing allows "top-down" information from higher levels of linguistic representation to rapidly constrain the processing of the input at lower levels.[7]

From the viewpoint of the Now-or-Never bottleneck, prediction provides an opportunity to begin Chunk-and-Pass processing as early as possible: to constrain representations of new linguistic material as it arrives, and even incrementally to begin recoding predictable linguistic input *before* it arrives.

7. It is often difficult empirically to distinguish bottom-up and top-down effects. Bottom-up statistics across large corpora of low-level representations can mimic the operation of high-level representations in many cases; indeed, the power of such statistics is central to the success of much statistical natural language processing, including speech recognition and machine translation (e.g., Manning & Schütze, 1999). However, rapid sensitivity to background knowledge and non-linguistic context suggests that there is also an important top-down flow of information in human language processing (e.g., Altmann 2004; Altmann, & Kamide, 1999; Marslen-Wilson, Tyler, & Koster, 1993) as well as in cognition, more generally (e.g., Bar, 2004).

This viewpoint is consistent with recent suggestions that the production system may be pressed into service to anticipate upcoming input (e.g., Pickering & Garrod, 2007, 2013). Chunk-and-Pass processing implies that there is practically no possibility of going back once a chunk is created because such backtracking tends to derail processing (e.g., as in classic garden path phenomena). This imposes a *Right-First-Time* pressure on the language system in the face of linguistic input that is highly locally ambiguous.[8] The contribution of predictive modeling to comprehension and production is that it facilitates local ambiguity resolution while the stimulus is still available. Only by recruiting multiple cues and integrating these with predictive modeling is it possible to resolve local ambiguities quickly and correctly (see box 4.3 on how artificial

Box 4.3
Getting It Right-First-Time

From a computational perspective, if the language system uses background knowledge predictively to be, mostly, Right-First-Time, and the computational cost of creating each chunk is constant, then the overall computational cost will be roughly linear in the length of linguistic material (more accurately, slightly slower than linear, as the number of hierarchical levels will increase with the log of input length). A system unable immediately to apply background knowledge to guess Right-First-Time will need to explore a very large number of possible parses because the pervasive local ambiguity of natural language produces a rapidly increasing set of forking paths to be explored, leading to computationally explosive costs in input length (Pratt-Hartmann, 2010). Thus, the rapid predictive application of background knowledge may be essential to the computational feasibility of language processing.

Interestingly, whereas much of computational linguistics has tended to work with traditional parsers that build syntactic trees in one form or another (e.g., Clark, Fox, & Lappin, 2010), engineers working on systems that need to respond in real-time to human speech have generally chosen a different approach involving probabilistic pattern matching. The latter systems face the same problems as humans when interacting with another person: language needs to be processed here-and-now so that responses can be made within a reasonably short amount of time (e.g., there are on average about 200 msec between turns in human conversation; Stivers et al., 2009). For example, so-called *chatbots* (e.g., Wallace, 2005) receive human language input (in text or voice) and produce language output (in text or synthesized voice) in real time. Apple's Siri is a kind of chatbot that responds to user commands or queries either through synthetic voice feedback or text. As no one is willing to wait even a few seconds for a response, these chatbots need to process language in the here-and-now, just like people. The strategies they employ to do this are revealing: they rely on probabilistic pattern matching with respect to individual words, multiword strings, or parts of strings with "wildcards" (e.g., *what's your X*, where *X* can be instantiated by *name* to get somebody's name), but not syntactic trees. They also incorporate as much prior context or other background knowledge (e.g., content area being discussed) as possible to ensure that they are Right-First-Time (or at least most of the time). Thus, language processing in the here-and-now can get by without hierarchical syntactic structure (Frank, Bod, & Christiansen 2012).

8. Strictly speaking, "good-enough first time" may be a more appropriate description. As may be true across cognition (e.g., Simon, 1956), the language system may be a satisficer rather than a maximizer (Ferreira et al., 2002; Ferreira & Patson, 2007).

language systems, such as Apple's Siri, adopt similar strategies to deal with language in the here-and-now).

Right-First-Time parsing fits with proposals such as Marcus (1980), where local ambiguity resolution is temporarily delayed until later disambiguating information arrives, and models in which aspects of syntactic structure may be underspecified, thus not requiring the ambiguity to be resolved (e.g., Gorrell, 1995; Sturt & Crocker, 1996). As we have noted, it also parallels Marr's (1976) Principle of Least Commitment, according to which the perceptual system should, as far as possible, only resolve perceptual ambiguities when sufficiently confident that they will not need to be undone. Moreover, it is compatible with the fine-grained weakly parallel interactive model (Altmann & Steedman, 1988) in which possible chunks are proposed, word-by-word, by an autonomous parser and one chunk is rapidly selected using top-down information.

To facilitate chunking across multiple levels of representation, prediction takes place in parallel across the different levels but at varying timescales. Predictions for higher-level chunks may "run ahead" of those for lower-level chunks. For example, most people simply answer *two* in response to the question *How many animals of each kind did Moses take on the Ark?* failing to notice the semantic anomaly (i.e., it was Noah's Ark, not Moses' Ark) even in the absence of time pressure and when made aware that the sentence may be anomalous (Erickson & Matteson, 1981). That is, anticipatory pragmatic and communicative considerations relating to the required response appear to trump lexical semantics. More generally, the time course of normal conversation may lead to an emphasis on the more temporally extended higher-level predictions at the expense of lower-level ones. This may facilitate the rapid turn-taking that has been observed cross-culturally (Stivers et al., 2009) and which seems to require that listeners make quite specific predictions about turn-ends (Magyari & De Ruiter, 2012) as well as being able to quickly adapt their expectations to specific linguistic environments (Fine, Jaeger, Farmer, & Qian, 2013).

We view the anticipation of turn-taking as one instance of the broader alignment that takes place between dialogue partners across all levels of linguistic representation (for a review, see Pickering & Garrod, 2004). This dovetails with fMRI analyses indicating that while there are some comprehension- and production-specific brain areas, spatiotemporal patterns of brain activity are in general closely coupled between speakers and listeners (e.g., Silbert, Honey, Simony, Poeppel, & Hasson, 2014). In particular, Stephens, Silbert, and Hasson (2010) found that there was a close synchrony between neural activations in speakers and listeners in early auditory areas; speaker activations preceded those of the listeners in posterior brain regions (including parts of

Wernicke's area), while listeners preceded the speakers in activation patterns in the striatum and anterior frontal areas. In the Chunk-and-Pass framework, the listener lag primarily derives from delays caused by the chunking process across the various levels of linguistic representation, whereas the speaker lag predominantly reflects the listener's anticipation of upcoming input, especially at the higher levels of representations (e.g., pragmatic and discourse levels). Strikingly, the extent of the listener's anticipatory brain responses was strongly correlated with successful comprehension, further underscoring the importance of prediction-based alignment for language processing. Indeed, analyses of real-time interactions show that alignment increases when the communicative task becomes more difficult (Louwerse, Dale, Bard, & Jeuniaux, 2012). By decreasing the impact of potential ambiguities, alignment thus makes processing as well as production easier in the face of the Now-or-Never bottleneck (see also Goldstein et al., 2010, for discussion of the importance of social feedback in such alignment).

4.3 Acquisition Is Learning to Process

If speaking and understanding language involve Chunk-and-Pass processing, then acquiring a language requires *learning* how to create and integrate the right chunks rapidly, before current information is overwritten by new input. Indeed, the ability to process linguistic input quickly—which has been proposed as an indicator of chunking ability (Jones, 2012)—is a strong predictor of language acquisition outcomes from infancy to middle childhood (Marchman & Fernald, 2008). The importance of this process is also introspectively evident to anyone acquiring a second language; initially, even segmenting the speech stream into recognizable sounds can be challenging, let alone parsing it into words, or processing morphology and grammatical relations rapidly enough to build a semantic interpretation. The ability to acquire and quickly deploy a hierarchy of chunks at different linguistic scales is parallel to the ability to chunk sequences of motor movements, numbers, or chess positions: it is a *skill*, built up by continual practice.

Viewing language acquisition as continuous with other types of skill learning is very different from the standard formulation of the problem of language acquisition in linguistics. There, the child is viewed as a *linguistic theorist*, with the goal of inferring an abstract grammar from a corpus of example sentences (e.g., Chomsky, 1957, 1965) and only secondarily learning the *skill* of generating and understanding language. But perhaps the child is not a mini-linguist; rather, we suggest that *language acquisition is nothing more than learning to process*: to turn meanings into streams of sound or sign (when generating language), and to turn streams of sound or sign back into meanings (when understanding language).

If linguistic input is available only fleetingly, then any learning must occur while that information is present, i.e., learning must occur in real-time, as the Chunk-and-Pass process takes place. That is, any modifications to the learner's cognitive system in light of processing must, according to the Now-or-Never bottleneck, occur *at the time of processing*. The learner must learn to "chunk" the input appropriately—to learn to recode the input at increasingly abstract linguistic levels; and to do this requires, of course, learning the structure of the language being spoken. But how is this structure learned?

We suggest that, in language acquisition, as in other areas of perceptual-motor learning, people learn *by processing,* and that past processing leaves traces that can facilitate future processing. What, then, is retained, so that language processing gradually improves? Various possibilities can be considered: for example, the weights of a connectionist network can be updated online in light of current processing (Rumelhart & McClelland, 1986); in an exemplar-based model, traces of past examples can be reused in future processing (e.g., Hintzman, 1988; Logan, 1988; Nosofsky, 1986). Here, we focus on the Chunk-Based Learner (CBL; McCauley, & Christiansen, 2011, 2014a), which provides a comprehensive computational implementation of children's use of chunking in language acquisition. The model gradually builds an inventory of chunks consisting of one or more words—a "*chunkatory*"—used in both language comprehension and production. CBL processes linguistic input word-by-word as it is encountered and learns incrementally, relying only on the input seen so far. Learning is based on computing backward transitional probabilities between words, which even eight-month-olds can track (Pelucchi, Hay, & Saffran, 2009). When the transitional probabilities between words dips below the running average up to that point in time, a multiword unit is formed.

CBL simulates language learning by improving its ability to perform two tasks: "comprehension" of child-directed speech through the statistical discovery and use of chunks as building blocks, and "sentence production" utilizing the same chunks and statistics as in comprehension. Comprehension is approximated in terms of the model's ability to use its chunkatory to segment a corpus into phrasal units—a process comparable to *shallow parsing* in the NLP literature (e.g., Hammerton, Osborne, Armstrong, & Daelemans, 2002) and consistent with evidence on the relatively underspecified nature of human sentence comprehension (e.g., Ferreira et al., 2002; Frank & Bod, 2011; Sanford & Sturt, 2002). Production is approximated in terms of the model's ability to reconstruct utterances produced by the child using its chunkatory. The model has broad cross-linguistic coverage and captures a variety of developmental data regarding children's use of multiword chunks (e.g., Arnon & Clark, 2011; Bannard & Matthews, 2008). CBL thus provides a computational demonstration that key aspects of children's early syntactic acquisition can be captured by a chunking model, implemented with the Now-or-Never bottleneck in mind. Importantly, the model has no separate component for grammar acquisition; rather, it learns by gradually becoming better at processing, both in comprehension and production.

More generally, the Now-or-Never bottleneck requires that language acquisition be viewed as a type of skill learning, such as learning to drive, juggle, play the violin, or play chess. Such skills appear to be learned through *practicing* the skill, using online feedback during the practice itself, although consolidation of learning occurs subsequently (Schmidt & Wrisberg, 2004). In chapter 6, we discuss how differences in experience with language can result in differences in the ability to process language. Thus, the challenge of language acquisition is to learn a dazzling sequence of rapid processing operations, rather than conjecturing a correct "linguistic theory."

4.3.1 Online Learning

The Now-or-Never bottleneck implies that learning can only depend on material currently being processed. As we have seen, this requires a processing strategy according to which modification to current representations (in this context, learning) occurs right away; in machine learning terminology, learning is "online." If learning does not occur at the time of processing, the representation of linguistic material will be obliterated, and the opportunity for learning will be gone forever. To facilitate such online learning, the child must learn to utilize all available information to help constrain processing, as discussed further in chapter 5. The integration of multiple constraints—or "cues"—is a fundamental component of many current theories of language acquisition (see e.g., contributions in Golinkoff et al., 2000; Morgan & Demuth, 1996; Weissenborn & Höhle, 2001; for a review, see Monaghan & Christiansen, 2008). For example, second-graders' initial guesses about whether a novel word refers to an object or an action is affected by that word's phonological properties (Fitneva, Christiansen, & Monaghan, 2009), 7-year-olds use visual context to constrain on-line sentence interpretation (Trueswell, Sekerina, Hill, & Logrip, 1999), and preschoolers' language production and comprehension is constrained by pragmatic factors (Nadig & Sedivy, 2002). Thus, children learn rapidly to apply the multiple constraints used in incremental adult processing (Borovsky, Elman, & Fernald, 2012).

Nonetheless, online learning contrasts with traditional approaches in which the structure of the language is learned "off-line" by the cognitive system acquiring a "corpus" of past linguistic inputs and choosing the grammar or other model of the language that best fits with those inputs. For example, in both mathematical and theoretical analysis (e.g., Gold, 1967; Hsu, Chater & Vitányi, 2011; Pinker, 1984) and in grammar-induction algorithms in machine learning and cognitive science, it is typically assumed that a corpus of language can be held in "memory," and that the candidate "grammar" is successively adjusted to fit the corpus as well as possible (e.g., Manning & Schütze, 1999; Pereira & Schabes, 1992; Redington, Chater, & Finch, 1998; Solan, Horn, Ruppin, & Edelman, 2005). However, this approach involves learning linguistic regularities (at, say, the morphological level), by storing and later surveying relevant linguistic input at a lower level of analysis (e.g., involving strings of phonemes), and then attempting to determine which higher-level regularities best fit the "database" of lower-level examples. There are a number of difficulties with this type of proposal, for example, that only a very rich lower-level representation (perhaps combined with annotations concerning relevant syntactic and semantic context) are likely to be a useful basis for later analysis. But more fundamentally, the Now-or-Never bottleneck requires that only if information is recoded *at processing time* will it be retained at all; phonological information that is not chunked at the morphological level and beyond will be obliterated by on-coming phonological material.[9]

So if learning is shaped by the Now-or-Never bottleneck, linguistic input must, when it is encountered, be recoded successively at increasingly abstract linguistic levels, if it is to be retained at all—a constraint imposed, we argue, by basic principles of memory. Crucially, such information is not, therefore, in a suitably "neutral" format to allow for the discovery of previously unsuspected linguistic regularities. In a nutshell: the lossy compression of the linguistic input is achieved by applying the learner's *current* model of the language. But information that would point toward a better model of the language (if examined in retrospect) will typically be lost (or, at best, badly obscured) by this compression, precisely because those regularities are *not* captured by the current model of the language. Suppose, for example, that we create a lossy encoding of language using a simple context-free phrase structure grammar that cannot handle, say, noun-verb agreement. This lossy

9. Some classes of learning algorithm can be converted from 'batch-learning' to 'incremental' or 'on-line' form, including connectionist learning (Saad, 1998) and the widely used Expectation-Maximization (EM) algorithm (Neal & Hinton, 1998), typically with diminished learning performance. How far it is possible to create viable on-line versions of existing language acquisition algorithms is an important question for future research.

encoding of the linguistic input produced using this grammar will provide a poor basis for learning a more sophisticated grammar that includes agreement precisely because agreement information will have been thrown away. Thus, the Now-or-Never bottleneck rules out the possibility that the learner can survey a neutral database of linguistic material, to optimize its model of the language.

The emphasis on online learning does not, of course, rule out the possibility that any linguistic material that *is* remembered may subsequently be used to inform learning. But according to the present viewpoint, any further learning requires *re*-processing that material. So if a child comes to learn a poem, song, or story verbatim, the child might extract more structure from that material by mental rehearsal (or, indeed, saying it aloud). The online learning constraint is that material is learned *only* when it is being processed—ruling out any putative learning processes that involve carrying out linguistic analyses or compiling statistics over a stored corpus of linguistic material.

If this general picture of acquisition as learning-to-process is correct, then we might expect the exploitation of memory to require "replaying" learned material, so that it can be re-processed. Thus, the application of memory itself requires passing through the Now-or-Never bottleneck—there is no way of directly interrogating an internal database of past experience. Indeed, this viewpoint fits with our subjective sense that we need to "bring to mind" past experiences or rehearse verbal material, in order to process it further. Interestingly, there is now also substantial neuroscientific evidence that replay does occur (e.g., in rat spatial learning; Carr, Jadhav, & Frank, 2011). Moreover, it has long been suggested that dreaming may have a related function (here using "reverse" learning over "fictional" input to eliminate spurious relationships identified by the brain, Crick & Mitchison, 1983; see Hinton & Sejnowski, 1985, for a closely related computational model). Deficits in the ability to replay material would, on this view, lead to consequent deficits in memory and inference. Consistent with this viewpoint, Martin and colleagues have argued that rehearsal deficits for phonological pattern and semantic information may lead to difficulties in the long-term acquisition and retention of word forms and word meanings, respectively, and their use in language processing (e.g., Martin & He, 2004; Martin, Shelton, & Yaffee, 1994). In summary, then, language acquisition involves learning to process, and generalizations can only be made over past processing episodes (see also chapter 6).

4.3.2 Local Learning

Online learning faces a particularly acute version of a general learning problem: the *stability-plasticity dilemma* (e.g., Mermillod, Bugaïska, & Bonin, 2013).

How can new information be acquired without interfering with prior information? The problem is especially challenging because reviewing prior information is typically difficult (because recalling earlier information interferes with new input) or impossible (where prior input has been forgotten). Thus, to a good approximation, the learner can only update its "model" of the language in a way that responds to current linguistic input, without being able to review whether any updates are inconsistent with prior input. Specifically, if the learner has a global model of the entire language (e.g., a traditional grammar), it runs the risk of over-fitting that model to capture regularities in the momentary linguistic input at the expense of damaging the match with past linguistic input.

Avoiding this problem, we suggest, requires that learning is highly *local,* consisting of learning about specific relationships between particular linguistic representations. New items can be acquired, with implications for later processing of similar items, but learning current items does not thereby create changes to the entire model of the language, thus potentially interfering with what was learned from past input. One way to learn in a local fashion is to store individual examples—this requires, in our framework, that those examples have been abstractly recoded by successive Chunk-and-Pass operations, of course—and then to generalize piecemeal from these examples. This standpoint is consistent with the idea that the "priority of the specific," as observed in other areas of cognition (e.g., Jacoby, Baker, & Brooks, 1989), also applies to language acquisition. For example, children seem to be highly sensitive to multiword chunks (Arnon & Clark, 2011; Bannard & Matthews, 2008; see Arnon & Christiansen, submitted, for a review[10]). More generally, learning based on past processing traces will typically be sensitive to details of that processing, as is observed across phonetics, phonology, lexical access, syntax, and semantics (e.g., Bybee, 2006; Goldinger, 1998; Pierrehumbert, 2002; Tomasello, 1992).

That learning is local provides a powerful constraint, incompatible with typical computational models of how the child might infer the grammar of the language, because these models typically do not operate incrementally, but range across the input corpus, evaluating alternative grammatical hypotheses (so-called "batch" learning). But, given the Now-or-Never bottleneck, the "unprocessed" corpus, so readily available to the linguistic theorist, or to a computer model, is lost to the human learner almost as soon as it is encoun-

10. Nonetheless, as would be expected from a statistical model of learning, some early productivity is observed at the word level, where words are fairly independent and may not form reliable chunks (e.g., children's determiner-noun combinations; Valian, Solt, & Stewart 2009; Yang 2013; though see McCauley & Christiansen, 2014c; Pine, Freudenthal, Krajewski, & Gobet, 2013, for evidence that such productivity is not driven by syntactic categories).

tered. Where such information has been memorized (as in the case of SF's encoding of streams of digits), recall and processing is slow and effortful. Moreover, because information is encoded in terms of the current encoding, it becomes difficult to review that input to create a better encoding and cross-check past data to test wide-ranging grammatical hypotheses.[11] So, as we have already noted, the Now-or-Never bottleneck seems incompatible with the view of the child as a mini-linguist.

By contrast, the principle of local learning fits with several other approaches. For example, item-based (Tomasello, 2003), connectionist (e.g., Chang, Dell, & Bock, 2006; Christiansen & Chater, 1999; Elman, 1990; MacDonald & Christiansen, 2002),[12] exemplar-based (e.g., Bod, 2009), chunk-based (such as CBL, McCauley & Christiansen, 2011, 2014a), and other usage-based (e.g., Arnon & Snider, 2010; Bybee, 2006) accounts of language acquisition tie learning and processing together, and assume that language is acquired piecemeal, in the absence of an underlying *bauplan*. Such accounts, based on local learning, provide a possible explanation for the frequency effects that are found at all levels of language processing and acquisition (e.g., Bybee, 2007, Bybee & Hopper, 2001; Ellis, 2002; Tomasello, 2003), analogous to exemplar-based theories of how performance speeds up with practice (Logan, 1988).

The local nature of learning need not, though, imply that language has no integrated structure. Just as in perception and action, local chunks can be defined at many different levels of abstraction, including highly abstract patterns, e.g., governing subject, verb, and object; and generalizations from past processing to present processing will operate across all of these levels. Thus, in generating or understanding a new sentence, the language user will be influenced by the interaction of multiple constraints from innumerable traces of past processing across different linguistic levels. This view of language processing involving the parallel interaction of multiple local constraints is embodied in a variety of influential approaches to language (e.g., Jackendoff, 2007; Seidenberg, 1997; see also chapter 5).

11. Interestingly, though, the notion of 'triggers' in the Principles and Parameters model (Chomsky, 1981) potentially fits with the on-line learning framework outlined here (Berwick, 1985; Fodor, 1998; Lightfoot, 1989): parameters are presumed to be set when crucial 'triggering' information is observed in the child's input (see Gibson & Wexler, 1994; Niyogi & Berwick, 1996; Yang, 2002, for discussion). However, these models are very difficult to reconcile with incremental processing and, moreover, do not provide a good fit with empirical linguistic data (Boeckx & Leivada, 2013).

12. Note that the stability-plasticity dilemma arises in connectionist modeling: models that globally modify their weights in response to new items, often only learn very slowly to avoid "catastrophic interference" with prior items (e.g., French, 1999; McCloskey & Cohen, 1989; Ratcliff, 1990). Notably, though, catastrophic interference tends to occur only if the old input rarely reappears later in learning.

4.3.3 Learning to Predict

If language processing involves prediction, in order to make the encoding of new linguistic material sufficiently rapid, then a critical aspect of language acquisition is *learning* to make such predictions successfully (Altmann & Mirkovic, 2009). Perhaps the most natural approach to predictive learning is to compare predictions with subsequent reality, thus creating an "error signal," and then to modify the predictive model in order to reduce this error systematically. Throughout many areas of cognition, such error-driven learning has been widely explored in a range of computational frameworks (e.g., from connectionist networks, to reinforcement learning, to support vector machines) and has considerable behavioral (e.g., Kamin, 1969) and neurobiological support (e.g., Schultz, Dayan, & Montague, 1997).

Predictive learning can, in principle, take a number of forms: for example, predictive errors can be accumulated over many samples, and then modifications are made to the predictive model to minimize the overall error over those samples (i.e., batch learning, as mentioned above). But this is ruled out by the Now-or-Never bottleneck: linguistic input, and the predictions concerning it, is present only fleetingly. However, error-driven learning can also be online—each prediction error leads to an immediate, though typically small, modification of the predictive model, and the aggregate of these small modifications gradually reduces prediction errors on future input.

A number of computational models adhere to these principles; learning involves creating a predictive model of the language, using online error-driven learning. Such models, limited though they are, may provide a starting point for creating an increasingly realistic account of language acquisition and processing. For example, a connectionist model that embodies these principles is the Simple Recurrent Network (Altmann, 2002; Christiansen & Chater, 1999; Elman, 1990), which learns to map from the current input on to the next element in a continuous sequence of linguistic (or other) inputs, and which learns online by adjusting its parameters (the "weights" of the network) to reduce the observed prediction error via the back-propagation learning algorithm (we discuss specific instantiations of such networks in chapters 5–7). Using a very different framework, the CBL model mentioned above incorporates prediction to generalize to new chunk combinations (McCauley & Christiansen, 2011, 2014a). And exemplar-based analogical models of language acquisition and processing may also be constructed that build and predict language structure online, by incrementally creating a database of possible structures, and dynamically using online computation of similarity to recruit these structures to process and predict new linguistic input.

We have argued that prediction allows for top-down information to influence current processing across different levels of linguistic representation, from phonology to discourse, and at different temporal windows (as indicated by figure 4.1). Importantly, though, the Now-or-Never bottleneck suggests that the ability to use such top-down information will emerge gradually, building on bottom-up information. That is, children gradually learn to apply top-down knowledge to facilitate processing via prediction, as higher-level information becomes more entrenched and allows for anticipatory generalizations to be made. Next, we discuss how chunk-based language acquisition and processing have shaped linguistic change and, ultimately, the evolution of language.

4.4 Language Change Is Item-Based

In chapter 2, we argued that language has been shaped through processes of cultural evolution to fit pre-existing constraints deriving from the human brain (and, indeed, body: for example, our speech articulators and sensory apparatus). Thus, language is construed as a complex evolving system in its own right (see also e.g., Arbib, 2005; Beckner et al., 2009; Christiansen & Chater, 2008; Hurford, 1999; Niyogi, 2006; Smith & Kirby, 2008; Tomasello, 2003; for a review, see Dediu et al., 2013); linguistic forms that are easier to use and learn, or are more communicatively efficient, will tend to proliferate, whereas those that are not will be prone to die out. Over time, processes of cultural evolution involving repeated cycles of learning and use have in this way shaped the languages we observe today.

If aspects of language survive only when they are easy to produce and understand, then moment-by-moment processing will not only shape the structure of language (see also Hawkins, 2004; O'Grady, 2005) but also the learning problem that the child faces, as we noted in chapter 3. From the perspective of language as an evolving system, language processing at the time-scale of seconds has implications for the longer timescales of language acquisition and evolution. Revisiting figure 1.1, we can see how the effects of the Now-or-Never bottleneck flow from the timescale of processing to those of acquisition and evolution.

Chunk-and-Pass processing carves the input (or output) into chunks at different levels of linguistic representation at the timescale of the utterance (seconds). These chunks constitute the comprehension and production events from which children and adults learn and update their ability to process their native language over the timescale of the individual (tens of years). Each learner, in turn, is part of a population of language users that shape the cultural evolution of language across a historical timescale (hundreds or thousands of years): language will be shaped by the linguistic patterns that learners find easiest to acquire and process. And the learners will, of course, be strongly constrained by the basic cognitive limitation that is the Now-or-Never bottleneck—and hence, through cultural evolution, linguistic patterns that can be processed through that bottleneck will be strongly selected. Moreover, if acquiring a language is learning to process, and processing involves incremental Chunk-and-Pass operations, then language change will operate through changes driven by Chunk-and-Pass processing, both within and between individuals. But this, in turn, implies that processes of language change should be *item-based*, and driven by processing/acquisition mechanisms defined over Chunk-and-Pass type representations (rather than, for example, being defined over abstract linguistic parameters, with diverse structural consequences across the entire language).

We noted earlier that a consequence of Chunk-and-Pass processing for production is a tendency toward reduction, especially of more frequently used forms, and this constitutes one of several pressures on language change (see also MacDonald, 2013). In extreme cases, this pressure can lead to a communication collapse, as evidenced by a lab-based analogue of the game of telephone in which participants were exposed to a miniature artificial language consisting of simple form-meaning mappings (Kirby, Cornish, & Smith, 2008). The initial "language" that people were taught contained random mappings between syllable strings and pictures of moving geometric figures in different colors. After exposure to the language, they were asked to produce linguistic forms corresponding to specific pictures. Importantly, the participants only saw a subset of the language but nonetheless had to generalize to the full language. The productions of the initial learner were then used as the input language for the next learner, and so on for a total of ten "generations." In the absence of other communicative pressures (such as to avoid ambiguity; Grice, 1967), the language collapsed into just a few different forms that allowed for systematic, albeit semantically underspecified, generalization to unseen items. In natural languages, however, the pressure toward reduction is normally kept in balance by the need to maintain effective communication. Consider, for example, the ASL sign for HOME, which was originally an iconic compound of EAT (O-handshape at the mouth) and BED (flat hand to the head). Over time, the BED handshape assimilated to the O-handshape, and the sign is now made with one contact next to the mouth and a second contact at the back of the jaw with an O-handshape (Frishberg, 1975). Thus, reduction to facilitate Chunk-and-Pass processing gradually made the HOME sign less iconic but more easy to articulate.

Expanding on the notion of reduction and erosion, we suggest that constraints from Chunk-and-Pass processing can provide a cognitive foundation for grammaticalization (Hopper & Traugott, 2003). Specifically, chunks at each level of linguistic structure—discourse, syntax, morphology, and phonology—are potentially subject to reduction. Consequently, we can distinguish between four different types of grammaticalization, from discourse syntacticization and semantic bleaching to morphological reduction and phonetic erosion. Repeated chunking of loose discourse structures may result in their reduction into more rigid syntactic constructions, reflecting Givón's (1979) previously mentioned hypothesis that today's syntax is yesterday's discourse. For example, the resultative construction *He pulled the window open* might derive from syntactization of a loose discourse sequence such as *He pulled the window and it opened* (Tomasello, 2003). As a further byproduct of chunking, some words that occur frequently in certain kinds of construction may gradually become bleached of meaning and ultimately signal only general syntactic properties. Recall our example of the construction *be going to,* which was originally used exclusively to indicate movement in space (e.g., *I'm going to Ithaca*) but which is now also used as an intention or future marker when followed by a verb (as in *I'm going to eat at seven;* Bybee, Perkins, & Pagliuca, 1994). Additionally, a chunked linguistic expression may further be subject to morphological reduction, resulting in further loss of morphological (or syntactic) elements. For instance, the demonstrative *that* in English (e.g., *that window*) lost the grammatical category of number ($\textit{that}_{\text{sing}}$ vs. $\textit{those}_{\text{plur}}$) when it came to be used as a complementizer, as in *the window/windows that is/are dirty* (Hopper & Traugott, 2003). Finally, as noted earlier, frequently chunked elements are likely to become phonologically reduced, leading to the emergence of new shortened grammaticalized forms, such as the phonetic erosion of *going to* into *gonna* (Bybee et al., 1994). Thus, the Now-or-Never bottleneck provides a constant pressure toward reduction and erosion across the different levels of linguistic representation, explaining why grammaticalization tends to be a largely unidirectional process (e.g., Bybee et al., 1994; Haspelmath, 1999; Heine & Kuteva, 2002b; Hopper & Traugott, 2003).

Beyond grammaticalization, we suggest that language change, more broadly, will be local at the level of individual chunks. At the level of sound change, our perspective is consistent with lexical diffusion theory (e.g., Wang, 1969, 1977; Wang & Cheng, 1977), suggesting that sound change originates with a small set of words and then gradually spreads to other words with a similar phonological make-up. The extent and speed of such sound change is affected by a number of factors, including frequency, word class, and phonological environment (e.g., Bybee, 2002; B. Phillips, 2006). Similarly, morpho-syntactic change is also predicted to be local in nature—what we might call "constructional diffusion." Thus, we interpret the cross-linguistic evidence indicating the effects of processing constraints on grammatical structure (e.g., Hawkins, 2004; Kempson, Meyer-Viol, & Gabbay, 2001; O'Grady, 2005; see Jaeger & Tily, 2011, for a review) as a process of gradual change over individual constructions, instead of wholesale changes to grammatical rules. Note, though, that because chunks are not independent of one another, but form a system within a given level of linguistic representation, a change to a highly productive chunk may have cascading effects to other chunks at that level (and similarly for representations at other levels of abstraction). For example, if a frequent construction changes, then constructional diffusion could in principle lead to rapid and far-reaching change throughout the language.

Our account further suggests that regularization—the ubiquitous process of language change whereby representations at a particular level of linguistic representation become more patterned—should also be a piecemeal process. This is exemplified by another of Kirby et al.'s (2008) game-of-telephone experiments (previously mentioned in chapter 2), showing that when ambiguity is avoided, a highly-structured linguistic system emerges across generations of learners, with morpheme-like substrings indicating different semantic properties (color, shape, and movement). That is, the gradual change from a set of random form-meaning mappings to a partially regularized system takes place through piecemeal variation of individual items. A further lab-based cultural evolution experiment along the same lines showed that this process of regularization does not result in the elimination of variability but, rather, in increased predictability through lexicalized patterns (Smith & Wonnacott 2010). Thus, whereas the initial language contained unpredictable pairings of nouns with plural markers, each noun became chunked with a specific marker in the final languages.

These examples illustrate how Chunk-and-Pass processing over time may lead to so-called obligatorification, where a pattern that was initially flexible or optional becomes obligatory (e.g., Heine & Kuteva, 2007). This is one of the ways in which new chunks may be created. So, while chunks at each linguistic level can lose information through grammaticalization, and cannot regain it, there is a countervailing process that constructs complex chunks by "gluing together" existing chunks.[13] That is, in Bybee's (2002) phrase, "items

13. Apparent counterexamples to the general unidirectionality of grammaticalization—such as the compound verb *to up the ante* (e.g., Campbell, 2000)—are entirely compatible with the present approach: they correspond to the creation of new idiomatic chunks, from other pre-existing chunks, and thus do not violate our principle that chunks generally decay.

that are used together fuse together." For example, auxiliary verbs (e.g., *to have*, *to go*) can become fused with main verbs to create new morphological patterns, as in many Romance languages, where the future tense is signaled by an auxiliary "tacked on" as a suffix to the infinitive. As we touched on briefly in chapter 3, in Spanish the future tense endings: *-é*, *-ás*, *-á*, *-emos*, *-éis*, *-án*, derive from the present tense of the auxiliary *haber*, namely, *he*, *has*, *ha*, *hemos*, *habéis*, *han*; and in French, the corresponding endings *-ai*, *-as*, *-a*, *-on*, *-ez*, *-on*, derive from the present tense of the auxiliary *avoir*, namely, *ai*, *as*, *a*, *avon*, *avez*, *ont* (Fleischman, 1982). Such complex new chunks are then subject to erosion. For example, in the example above, the Spanish for $you_{\text{informal, plural}}$ *will eat* is *comeréis*, rather than **comerhabéis*; the first syllable of the auxiliary has been stripped away.

Importantly, the present viewpoint is neutral regarding how far children, rather than adults, are the primary source of innovation (e.g., Bickerton, 1984) or regularization (e.g., Hudson Kam & Newport, 2005) of linguistic material, although constraints from child language acquisition are likely to play some role (e.g., in the emergence of regular subject-object-verb word order in the Al-Sayyid Bedouin Sign Language; Sandler, Meir, Padden, & Aronoff, 2005). In general, we would expect that multiple forces influence language change in parallel (for reviews, see Dediu et al., 2013; Hruschka et al., 2009), including sociolinguistic factors (e.g., Trudgill, 2011), language contact (e.g., Mufwene, 2008), and use of language as an ethnic marker (e.g., Boyd & Richerson, 1987).

Because language change, like processing and acquisition, is driven by multiple competing factors, amplified by cultural evolution, linguistic diversity will be the norm. Accordingly, as discussed in chapter 3, we would expect few, if any, "true" language universals to exist in the sense of constraints that can only be explained in purely linguistic terms. This picture is consistent with linguistic arguments suggesting that there may be no "strict" language universals (Bybee, 2009; Evans & Levinson, 2009). For example, computational phylogenetic analyses indicate that word order correlations are lineage-specific (Dunn, Greenhill, Levinson, & Gray, 2011), and thus shaped by specific histories of cultural evolution, rather than following universal patterns, as would be expected if they were the result of innate linguistic constraints (e.g., Baker, 2001) or language-specific performance limitations (e.g., Hawkins, 2009). Thus, the process of piecemeal tinkering that drives item-based language change is subject to constraints deriving not only from Chunk-and-Pass processing and multiple-cue integration but also from the specific trajectory of cultural evolution that a language follows. From this perspective, there is no sharp distinction between language evolution and language change: language evolution is just the result of language change over a long timescale (see also Heine & Kuteva, 2007), obviating the need for separate theories of language evolution and change (e.g., as proposed by Berwick, Friederici, Chomsky, & Bolhuis, 2013; Hauser, Chomsky, & Fitch, 2002; Pinker 1994).

4.5 Explaining Key Properties of Language

The Now-or-Never bottleneck implies, we have argued, that language comprehension involves incrementally chunking linguistic material and immediately passing the result to further processing, and production involves a similar cascade of processing operations in the opposite direction. And language will be shaped through cultural evolution to be easy to learn and process by generations of speakers/hearers, who are forced to chunk and pass the on-coming stream of linguistic material. What are the resulting implications for the structure of language and its mental representation? Here, we show that certain key properties of language follow naturally from this framework.

4.5.1 The Bounded Nature of Linguistic Units

In non-linguistic sequential tasks, memory constraints are so severe that chunks of more than a few items are rare. People typically encode phone numbers, number plates, postal codes, and social security numbers into sequences of three or four digits or letters; memory recall deteriorates rapidly for un-chunked item-sequences longer than about four (Cowan, 2000), and memory recall typically breaks into short chunk-like phrases. Similar chunking processes are thought to govern non-linguistic sequences of actions (e.g., Graybiel, 1998). As we have argued above, the same constraints apply at all levels of language processing.

Across different levels of linguistic representation, units also tend to have only a few component elements. Even though the nature of sound-based units in speech is theoretically contentious, all proposals capture the sharply bounded nature of such units. For example, a traditional perspective on English phonology would postulate phonemes, short sequences of which are grouped into syllables, with multisyllabic words being organized by intonational or perhaps morphological groupings. Indeed, the tendency toward few-element units is so strong that a long nonsense word with many syllables such as *supercalifragilisticexpialidocious* is chunked successively, for example, as tentatively indicated:

[[[Super][cali]] [[fragi][listic]] [[expi] [ali]] [docious]]

Similarly, agglutinating languages, such as Turkish, chunk complex multimorphemic words via local grouping mechanisms that include formulaic morpheme expressions (Durrant, 2013). Likewise, at higher levels of linguistic representation, verbs normally only have two or three arguments at most. Within both generative and usage-based linguistic theories, syntactic phrases typically consist of only a few constituents. Thus, the Now-or-Never bottleneck provides a strong bias toward bounded linguistic units across various levels of linguistic representations.

4.5.2 The Local Nature of Linguistic Dependencies

Just as Chunk-and-Pass processing leads to bounded linguistic units with only a small number of components, we suggest that it also produces a powerful pressure toward local dependencies. Dependencies between linguistic elements will primarily be adjacent or separated only by a few other elements. For example, at the phonological level, processes are highly local, as reflected by data on coarticulation, assimilation, and phonotactic constraints (e.g., Clark, Yallop, & Fletcher, 2007). Similarly, we expect word formation processes to be highly local in nature, which is in line with a variety of different linguistic perspectives on the prominence of adjacency in morphological composition (e.g., Carstairs-McCarthy, 1992; Hay, 2000; Siegel, 1978). Strikingly, adjacency even appears to be a key characteristic of multi-morphemic formulaic units in an agglutinating language such as Turkish (Durrant, 2013).

At the syntactic level, there is also a strong bias toward local dependencies. For example, when processing the sentence *"The key to the cabinets was ..."* comprehenders experience local interference from the plural *cabinets* when the verb *was* needs to agree with the singular *key* (Nicol, Forster, & Veres; 1997; Pearlmutter, Garnsey, & Bock, 1999). Indeed, individuals who are good at picking up adjacent dependencies among sequence elements in a serial-reaction time task also experience greater local interference effects in sentence processing (Misyak & Christiansen, 2010). Moreover, similar local interference effects have been observed in production when people are asked to continue the above sentence after *cabinets* (Bock & Miller, 1991).

More generally, analyses of Romanian and Czech (Ferrer-i-Cancho, 2004) as well as Catalan, Basque, and Spanish (Ferrer-i-Cancho & Liu, 2014) point to a pressure toward minimization of the distance between syntactically related words. This tendency toward local dependencies seems to be particularly strong in strict word-order languages such as English, but somewhat less so for more flexible languages like German (Gildea & Temperley, 2010). However, the use of case marking in German may provide a cue to counteract this pressure by providing an alternative way, not dependent on word order, of indicating who does what to whom, as suggested by simulations of the learnability of different word orders with or without case markings (e.g., Lupyan & Christiansen, 2002; Van Everbroeck, 1999). This highlights the use of not just simple distributional information (e.g., regarding word order) but also of other types of cues (e.g., involving phonological, semantic, or pragmatic information), as discussed previously.

We want to stress, however, that we are not denying the existence of long-distance syntactic dependencies; rather, we are suggesting that our ability to process such dependencies will be bounded by the number of chunks that can be kept in memory at a given level of linguistic representation. In many cases, chunking may help to minimize the distance over which a dependency has to remain in memory. For example, as discussed further in chapter 6, the use of personal pronouns can facilitate the processing of otherwise difficult object relative clauses, because they are more easily chunked (e.g., *People [you know] are more fun*; Reali & Christiansen, 2007a). Similarly, the processing of long-distance dependencies is eased when they are separated by highly frequent word combinations that can be readily chunked (e.g., Reali & Christiansen, 2007b).

The Chunk-and-Pass account is in line with other approaches that assign processing limitations and complexity as primary constraints on long-distance dependencies, thus potentially providing explanations for linguistic phenomena, such as subjacency (e.g., Berwick & Weinberg, 1984; Kluender & Kutas, 1993), island constraints (e.g., Hofmeister & Sag, 2010), referential binding (e.g., Culicover, 2013b), and scope effects (e.g., O'Grady, 2013). However, the impact of these processing constraints may be lessened to some degree by the integration of multiple sources of information (e.g., from pragmatics, discourse context, and world knowledge) to support the ongoing interpretation of the input (e.g., Altmann & Steedman, 1988; Heider, Dery, & Roland, 2014; Tanenhaus et al., 1995), as we discuss further in chapter 6.

4.5.3 Multiple Levels of Linguistic Representation

Speech involves transmitting a digital, symbolic code over a serial, analog "channel" using time-variation in sound-pressure. How might we expect this digital-analog-digital conversion to be tuned, to optimize the amount of information transmitted?

The problem of encoding and decoding digital signals over an analog serial channel is well-studied in communication theory (Shannon, 1948)—and, interestingly, the solutions typically adopted look very different from those employed by natural language. To maximize the rate of transfer of information it is, in general, best to transform the message to be conveyed across the analog signal in a very non-local way. That is, rather than matching up portions of the information to be conveyed (e.g., in an engineering context, these might be the contents of a database) to particular portions of the analog signal, the best strategy is to encrypt the whole digital message using the entire analog signal, so that the message is coded as a "block" (e.g., MacKay, 2003)—for example, using so-called "block-codes." But why is the engineering solution to information transmission so very different from that used by natural language, in which distinct portions of the analog signal correspond to meaningful units in the digital code (e.g., phonemes, words)? The Now-or-Never bottleneck provides a natural explanation.

A block-based code requires decoding a stored memory trace for the *entire* analog signal (for language, typically, acoustic)—i.e., the whole block. This is straightforward for artificial computing systems, where memory interference is no obstacle. But this is, of course, precisely what the Now-or-Never bottleneck rules out. The human perceptual system must turn the acoustic input into a (lossy) compressed form right away, or else the acoustic signal is lost forever. Similarly, the speech production system cannot decide to send a single lengthy analog signal, and then successfully "reel off" the corresponding long sequence of articulatory instructions, because this will vastly exceed our memory capacity for sequences of actions. Instead, the acoustic signal must be generated and decoded *incrementally*: so that the symbolic information to be transmitted maps, fairly locally, to portions of the acoustic signal. Thus, to an approximation, while individual phonemes acoustically exhibit enormous contextual variation, diphones (pairs of phonemes) are a fairly stable acoustic signal, as is evident from their use in tolerably good speech synthesis and recognition (e.g., Jurafsky, Martin, Kehler, Vander Linden, & Ward, 2000). Overall, then, each successive segment of the analog acoustic input must correspond to a part of the symbolic code being transmitted. This is not because of considerations of informational efficiency, but because of the brain's processing limitations in encoding and decoding: specifically, by the Now-or-Never bottleneck.

The need to rapidly encode and decode implies that spoken language will consist of a sequence of short sound-based units (the precise nature of these units may be controversial, and may even differ between languages, but could correspond to diphones, phonemes, mora, syllables, etc.). Similarly, in speech production, the Now-or-Never bottleneck rules out planning and executing a long articulatory sequence (as in the block-codes used in communication technology); rather, speech must be planned incrementally, in a Just-in-Time fashion, requiring that the speech signal corresponds to sequences of discrete sound-based units.

4.5.4 Duality of Patterning

Our perspective has yet further intriguing implications. Because the Now-or-Never bottleneck requires that symbolic information must be read off the analog signal rapidly, the number of such symbols will be severely limited—and, in particular, may be much smaller than the vocabulary of a typical speaker (many thousands or tens of thousands of items). This implies that the short symbolic sequences into which the acoustic signal is initially recoded cannot, in general, be bearers of meaning; instead, the primary bearers of meaning, lexical items and morphemes, will be composed out of these smaller units.

Thus, the Now-or-Never bottleneck provides a potential explanation for a puzzling but ubiquitous feature of human languages, including signed languages. This is *duality of patterning*: the existence of one or more levels of symbolically encoded sound structure (whether described in terms of phonemes, mora, or syllables) from which the level of words and morphemes, over which meanings are defined, are composed. Such patterning arises, in the present analysis, as a consequence of rapid online multilevel chunking in both speech production and perception. In the absence of duality of patterning, the acoustic signal corresponding, say, to a single noun, could not be recoded incrementally as it is received (Warren & Marslen-Wilson, 1987) but would have to be processed as a whole, thus dramatically overloading sensory memory.

It is, perhaps, also of interest to note that the other domain in which people process enormously complex acoustic input, music, also typically consists of multiple layers of structure (notes, phrases, and so on). We may conjecture that Chunk-and-Pass processing operates for music as well as language, thus helping to explain why our ability to process musical input spectacularly exceeds our ability to process arbitrary sequential acoustic material (Clément, Demany, & Semal, 1999).

4.5.5 The Quasi-Regularity of Linguistic Structure

We have argued that the Now-or-Never bottleneck implies that language processing involves applying highly local Chunk-and-Pass operations across a range of representational levels; and that learning language involves learning to perform such operations. But, as in the acquisition of other skills, learning from such specific instances does not operate by rote, but leads to generalization and hence modification from one instance to another (Goldberg, 2006). Indeed, such processes of local generalization are ubiquitous in language change, as we have noted above. From this standpoint, we should expect the rule-like patterns in language to emerge from generalizations across specific instances (see, e.g., Hahn & Nakisa, 2000, for an example of this approach to inflectional morphology in German); once entrenched, such rule-like patterns can, of course, be applied quite broadly to newly encountered cases. Thus, patterns of regularity in language will emerge locally and bottom-up, from generalizations across individual instances, through processes of language use, acquisition, and change.

From this point of view, we should expect language to be quasi-regular across phonology, morphology, syntax, and semantics—to be an amalgam of overlapping and partially incompatible patterns, involving generalizations from the variety of linguistic forms that successive language learners encounter. For example, English past-tense morphology has, famously, the regular *–ed* ending, a range of subregularities (*sing* → *sang*, *ring* → *rang*, *spring* → *sprang*, but *fling* → *flung*, *wring* → *wrung*; and even *bring* → *brought*; with some verbs having the same present and past tense forms, e.g., *cost* → *cost*, *hit* → *hit*, *split* → *split*; while others differ wildly, e.g., *go* → *went*, *am* → *was*; see, e.g., Bybee & Slobin, 1982; Pinker & Prince, 1988; Rumelhart & McClelland, 1986). This quasi-regular structure does indeed seem to be widespread throughout many aspects of language (e.g., Culicover, 1999; Goldberg, 2006; Pierrehumbert, 2002).

From a traditional generative perspective on language, such quasi-regularities are puzzling: language is assimilated, somewhat by force, to the structure of a formal language, with a precisely defined syntax and semantics—the ubiquitous departures from such regularities are mysterious. From the present standpoint, by contrast, the quasi-regular structure of language arises in rather the same way that a partially regular pattern of tracks comes to be laid down through a forest, through the overlaid traces of endless animals finding the path of local least resistance; and where each language processing episode tends to facilitate future, similar, processing episodes, just as an animal's choice of path facilitates the use of that path for animals that follow (see also Culicover & Nowak, 2003, for a similar metaphor).

4.6 Summary

The perspective developed in this chapter sees language as composed of a myriad of specific processing episodes, where particular messages are conveyed and understood. Like other action sequences, linguistic acts have their structure by virtue of the cognitive mechanisms that produce and perceive them. We have argued that the structure of language is, in particular, strongly affected by a severe limitation on human memory: the Now-or-Never bottleneck, that sequential information, at many levels of analysis, must rapidly be recoded to avoid being interfered with, or overwritten, by the deluge of subsequent material. To cope with the Now-or-Never bottleneck, the language system chunks new material as rapidly as possible at a range of increasingly abstract levels of representation. As a consequence, Chunk-and-Pass processing induces a multi-level structure over linguistic input. As we detail further in chapter 8, the history of this chunk-building process can be viewed as analogous to a shallow surface structure in linguistics, and thus the repertoire of possible chunking mechanisms and the principles by which they can be combined can be viewed as defining a grammar. Indeed, we suggest that chunking procedures may be one interpretation of the constructions that are at the core of construction grammar in linguistics.

Given the wide-reaching and fundamental implications of the Now-or-Never bottleneck, we believe that the theoretical perspective that we have outlined in this chapter may serve as a framework within which to reintegrate the language sciences, from the psychology of language comprehension and production, to language acquisition, language change and evolution, to the study of language structure. In the second part of this book, we apply this framework to explain specific aspects of language, including the structure of the vocabulary (chapter 5), the role of experience in the processing of relative clauses (chapter 6), and the nature of complex recursive structure (chapter 7).

II Implications for the Nature of Language

5 Language Acquisition through Multiple-Cue Integration

Le signe est arbitraire.
—Ferdinand de Saussure, 1916, *Cours de Linguistique Générale*

In the previous chapters, we argued that the evolution, acquisition, and processing of language is fundamentally shaped by multiple constraints. Importantly, rather than being rigid rules and principles, these constraints provide "soft," partially overlapping, probabilistic cues to linguistic structure and its associated meaning. The integration of multiple cues in the service of language processing applies across all aspects of language from morphemes and words to constructions and full utterances. But whereas such multiple-cue integration is now well accepted by most theories of syntactic processing (e.g., MacDonald, Pearlmutter, & Seidenberg, 1994; Tanenhaus & Trueswell, 1995), at the word level it seems to run head-long into the century-old assumption of linguistic theory illustrated by this chapter's opening quote: the arbitrariness of the sign (de Saussure, 1916), celebrated as a defining feature of human language (Hockett, 1960). That is, if words were to provide cues to their meanings, it would imply that the relationship between the form of a word (sound or sign) and its meaning is not entirely arbitrary but subject to nontrivial constraints.

The suggestion that words may contain cues within them to discover their meaning and use might seem like a non-starter. For example, when considering that the perennial woody plant that we refer to in English as *tree* is called *Baum* in German, *arbre* in French, and *shù* (樹) in Mandarin Chinese, the arbitrariness of the sign seems obvious. The sound of a word does not seem to tell us much about what it means, and the same meaning can be conveyed by wildly different sound combinations. Even onomatopoeia can appear surprisingly arbitrary: the sound that pigs make is called *oink oink* in English, *ut it* in Vietnamese, *kvik kvik* in Czech, and *øf øf* in Danish. It is therefore not surprising that the assumption of form-meaning arbitrariness has become fundamental to modern linguistic theories on both sides of the Chomskyan divide. For example, Pinker (1999, p. 2) states that "onomatopoeia and sound symbolism certainly exist, but they are asterisks to the far more important principle of the arbitrary sign—or else we would understand the words in every foreign language instinctively, and never need a dictionary for our own!" In a similar vein, Goldberg (2006, p. 217) notes that "... the particular phonological forms that a language chooses to convey particular concepts [...] generally are truly arbitrary, except in relative rare cases of phonaesthemes."

However, the last century has seen a growing body of evidence suggesting that there is, nonetheless, a substantial amount of *systematicity* in form-meaning mappings (see Dingemanse, Blasi, Lupyan, Christiansen, & Monaghan, 2015; Perniss, Thompson, & Vigliocco, 2010; Schmidtke, Conrad, & Jacobs, 2014, for reviews)—while form-meaning mappings are clearly very flexible, they are by no means completely unconstrained. For example, consider the two nonsense shapes in figure 5.1. If you were told that one of them was called *bouba* and the other *kiki*, which name would you say goes with which shape? If you matched the rounded shape on the right with *bouba* and selected the spikey shape on the left to go with *kiki*, then you made the same choices that most people do. This type of systematic sound-meaning mapping was first demonstrated by Köhler (1929) using native Spanish speakers. Later, the *bouba/kiki* effect was replicated with undergraduate students in the US and Tamil speakers in India (Ramachandran & Hubbard, 2001) as well as with Himba speakers in Namibia (Bremner et al., 2013). Subsequent studies have shown that similar sound-meaning preferences are observed in two-and-a-half-year-olds (Maurer, Pathman, & Mondloch, 2006) and are present in infants as young as four months (Ozturk, Krehm, & Vouloumanos, 2013).

More generally, the focus of Western scholars across all but the twentieth century has been on the degree to which form-meaning mappings are systematic, rather than arbitrary (see box 5.1). A systematic relationship may appear either as *absolute iconicity*, where a linguistic feature directly imitates some

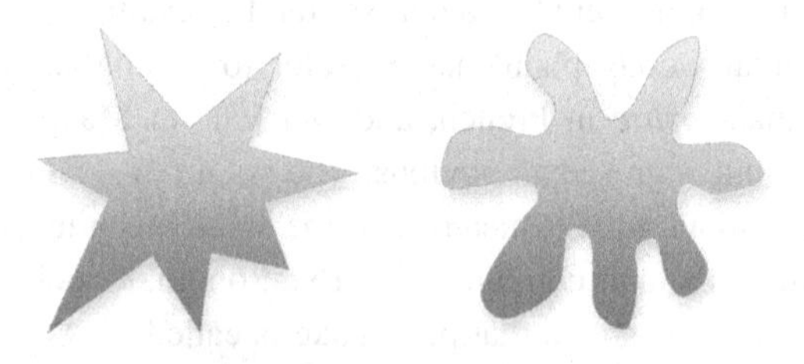

Figure 5.1
Illustrations of the kind of spikey and rounded nonsense shapes used to test the *bouba/kiki* effect.

Box 5.1
Sound Symbolism as the Basis for the Perfect Language

Historically, the idea of the arbitrary sign was far from obvious. For example, to some Greek philosophers, it seemed plausible that words were intrinsically linked to their meaning, as evidenced by a 2,300-year-old debate between Hermogenes and Cratylus (Hamilton & Cairns, 1961) on naturalism vs. conventionalism in meaning. The old testament, too, has been seen to support the notion that meaning can be uncovered from the sound of words: "And out of the ground the LORD God formed every beast of the field, and every fowl of the air; and brought them unto Adam to see what he would call them: and whatsoever Adam called every living creature, that was the name thereof" (*King James Bible*, Genesis 2:19). In this way, Adam was thought to create *the perfect language*, common to all humankind, in which the sound of a word directly expresses its inherent meaning. Of course, with the building of the Tower of Babel, humans were subsequently cursed with arbitrariness, resulting in a multitude of different tongues that no longer expressed the direct sound-meaning relationship of Adam's perfect language.

For much of Western intellectual history, it has been a goal of many scholars up through the Renaissance and Enlightenment to uncover the lost pre-Babel language that Adam created (for a review, see Eco, 1995). These efforts took on many forms, but all aimed to recover a more systematic relationship between words and their meaning. The English natural philosopher, John Wilkins (1668), provided an illustrative example of such an endeavor in his treatise, "An Essay Towards a Real Character and a Philosophical Language." He devised a version of a perfect language in which there were systematic relations between letters and particular meaning classes. For example, all plants begin with the letter "g," whereas all animals begin with the letter "z." Plants are then further divided up into several subcategories depending on their primary characteristics. Thus, leafy plants start with "gα," flowers with "ga," seed-vessels with "ge," shrubs with "gi," and trees with "go." However, although his system was ingenious, Wilkins unwittingly demonstrated its inherent problems: closely related words are easily confused with one another. Indeed, as noted by Eco (1995), on the very same page that he introduced the scheme, he wrote *gαde* (barley) in place of the intended *gape* (tulip). Ultimately, Wilkins' perfect language was doomed, but his classification scheme survived as a key inspiration to Peter Mark Roget when writing his *Thesaurus* (Hüllen, 2003).

aspect of semantics (as in onomatopoeia) or *relative iconicity*, where there may be statistical regularities between similar sounds and similar meanings in the absence of imitation (Gasser, Sethuraman, & Hockema, 2011; for further discussion of different kinds of non-arbitrariness, see Dingemanse et al., 2015). The *bouba/kiki* effect can be seen as an instance of relative iconicity, which might be rooted in cross-modal neurobiological connections given that the effect is not only observed across different languages and cultures but also present very early on in development (e.g., Spector & Maurer, 2009). Very roughly, there may be basic perceptual reasons why a rounded shape is represented more similarly to the more rounded-sounding *bouba* than the more spiky-sounding *kiki*. By contrast, an example of relative iconicity that is likely to be acquired during development, rather than arising from basic perceptual processes, is the systematic form-meaning mapping found in phonaesthemes, e.g., the tendency for English words starting with *sn-* to refer to something related to the nose and mouth (Bloomfield, 1933) such as *snout, sneeze, snuffle, sniff, snore, snooze, snort, snarl, sneer,* and *snack* (an effect that can even be primed in native English speakers; Bergen, 2004). Moreover, preferences for certain sound-meaning mappings have been shown to facilitate the learning of new words (e.g., Imai, Kita, Nagumo, & Okada, 2008; Kovic, Plunkett, & Westermann, 2010; Nygaard, Cook, & Namy, 2009). Thus, it seems that, after all, the sign is not entirely arbitrary and that some, perhaps even considerable, systematicity does seem to exist in form-meaning mappings.

So there are both pre-existing biological constraints on learning and processing (which may shape the *bouba/kiki* results), and there is a tendency for learners to form local regularities (as with the *sn-* words of English). Viewing language as a culturally evolving system, we might plausibly expect that vocabularies across the world's languages will not, after all, be wholly arbitrary; instead, systematic patterns across vocabulary items will tend to be favored because they make words that follow these patterns easier to learn and use. We should therefore expect languages to evolve to have some degree of systematicity in their mappings from form to meaning.

If languages do evolve to have a nontrivial amount of systematicity in form-meaning mappings, and the child is sensitive to such systematicity, then this may provide a useful source of information with which to guide vocabulary learning. More generally, if other aspects of language may similarly be supported by different types of cues available in the input to children, then the integration of such cues during language acquisition could provide crucial constraints on the kind of linguistic structure that children learn.

In this chapter, we explore the proposal that pre-existing neural constraints not only provided important restrictions on cultural evolution (as discussed in chapter 2), but also made available multiple sources of information that can facilitate both the acquisition and use of language. By recruiting such cues, some of which may be partially overlapping and redundant, language could evolve culturally to become more expressive, while still being learnable and processable by mechanisms not dedicated to language. Consequently, as a product of cultural evolution, every language today incorporates its own unique constellation of probabilistic cues to signal different aspects of linguistic structure from word boundaries (e.g., Mattys, White, & Melhorn, 2005) to syntactic relations (Monaghan & Christiansen, 2008), and beyond (Evans & Levinson, 2009). Because individual cues are probabilistic in nature, and therefore unreliable when considered in isolation, we suggest that the integration of multiple cues is likely to be a necessary component of language acquisition.

The hypothesis put forward here is that cultural evolution has shaped languages to depend on multiple-cue integration for their acquisition and processing. But how might this work? We present an extended case study in multiple-cue integration, focusing on one possible source of information about syntactic structure—phonological cues—and how these may be integrated with distributional information during language acquisition and processing. We start by considering a possible functional explanation for de Saussure's (1916) notion of the arbitrary sign, suggesting that lexical items may have been adapted by cultural evolution to handle the Now-or-Never bottleneck that we discussed in chapter 4. But, looking more closely, we then re-examine the question of how much systematicity natural language vocabulary retains and why. While processing considerations provide a drive toward an arbitrary relation between form and meaning, it turns out that other psychological forces, including ease of acquisition, may push in the opposite direction; and we ask how far mechanisms for processing and acquisition are able to exploit such systematic relationships where they are present. Thus, the mixture of arbitrariness and systematicity in the relation between form and meaning provides a rich domain within which to explore the interplay of language processing, acquisition, and evolution.

5.1 A Pressure for Arbitrary Signs: A Cultural Evolution Perspective

Below we explore some of the systematic departures from complete arbitrariness in the mapping between form and meaning. But first we want to address the natural question: why does de Saussure's (1916) general observation about the arbitrariness of signs hold good to a first approximation? Interestingly, the demands of incremental Chunk-and-Pass processing provide a simple explanation. Recall that the onslaught of oncoming linguistic material demands that the language system recodes linguistic input as rapidly as possible. A particularly time-critical aspect of this chunking process involves the recoding of sound-based input as a sequence of words. The identification of words is known to be extremely rapid, deploying both contextual factors and acoustic information to identify words, typically well before the entire word has been heard (e.g., Marslen-Wilson, 1987). Thus, there are two key sources of information that can be used by language system to identify words as rapidly as possible: the speech stream itself provides information about the *form* of the word (typically, at the point of recognition, the speech stream will have revealed the form of the beginning of the word, of course); and contextual information, drawing on what has just been said, background knowledge, and the current perceptual environment, which will provide information about the *meaning* of the word (and, indeed, correlated syntactic properties, which we discuss further below).

An elementary observation in information theory and statistics (e.g., Cover & Thomas, 2006) is that two sources of information provide maximal constraint when they are *independent*. For word recognition to be achieved as early as possible requires, then, that form and meaning/syntactic information should be as independent as possible. To the extent that the sources of information are systematically related, they are redundant, and hence each adds little to the other; obtaining two highly correlated sources of information is rather like checking a story by buying two copies of the same newspaper. More concretely, suppose that all words relating to animals began with a particular phoneme, say, /d/. Then contextual information (being located at a zoo or a farm), including linguistic context such as *Can you see the …?* may strongly suggest that an animal will be referred to, and as the word unfolds, the initial /d/ conveys the same information, and hence provides little or no extra guidance about which lexical item is intended. By contrast, with an arbitrary mapping, the semantic information given by the context is uncorrelated with the phonological constraint (beginning with a /d/), so after the initial phoneme, the listener can infer that the word is likely to be one of a relatively small class of animals words beginning with a /d/, rather than that it is merely some animal word or other[1]. So the approximate arbitrariness of form-meaning mappings need not be taken as primitive, rather, it can be seen as an adaptation of the vocabulary, through cultural evolution, to be as easy as possible to handle using Chunk-and-Pass processing, in the face of the Now-or-Never bottleneck we encountered in chapter 4.

As stated, though, this argument may appear to apply globally to the "optimal design" of the entire vocabulary, yet the item-based perspective of language processing, acquisition, and evolution that, as we saw, arises from Chunk-and-Pass processing implies that it is individual lexical items, rather than the entire vocabulary, that will be optimized through cultural evolution to be as easy to process and acquire as possible. But the pressure for independence between form and meaning to support successful chunking applies equally at the level of individual lexical items. For example, if all animal words did indeed begin with /d/, then an animal word that "breaks the pattern" and begins with, say, a /p/, will be especially rapidly recognized when semantics is clearly signaled by context. Once we know that an animal is being talked about, the initial phoneme will instantly give away which animal word is

1. This argument was suggested to us by Padraic Monaghan. It is also consistent with the processing-based explanation of the suffixing preference by Cutler, Hawkins, and Gilligan (1985).

relevant, allowing chunking to proceed immediately. Animal words beginning with /d/ can only be identified by waiting for subsequent speech input, making chunking more difficult and fragile. If lexical items that are more readily chunked are favored by cultural evolution, then we should expect pressure on individual lexical items to violate strong interdependencies between form and meaning, resulting in an approximate "arbitrariness of the sign" across the entire vocabulary.

Notice, though, that the pressure for arbitrariness will be opposed to other selectional pressures on the vocabulary. We have already suggested that innate neural connections (e.g., perhaps more naturally associating "rounded" sounding words such as *bouba* with rounded objects, and "sharper" sounding words such as *kiki* with pointed objects), and pressures from learning—e.g., a bias toward systematic patterns linking form and meaning—will make acquiring new words much easier. Yet another possible origin of systematicity arises from processes of morphological erosion or other modifications, as we discussed in chapter 4: traces of the systematic form-meaning mappings in morphology will be retained as morphemes themselves are gradually obliterated by processes such as reduction—we shall consider such "emergent" form-meaning systematicity below. More generally, we suggest that the sometimes conflicting pressures from learning and processing lead to a trade-off between arbitrariness and systematicity in the development of the vocabulary. Thus, viewing the vocabulary as a product of cultural evolution, shaped by potentially opposing selectional pressures from a myriad of processing episodes across generations of speakers, suggests that we should anticipate it to comprise a complex mixture of arbitrary and systematic form-meaning mappings.

5.2 How Systematic Is Spoken Language?

As a first step toward quantifying the amount of form-meaning systematicity present in spoken language, and thus its potential usefulness as a cue to vocabulary learning, Monaghan, Shillcock, Christiansen, and Kirby (2014) conducted a series of corpus analyses (see box 5.2 for an outline of the long history of corpus usage in the language sciences). They extracted phonological forms for all the English monosyllabic words found in the CELEX database (Baayen, Pipenbrock, & Gulikers, 1995), accounting for about 70% of words used in everyday English (Baayen et al., 1995). To ensure that potential systematicity in form-meaning mappings was not due to the specific ways in which either form or meaning similarity was represented, Monaghan et al. employed several different methods. They computed sound similarity between word forms in three different ways: (a) *phonological feature edit distance*: the

Box 5.2
Big Data in the Language Sciences: Linguistic Corpora

The use of "big data" in the form of corpora has a long historical pedigree in the language sciences. Initially, corpora were employed primarily to determine various distributional properties of language (see McEnery & Wilson, 1996, for a review). For example, Kading (1897) used a corpus consisting of 11 million words to determine the frequency of letters and sequences thereof in German. Corpora were subsequently collected by field linguists (e.g., Boas, 1940) and structuralist linguists (e.g., Harris, 1951; Hockett, 1954) to inform their linguistic theories. However, this work was limited by their treatment of corpora simply as collections of utterances that could be subjected only to relatively simple bottom-up analyses. A more comprehensive approach to corpora was proposed by Quirk (1960) in his introduction to the still ongoing project, *The Survey of English Usage*. From the viewpoint of psychology, the compilation of the one million word *Brown Corpus* (Kucera & Francis, 1967) in the 1960s was a major milestone. Computerized analyses of this corpus resulted in word-frequency statistics that were used to control psycholinguistic studies until quite recently. Currently, word frequency is generally assessed using much larger corpora, such as the British National Corpus (Burnard & Aston, 1998), the Corpus of Contemporary American English (Davies, 2010), or the Google Terabyte Corpus (Brants & Franz, 2006), which provides a snapshot of the World Wide Web in 2006.

There is also a long history of using corpora in the study of language acquisition (for reviews, see Behrens, 2008; Ingram, 1989). Much of this history is characterized by diary studies of children, many of which focused on development in general rather than on language per se. Indeed, even Charles Darwin (1877) wrote a paper on the development of his infant son. Modern diary studies have generally concentrated more on children's production of specific aspects of language such as errors in argument structure (Bowerman, 1974) or verb use (Tomasello, 1992). Other studies have approached children's productions by collecting extended samples longitudinally from multiple children (e.g., Bloom, 1970; Braine, 1963). Brown's (1973) longitudinal study of three children—Adam, Eve, and Sarah—constitutes an important milestone because of the use of tape recordings. Importantly, the transcriptions of the three children's language samples across their early linguistic development eventually became part of the *Child Language Data Exchange System* (CHILDES; MacWhinney, 2000), an electronic depository of language acquisition corpus data (MacWhinney & Snow, 1985). The CHILDES database, as a source of (relatively) big data, has become the most prominent source of language acquisition data for both sophisticated statistical analyses and computational modeling. As such, corpus data has become a key component of both developmental and adult psycholinguistics.

minimum number of phonological feature changes required to convert one word to another (e.g., *cat* and *dog* differ by eight features, associated with manner and place of articulation); (b) *phoneme edit distance*: the number of phoneme changes required to convert one word to another (e.g., *cat* and *dog* differ by three phonemes); and (c) *phonological feature Euclidean distance*: the Euclidean distance between phonological feature representations of words (e.g., *cat* and *dog* turn out to differ by a distance of 0.881). Two different representations of meaning[2] were used to compute meaning similarity: (a)

2. Both types of semantic representations have been used extensively in computational linguistics, in part reflecting behavioral responses to meaning similarity (e.g., Huettig, Quinlan, McDonald, & Altmann, 2006). Contextual co-occurrence vectors capture the tendency for words with similar meanings to occur in similar contexts, thus resulting in similar vectors (see Riordan & Jones,

contextual co-occurrence vectors generated by counting how often each of 446 context words occur within a three-word window before or after the target word in the British National Corpus (Burnard & Aston, 1998); and (b) *semantic features derived from WordNet* (Miller, 2013), in which words are grouped together according to hierarchical relations and grammatical properties (e.g., *cat* and *dog* share thirteen features [*entity, organism, animal, vertebrate, mammal, placental, carnivore, has paws, has tail, has ribs, has thorax, has head, has face*] and differ by eight features, including *feline* versus *canine*).

Using these representations, Monaghan et al. (2014) generated separate similarity spaces for sound and meaning by comparing the representation for a given word to all other words. For example, for the word *dog*, similarity measures would be computed between its phonological representation and the sound representations of all other words. Likewise, the meaning representation of *dog* would be compared with the semantic representations of all the other words in the meaning similarity space. This produces the same number of similarity pairs (such as *dog—cat, dog—cog*, etc.) across all similarity spaces (whether phonological or semantic). The degree of cross-correlation between any two similarity spaces can then be computed by aggregating the correlations between all matching similarity pairs across the two similarity spaces. Thus, these analyses determine whether the phonological similarity of any two words *Phon*(x,y) correlate with the semantic similarity of those words *Sem*(x,y) (e.g., comparing *Phon(dog,cat)* with *Sem(dog,cat)*, *Phon(dog,cog)* with *Sem(dog,cog)*, and so on).

When computing such cross-correlations between the phonological and semantic similarity spaces, Monaghan et al. (2014) found a small, but highly significant, positive correlation ($r^2 \approx 0.002$) suggesting that there is some overall systematicity in English sound-meaning correspondences. To make sure that the positive correlation was not a trivial property of the high dimensionality of the similarity spaces, they conducted a set of Monte Carlo analyses. These involved a Mantel (1967) test in which every word's meaning was randomly reassigned to a different word (e.g., the meaning for *dog* might be that of *cat*)—thereby "breaking" any potential correlation between form and meaning—and the sound-meaning cross-correlation was then recomputed. This process was repeated 10,000 times, revealing that English words contain more sound-meaning systematicity than would be expected by chance ($p < 0.0001$). This result was replicated across the different phonological and

2011, for a review of different ways of computing such vectors). In contrast, WordNet aims to capture similarity between word meanings in terms of hyponymy, that is, words are defined in terms of so-called *is-a* relations; e.g., a *dog* is-a canine, which is-a carnivore, which is-a mammal, and so on (for an introduction, see Fellbaum, 2005).

semantic representations. Moreover, to control for possible effects of both inflectional and derivational morphology on form-meaning systematicity, Monaghan et al. further redid their analyses with only monomorphemic versions of the words (*dog* but not *dogs*), and again obtained significant positive correlations.

A final set of analyses was conducted to control for potential phonological and/or semantic relatedness due to shared historical origin. For example, words related to the phonaestheme *gl-*, such as *gleam, glitter, glow,* and *glisten*, are proposed either to derive from the Proto-Indo-European root **ghel-*, meaning "to shine, glitter, glow, be warm" (Klein, 1966) or the Old English root **glim-*, meaning "to glow, shine" (OED Online, 2013). Thus, a family of morphologically related words in a root language (with, in consequence, a transparent form-meaning mapping) may independently become distorted over centuries of language change while retaining traces of the original form-meaning systematicity. To make sure that the previously observed systematicity in form-meaning mappings was not due to such etymological vestiges, Monaghan et al. (2014) redid their analyses after omitting words with proposed common roots in Old English, Old French, Old Norse, Greek, Latin, Proto-Germanic, or Proto-Indo-European. Strikingly, they still found that English incorporates a small but significant degree of form-meaning systematicity.

Additional analyses of the contribution of an individual word's form-meaning mapping to the overall systematicity of the vocabulary suggested that the systematicity of English is a property of the language as a whole, and not due to small isolated pockets of words with highly systematic form-meaning mappings. Of course, given the local item-based nature of processing and acquisition that we discussed in chapter 4, we might nonetheless expect that some isolated pockets of words with form-meaning relations might exist (e.g., as in the group of so-called mimetic verbs in Japanese; see Imai & Kita, 2014, for a review). Importantly, this form-meaning systematicity is not due to morphology or common historical origin. And, as we shall see below, the results cannot be explained by the presence of, in Pinker's (1999) terms, the "asterisks" of sound symbolism to the general arbitrariness of the sign: it turns out that there are rich and subtle statistical relations across multiple linguistic uses relating form and meaning (and, indeed, between form and grammatical category). Thus, although the very small amount of variance accounted for in the correlation between form and meaning indicates that the mappings between them are largely arbitrary, language nonetheless incorporates a reliable amount of systematicity between the sound of words and their meaning (at least as exemplified by English). Indeed, analyses of a small set of words from nearly two-thirds of the world's languages have revealed that considerable sound symbolism exists in the basic vocabulary, independent of geographical location and linguistic lineage (Blasi, Wichmann, Stadler, Hammarström, & Christiansen, submitted).

5.2.1 The Role of Systematicity in Vocabulary Learning

The corpus analyses surveyed so far suggest that there are systematic correspondences between the sound of a word and its meaning at the level of individual words. But does such sound-meaning systematicity provide a useful cue to vocabulary learning? Further corpus analyses by Monaghan et al. (2014) lend support to the idea that systematicity might help children get a foothold in language. They first measured the systematicity of individual words by determining whether omitting each word increased or decreased the correlation between sound and meaning for the whole vocabulary. Then, they determined the degree of correlation between this measure of sound-meaning systematicity for each word and the average age at which this word is generally acquired according to data from age of acquisition norms (Kuperman, Stadthagen-Gonzalez, & Brysbaert, 2012). The results revealed that early-acquired words tended to be more systematic in their sound-meaning correspondences (when controlling for other factors such as frequency and word length).

Given experimental data showing that three- to four-month old infants are able to form cross-modal correspondences between sounds and visual properties of objects—such as between spatial height and angularity with auditory pitch (Walker, Bremner, Mason, Spring, Mattock, Slater, & Johnson, 2010)—the sound-meaning systematicity of early words may thus provide a scaffolding for word learning (see also Miyazaki et al., 2013). Indeed, the cross-modal correspondences of early words may even help young infants learn in the first place that there *are* mappings between sound and meaning (Spector & Maurer, 2009).

However, as the vocabulary grows across development, the form-meaning systematicity of later-acquired words decreases. Figure 5.2 illustrates Monaghan et al.'s (2014) findings that form-meaning mappings tend to become more arbitrary for later acquired words. If, as we have suggested, systematic form-meaning mappings are easier to *acquire*, but, due to the lack of independence between contextual and sound-based cues, they are less efficiently *processed* by the Chunk-and-Pass language system, then the degree of form-meaning systematicity in different parts of the vocabulary is likely to depend on the relative importance of acquisition and processing. We suggest that acquisition is the primary challenge early on, where new lexical items are acquired slowly; we should therefore expect form-meaning systematicity to be greater for early vocabulary. Later, of course, vocabulary is acquired by the

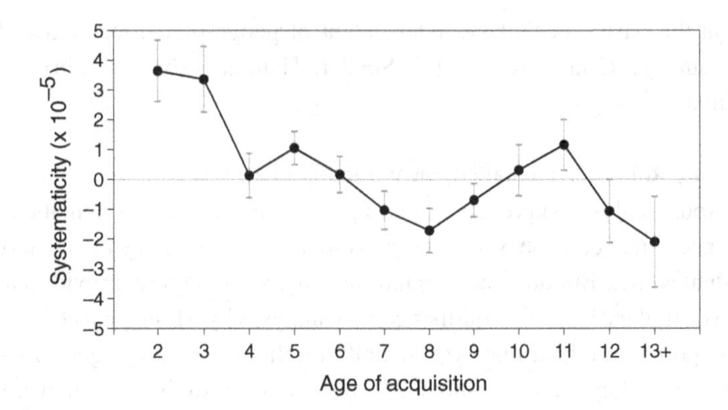

Figure 5.2
The form-meaning systematicity of words as a function of their age of acquisition. Early acquired words have a higher mean systematicity compared to later acquired words (based on Monaghan, Shillcock, Christiansen, & Kirby, 2014; error bars indicate standard error of the mean).

child at an astonishing speed (e.g., roughly ten new words per day), so that finding a foothold for acquisition may then be relatively less important, but processing the oncoming deluge of complex linguistic material becomes critical (and this may be particularly important after the onset of reading).

Intriguingly, computational simulations by Gasser (2004) showed that systematic form-meaning mappings were helpful for learning small vocabularies where perceptual representations of words can be kept sufficiently distinct from one another, while still allowing for sound symbolic systematicity to exist. As the vocabulary grows, however, it becomes increasingly hard for words with similar meanings to have sufficiently different word forms to avoid confusion because parts of the representational space become saturated with word forms. Eventually, with large vocabularies, arbitrary mappings rather than systematic ones end up resulting in better learning overall, as word forms can be more evenly distributed across representational space. Gasser's simulations suggest that, aside from the powerful pressure for independence of form and meaning arising from the need to deal with the Now-or-Never bottleneck, for some learning mechanisms, at least, there may be a pressure for arbitrariness arising purely from the acquisition process, with increasing vocabulary size.

5.3 The Sound of Syntax

The results of Monaghan et al. (2014) indicated that the English vocabulary contains a small but significant amount of systematicity in the mapping between the sound of an individual word and its meaning, which decreases with age of acquisition. Of course, children do not just learn words to use them in isolation; rather, they need to combine them with other words in sentences. So, perhaps it is possible to find stronger systematicity within categories of words? Here, we explore the hypothesis that the role of phonological cues changes from signaling specific *word meanings* in the service of initial vocabulary acquisition to indicating *word use* at the lexical category level to facilitate processing. That is, sound-meaning systematicity not only offers cues to the individual meaning of words, but also provides information about how words might be grouped together in terms of their broader meaning or syntactic usage.

Recent corpus analyses provide some initial support for this possibility, suggesting that nouns that differ in the semantic property of abstractness (measured in terms of imageability) also tend to differ along several phonological measures, including prosody, phonological neighborhood density, and rates of consonant clustering (Reilly & Kean, 2007). Subsequent psycholinguistic experiments have confirmed that adults are indeed sensitive to such phonological information when making semantic judgments about novel words or reading known words aloud (Reilly, Westbury, Kean, & Peelle, 2012). Thus, if abstract and concrete nouns may sound somewhat different from one another, then might there not also be phonological differences between lexical categories of words, such as nouns and verbs, which could be helpful for the acquisition of syntax? In other words, perhaps words contain within them the "sound of syntax"? If so, then the phonological forms of words would provide cues not only for early vocabulary learning but also for discovering how to use the acquired words in a syntactic context. We would further expect such phonological cues to be integrated with other types of cues to facilitate language acquisition and use.

5.3.1 Quantifying the Usefulness of Phonological Cues

Kelly (1992) was one of the earliest proponents of the idea that phonological cues may play an important role in syntactic acquisition and use. Potentially relevant phonological information includes lexical stress, vowel quality, and duration, and may help distinguish grammatical function words (e.g., determiners, prepositions, and conjunctions) from content words (nouns, verbs, adjectives, and adverbs) in English (e.g., Cutler, 1993; Gleitman & Wanner, 1982; Monaghan, Chater, & Christiansen, 2005; Monaghan, Christiansen, & Chater, 2007; Morgan, Shi, & Allopenna, 1996; Shi, Morgan, & Allopenna, 1998). As we demonstrate in detail below, phonological cues may also help separate nouns and verbs—implying, incidentally, a consequent statistical form-meaning systematicity, given that the semantics of verbs and nouns differ, of course. For example, English disyllabic nouns tend to receive initial-syllable (trochaic) stress whereas disyllabic verbs tend to receive final-syllable (iambic) stress, and adults are sensitive to this distinction (Kelly, 1988). Acoustic analyses have also shown that disyllabic words that are noun-verb ambiguous and have the same stress placement can still be differentiated by syllable duration and amplitude cue differences (Sereno & Jongman, 1995). Even three-year old children are sensitive to this stress cue, despite the fact that few multi-syllabic verbs occur in child-directed speech (Cassidy & Kelly, 1991, 2001). Additional noun/verb cues in English are likely to include differences in word duration, consonant voicing, and vowel types, and many of these cues may be cross-linguistically relevant (see Kelly, 1992; Monaghan & Christiansen, 2008, for reviews).

The question remains, though, as to just how useful such phonological cues might be for acquisition, and whether they can be integrated with other kinds of cues in an efficient and productive way. Monaghan, Chater, and Christiansen (2005) aimed to address this question by quantifying the contribution of phonological cues to lexical categories[3] through a series of corpus analyses of English child-directed speech. More than five million words were extracted from the CHILDES database (MacWhinney, 2000), comprising more than a million utterances spoken in the presence of children. Phonological forms and lexical categories were gleaned from the CELEX database (Baayen, Pipenbrock, & Gulikers, 1995) and results reported for the 5,000 most frequent words. As potential cues to lexical categories, Monaghan et al. used sixteen different phonological properties (listed in table 5.1) that have previously been proposed to be useful for separating nouns from verbs (and function words from content words).

Whereas previous studies of phonological cues to lexical categories had focused on the contribution of individual phonological cues, such as lexical stress (e.g., Cutler, 1993), Monaghan et al. *combined* the sixteen cues into a unified phonological representation for each word. Specifically, each word was represented by a 16-dimensional vector, with each dimension corresponding to the word's value for a particular phonological cue (as illustrated by the two examples in table 5.1), so that each word corresponds to a point in a 16-dimensional space defined by the sixteen phonological cues. Monaghan et al. used

3. We note here that these analyses of the degree to which the phonological forms of words carry information about their syntactic use as nouns or verbs do not presuppose the postulation of universal lexical categories. Instead, phonological and distributional cues provide probabilistic information about how words can be used in sentential contexts and this is what is assessed by the corpus analyses reported here.

Table 5.1
The 16 phonological cues used by Monaghan, Chater & Christiansen (2005)

	Examples	
Phonological Cue	*stretched*	*dog*
Word level		
Length in phonemes[a]	6	3
Length in syllables[a]	1	1
Presence of stress[b]	1	1
Syllable position of stress[c]	1	1
Syllable level		
Number of consonants in word onset[d]	3	1
Proportion of phonemes that are consonants[e]	0.8	0.67
Proportion of syllables containing reduced vowel[f]	3	0
Reduced 1st vowel[g]	0	0
-ed inflection[h]	0	0
Phoneme level		
Proportion of consonants that are coronal[e]	1	0.5
Initial /ð/ [i]	0	0
Final voicing[a]	2	1
Proportion of consonants that are nasals[a]	0	0
Position of stressed vowel[j] (1:front → 3:back)	1	3
Position of vowels[k] (1:front → 3:back)	1	3
Height of vowels[k] (0:high → 3:low)	2	3

Note. a. Kelly (1992); b. Gleitman & Wanner (1982); c. Kelly & Bock (1988); d. Shi et al. (1998); e. Morgan et al. (1996); f. Cutler (1993); g. Cutler & Carter (1987); h. Marchand (1969); i. Campbell & Besner (1981); j. Sereno & Jongman (1990); k. Monaghan et al. (2005).

discriminant analysis to determine whether the nouns and verbs formed separate clusters in this phonological space. Informally, this type of statistical analysis inserts a hyper-plane into the 16-dimensional phonological cue space to produce the most optimal separation of nouns and verbs into two different categories. Correct classification of nouns and verbs can then be computed, given how well the hyperplane splits the two categories (with chance = 50%). This statistical analysis resulted in well above chance classifications of both nouns (58.5%) and verbs (68.3%)—with an indication that phonological cues may be more useful for discovering verbs than nouns. The advantage of phonological cues for verbs was subsequently confirmed by further analyses in Christiansen and Monaghan (2006).

5.3.2 Quantifying the Usefulness of Distributional Cues

Although the phonological cues provide useful information about the lexical category of a word, they are probabilistic in nature, and therefore on their own cannot be expected to lead to perfect classification. But perhaps other cues can be brought to bear? One possible source of additional cues to grammatical category comes from the distributional characteristics of linguistic fragments at or below the word level. Morphological patterns across words will, of course, be informative—e.g., English words that are observed to have both *–ed* and *–s* endings are likely to be verbs (Maratsos & Chalkley, 1980; Onnis & Christiansen, 2008). In artificial language learning experiments, adults acquire grammatical categories more effectively when they are cued by such word-internal patterns (Brooks, Braine, Catalano, Brody, & Sudhalter, 1993; Frigo & McDonald, 1998). Corpus analyses reveal that word co-occurrence also provides useful cues to grammatical categories in child-directed speech (e.g., Monaghan et al., 2005, 2007; Redington, Chater & Finch, 1998; St. Clair, Monaghan, & Christiansen, 2010). Given that function words primarily occur at phrase boundaries (e.g., phrase-initially in English and French; phrase-finally in Japanese), they can also help the learner by signaling syntactic structure. This idea has received support from corpus analyses (e.g., Fries, 1952; Mintz, Newport, & Bever, 2002; St. Clair et al., 2010) and artificial language learning studies (Green, 1979; Morgan, Meier, & Newport, 1987; Valian & Coulson, 1988). Finally, artificial language learning experiments indicate that the duplication of morphological patterns across related items in a phrase (e.g., in Spanish: L<u>os</u> *Estad*<u>os</u> *Unid*<u>os</u>) facilitates learning (Meier & Bower, 1986; Morgan et al., 1987).

As an additional cue, Monaghan et al. (2005) therefore assessed the usefulness of distributional information using a simple, developmentally plausible approach. They selected the twenty most frequent words in the corpus (*are, no, there, this, your, that's, on, in, oh, do, is, and, I, that, what, to, a, it, the, you*) and recorded how often these preceded one of the target words (e.g., *you want*). The rationale was that even though the child may not know the meaning of the twenty context words, these word forms nonetheless constitute highly frequent acoustic events to which the child is likely to be sensitive. To determine the usefulness of the distributional patterns thus recorded, Monaghan et al. used an information-based measure—a modified version of the Dunning (1993) score—to assess the strength of the association between the context word and the target word. Informally, this measure provides an estimation of how surprising it is that the context and target words occur together (e.g., Freq[*you* want]) given how often each occurs on its own (e.g., Freq[*you*] and Freq[*want*]). Each word was then represented as a 20-dimensional vector, corresponding to the signed Dunning log-likelihood scores for each of the twenty context words (e.g., for *cats*, the scores for *are cats, no cats, there cats,* and so on). Classification of nouns and verbs given these distributional representations was then assessed using a discriminant analysis, similar to the phonological cue analysis. The results showed a very good classification of nouns (93.7%) but not of verbs (31.1%).

5.3.3 Integrating Phonological and Distributional Cues

Monaghan et al.'s (2005) results from the separate analyses of phonological and distributional cues suggest that their respective usefulness may differ across nouns and verbs. Perhaps integration across the two types of cues may improve classification? Monaghan et al. combined the phonological cues and the distributional cues into a single representation, by appending the 20-dimensional distributional vector to the 16-dimensional phonological cue vector, thus producing a 36-dimensional multiple-cue vector representation for each word. A discriminant analysis was then conducted on the 36-dimensional cue space defined by these word representations. The results yielded reliable classifications of both nouns (67.0%) and verbs (71.4%).

When considering correct classifications of nouns and verbs together, they further noted an interesting interaction of phonological/distributional cues with frequency, as shown in figure 5.3. Distributional cues (black bars) appear to work very well for high-frequency words but less so for low-frequency words. A natural explanation is that high-frequency words occur in more contexts and this provides for more accurate distributional information about their lexical categories. Phonological cues (white bars), on the other hand, seem to be more reliable for low-frequency words than for high-frequency words[4]. One explanation may be the tendency for high-frequency words to be shortened over generations of cultural evolution, to yield more rapid processing, but which may also lead to the omission of important phonological cues to their lexical category. In contrast, low-frequency words are not subject to the same processes of erosion, allowing the cues to remain in place. Crucially, though, when both phonological and distributional cues are integrated (gray bars), good classification can be found across all frequency bins, as illustrated in figure 5.3.

5.3.4 Multiple-Cue Integration across Languages

The results presented so far apply only to English. If language, in general, has evolved to rely on multiple-cue integration, then it should be possible to find

4. Note that low-frequency words tend to be acquired later than high-frequency ones, suggesting that phonological cues to lexical categories may not be available for the earliest vocabulary items. However, this might allow form-meaning systematicity at the individual word level to be more efficient for initial vocabulary items, before systematicity at the category level comes into play, as we discuss further below.

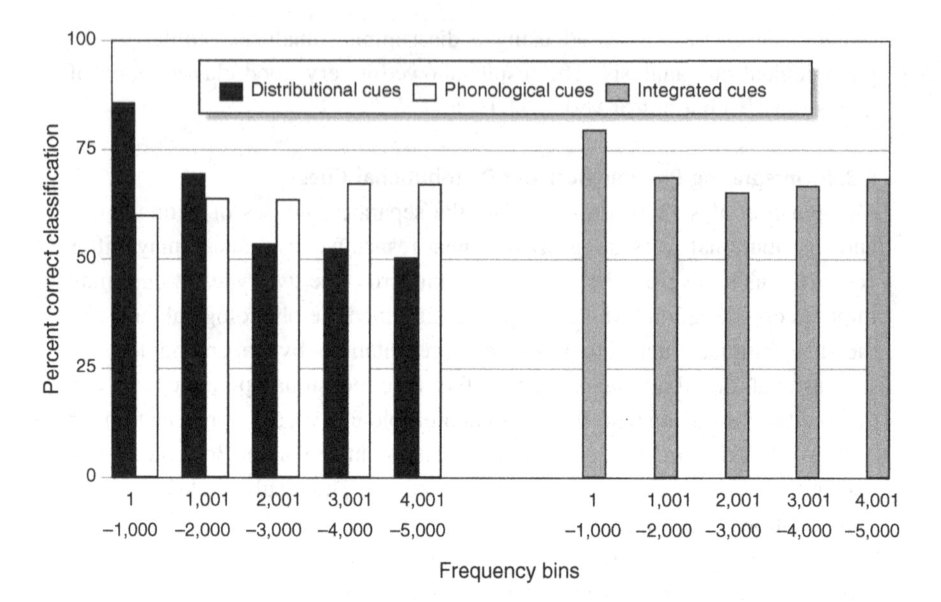

Figure 5.3
The percentage of nouns and verbs correctly classified according to their lexical category across different frequency bins for distributional (black bars) and phonological cues (white bars) treated separately, and when both cues are integrated with one another (grey bars) (based on Monaghan, Chater, & Christiansen, 2005).

evidence of similar kinds of cue information in other languages as well. However, many of the phonological cues used by Monaghan et al. (2005) were specific to English (table 5.1) and thus may not be readily applicable to other languages. Monaghan, Christiansen, and Chater (2007) therefore generated a set of fifty-three cross-linguistic phonological cues, including gross-level word cues (such as length), consonant cues relating to manner and place of articulation of phonemes in different parts of the words, and vowel cues relating to tongue height and position as well as whether the vowel was reduced. They then conducted analyses of child-directed speech in English, French, and Japanese. Using the new cues, they replicated the results of the previous study in terms of correct noun/verb classification (16 cues: 63.4% vs. 53 cues: 67.5%). These analyses were also replicated on a set of monomorphemic words, demonstrating that the results are not dependent on morphological information but stem from inherent phonological properties of the words. Noun/verb classification using phonological cues was also very good for both French (82%) and Japanese (82%). Classification performance was further improved across all three languages (English: 94%; French: 91.4%; Japanese: 93.4%) when the phonological cues were integrated with distributional cues (computed as before).

Together, the results of the corpus analyses show that across representatives of three different language genera—Germanic (English), Romance (French), and Japanese (Japanese)—child-directed speech contains useful cues for distinguishing between nouns and verbs (see also, Kelly, 1992). Note that this outcome does not appear to be dependent on the specific phonological representations used by Monaghan et al. (2007): the cross-linguistic results have been replicated using just the initial and final phoneme/mora of a word (Onnis & Christiansen, 2008). More generally, the results are consistent with the hypothesis that, as a result of the cultural evolution of language, and despite processing pressures for independence between the sound of a word (obtained from the immediate speech input) and its meaning and function (inferred from prior linguistic and environmental context), word forms nonetheless incorporate the sound of syntax: nouns and verbs differ in their phonology. Importantly, the specific cues differed considerably across languages, suggesting that each language has recruited its own unique set of cues to facilitate acquisition through multiple-cue integration. However, these analyses only demonstrate that there are probabilistic cues available for learning about aspects of syntax, not that such learning can be exploited effectively. Next, we shall see that a domain-general sequence learner can take advantage of both phonological and distributional cues to learn about syntax.

5.4 Multiple-Cue Integration by a Domain-General Sequence Learner

A potential concern regarding the viability of multiple-cue integration is the sheer number and variety of types of information that could potentially inform language acquisition. As noted by Pinker (1984, p. 49), "... in most distributional learning procedures there are vast numbers of properties that a learner could record, and since the child is looking for correlations among these properties, he or she faces a combinatorial explosion of possibilities. ... Adding semantic and inflectional information to the space of possibilities only makes the explosion more explosive." Pinker expresses a common intuition about the use of multiple, partially correlated sources of information by a domain-general learning device: that combining different kinds of partially reliable information can only result in unreliable outcomes. However, research in formal learning theory has shown that this intuition is incorrect. Mathematical analyses of neural network learning using the Vapnik–Chervonenkis (VC) dimension (Vapnik & Chervonenkis, 1971) have shown that multiple-cue integration with correlated information sources does not lead to a combinatorial explosion but instead to improved learning (Abu-Mostafa, 1993). The VC dimension establishes an upper bound for the number of examples needed by a learning process that starts with a set of hypotheses about the task solution. A cue (or "hint" in the engineering literature) may lead to a reduction in the VC dimension by weeding out bad hypotheses and reducing the number of examples needed to learn the solution. This holds even when one or more of the cues are either uncorrelated or otherwise uninformative with respect to the acquisition task, in which case they have no negative effect on learning (see Allen & Christiansen, 1996, for neural network applications, including the kind of network discussed below). In other words, the integration of multiple cues may speed up learning by reducing the number of steps necessary to find an appropriate function approximation, as well as reducing the set of candidate functions considered, potentially ensuring better generalization. Indeed, one of the most powerful classification methods in machine learning, support vector machines (Cristianini & Shawe-Taylor, 2000), project the classification into a space typically with a huge number of correlated dimensions, and then search for a hyperplane in that space. Dealing with very high dimensional spaces turns out to be entirely compatible with computationally efficient machine learning techniques. Thus, mathematically speaking, Pinker's intuitive concern about combinatorial explosion may not be well-founded.

Although the issue of combinatorial explosion may not be a problem in principle, it may nonetheless pose a considerable obstacle in practical terms. Christiansen and Dale (2001) sought to address this issue head-on by training simple recurrent networks (SRNs; Elman, 1990) to do multiple-cue integration. The SRN is a type of connectionist model that implements a domain-general learner with sensitivity to complex sequential structure in the input. This model is trained to predict the next element in a sequence and learns in a self-supervised manner from violations of its own expectations regarding what should come next. SRNs provide a small-scale computational model of key aspects of language processing, rather than attempting to capture the entire lexical, syntactic, and semantic complexity involved in human language processing. Importantly, though, this model processes and learns incrementally, in line with the Now-or-Never bottleneck described in chapter 4. Moreover, the SRN has been successfully applied to the modeling of both language processing (e.g., Elman, 1993)—including multiple-integration in speech segmentation (Christiansen, Allen, & Seidenberg, 1998) as well as syntax acquisition (Christiansen, Dale, & Reali, 2010)—and sequence learning (e.g., Botvinick & Plaut, 2004). As a model of human performance, the SRN has been shown to closely mimic the processing of different kinds of recursive linguistic constructions (Christiansen & Chater, 1999; Christiansen & MacDonald, 2009; see chapter 7) as well as the learning of sequentially-structured nonadjacent dependencies (Misyak, Christiansen, & Tomblin, 2010; see chapter 6). In addition, the SRN has also been applied to the modeling of potential coevolution between language and learners (Batali, 1994).

Christiansen and Dale trained their SRNs on a corpus of artificially generated child-directed speech, incorporating declarative, imperative, and interrogative sentences with subject-noun/verb agreement and variations in verb argument structure. In one simulation, the networks were provided with three partially reliable cues to syntactic structure (word length, lexical stress, and pitch change) and three cues not related to syntax (presence of word-initial vowels, word-final voicing, and relative speaker pitch). The results of the simulations indicated that the SRNs were able to ignore the three unrelated cues while taking full advantage of informative ones, as might be anticipated from the mathematical results (see Gogate & Hollich, 2010, for a discussion of how language learners may detect invariances in the input more generally).

The question remains, though, whether Christiansen and Dale's SRN model can scale up to deal with the kind of cues found in the corpus analyses described previously. To answer this question, Reali, Christiansen, and Monaghan (2003) trained SRNs on a full-scale corpus of natural speech directed at children between the ages of one year and one month to one year and nine months (Bernstein-Ratner, 1984). Each word in the input was encoded in terms of the sixteen phonological cues used in the Monaghan et al. (2005) corpus analyses (and shown in table 5.1). Given a word represented in terms of these phonological cues, the task of the networks was to predict the next lexical category in the utterance. The networks thus were provided with phonological cues as well as distributional information that could be learned from the co-occurrence of words in the input. A second set of networks was provided with distributional information only, by randomizing the phonological cues for each word (e.g., all instances of the word *dog* might be assigned the phonological cues for *walk*), thereby breaking any systematic relationship between phonological cues and lexical categories while maintaining the same input representations.

The simulation results revealed that the networks provided with systematic phonological cues as input were significantly better at learning to predict the next lexical category in a sentence, compared to the networks that learned from distributional information alone. Analyses of the networks' hidden unit activations—essentially their internal state at a particular point in a sentence given previous input—revealed that the networks used the phonological cues to place themselves in "noun state" when processing novel nouns and in a separate "verb state" when encountering new verbs.

These simulations suggest that a domain-general sequence learner can acquire aspects of syntactic structure via multiple-cue integration. Despite intuitions to the contrary, the feared combinatorial explosion does not prove to be a problem. Rather, the *right* cues are recruited to facilitate acquisition, because the language itself has evolved, through cultural evolution over many generations of learners, to be learnable by way of those very cues. For example, phonological cues promote better learning and better generalization to new words. But these very cues are informative precisely because the language has evolved to be learnable by generations of children who are attuned to the relevant cues in their native language during the first years of life, as we shall see next. The vocabulary, like the language in general, has come to reflect the conditions in which it is learned, processed, and used in communication.

5.5 Multiple-Cue Integration in Acquisition and Processing

The corpus analyses above indicated that there are useful phonological and distributional cues for language acquisition, and the SRN simulations demonstrated that a domain-general learner benefits from integrating them. But are children sensitive to such cues and able to use them during language acquisition? A growing body of evidence suggests that they are. After just one year of language exposure, the perceptual attunement of infants to various properties of the input likely allows them to make use of such probabilistic cues (for reviews, see Jusczyk, 1997, 1999; Kuhl, 1999; Pallier, Christophe, & Mehler, 1997; Werker & Tees, 1999). Through early learning experiences, infants already appear sensitive to the acoustic differences between function and content words (Shi, Werker, & Morgan, 1999) and the relationship between function words and prosody in speech (Shafer, Shucard, Shucard, & Gerken, 1998). Young infants are able to detect differences in syllable number among isolated words (Bijeljac, Bertoncini, & Mehler, 1993). In addition, infants exhibit rapid distributional learning (e.g., Gómez & Gerken, 1999; Saffran, Aslin, & Newport, 1996; see Gómez & Gerken, 2000; Misyak, Goldstein, & Christiansen, 2012; Saffran, 2003 for reviews), and perhaps most importantly, they are capable of multiple-cue integration (Mattys, Jusczyk, Luce, & Morgan, 1999; Morgan & Saffran, 1995).

However, although these studies showed sensitivity to different types of cues, they did not demonstrate that children actually use such information during acquisition. When it comes to exposure to new words, Storkel (2001, 2003) showed that preschoolers find it easier to learn novel words when these consist of phonotactically common sound sequences. Fitneva, Christiansen, and Monaghan (2009) sought to determine whether children could use the kind of phonological cues attested by the corpus analyses to learn about the syntactic role of words. Specifically, they conducted a word-learning study to investigate whether children implicitly use phonological information when guessing about the referents of novel words.

To create novel words that resembled either nouns or verbs in terms of their phonological cues, Fitneva et al. used a measure of phonological typicality, originally proposed by Monaghan, Chater, and Christiansen (2003). Phonological typicality measures how typical a word's phonology is relative to other words in its lexical category, and reliably reflects the phonological systematicity of nouns and verbs (Monaghan, Christiansen, Farmer, & Fitneva, 2010). Thus, what we refer to as "noun-like" nouns are typical in terms of their phonology of the category of nouns, and likewise "verb-like" verbs are phonologically typical of other verbs. As might be expected from its origins in the integration of multiple distinct cues, the difference between noun-like and verb-like is quite subtle, and not easy to discern intuitively. For example, the analysis shows that *fact* is a noun-like noun whereas it turns out that *myth* is a verb-like noun; on the other hand *learn* is a verb-like verb, whereas *thrive* is a noun-like verb.

Fitneva et al. (2009) created a set of novel words that were either noun-like or verb-like in their phonology and asked English monolingual second-graders to guess whether these words referred to a picture of an object or a picture of an action. The results showed that the children used the phonological typicality of the nonword in making their choices. Interestingly, and in line with the corpus analyses (Christiansen & Monaghan, 2006), verbs benefitted more from phonological cues than nouns.

Second-graders may be considered too experienced as language learners to serve as a suitable population with which to investigate the usefulness of phonological cues, especially if such cues are to be used to inform early syntactic acquisition. To address this objection, Fitneva et al. conducted a second study with another group of second-graders, who were enrolled in a French immersion program. The stimuli were the same as in the experiment with the monolingual children. However, whereas half the nonwords were verb-like and the other half noun-like with respect to English phonology, all the nonwords were noun-like according to French phonology. Two groups of the French-English bilingual children were tested, with the only difference being in the language used for the instructions. When given English instructions, the bilingual children behaved exactly like the monolingual English children, showing an effect of English phonological typicality. Notably, though, when the instructions were provided in French, the patterns of results changed, in line with French phonology. Hence, not only did the children seem to use phonological cues to make guesses about whether a novel word was a noun or a verb, but they were also able to do so after relatively short experience with the relevant phonology (less than two years of exposure for the children in the French immersion program).

The results of the word-learning study suggest that phonological cues may come into play early in syntax acquisition. Farmer, Christiansen, and Monaghan (2006) explored whether the ability to exploit the integration of multiple phonological cues might also extend into adulthood. Analyzing an existing database of word-naming latencies (Spieler & Balota, 1997), they found that the processing of words presented in isolation is affected by how typical their phonology is relative to their lexical category: noun-like nouns are read aloud faster, as are verb-like verbs. Similarly, Monaghan et al. (2010) analyzed a lexical decision database (Balota, Cortese, Sergent-Marshall, Spieler, & Yapp, 2004), revealing that people produce faster responses for words that are phonologically typical of their lexical category. Farmer et al. further showed that the phonological typicality of a word could even affect how easily it is processed in a sentence context. Indeed, for noun/verb homonyms (e.g., *hunts* as in *the bear hunts were terrible* ... versus *the bear hunts for food* ...), if the continuation of the sentence is incongruent with the phonological typicality of the homonym, then people both experience on-line processing difficulties and have problems understanding the meaning of the sentence.

Together, the results of the human experimental studies indicate that the use of multiple phonological cues during acquisition is sufficiently important that it becomes a crucial part of the developing language-processing system. The phonological properties of words facilitate lexical acquisition through multiple-cue integration and become an intrinsic part of lexical representations. As a consequence, adult language users cannot help but pay attention to phonological cues to syntactic structure when processing language. In this way, the impact of phonological cues on language acquisition can be observed in adulthood as the influence of phonological typicality on sentence comprehension.

5.6 The Cultural Evolution of Multiple Cues in Language

So far, we have shown that there are multiple probabilistic cues available in the input that can facilitate language acquisition and use, and that the specific constellations of cues differ across languages. We have suggested, in general terms, that languages end up with such intricate systems of multiple, partially overlapping, probabilistic cues through competing pressures from processing and acquisition, and indeed, historical language change. But to examine aspects of this hypothesis in more detail, Christiansen and Dale (2004) conducted a set of computational simulations to investigate whether cultural evolution could result in the recruitment of cues to facilitate the learning of more complex linguistic structure (see box 5.3 on the use of computational modeling to explore cultural evolution).

As a computational model of language learners, Christiansen and Dale employed the same type of connectionist model, the SRN, which we discussed above as capturing the integration of phonological and distributional cues in the service of language acquisition. The SRNs were trained on miniature languages generated by small context-free grammars, each derived from the grammar skeleton illustrated in table 5.2. The curly brackets indicate that the order of the constituents on the right-hand side of a rule can be either as is (head-first) or in the reverse order (head-final). The SRNs were expected to use the distributional information afforded by the order of words in the sentences as a cue to the underlying structure of the language. As additional cues to linguistic structure, the languages could recruit a constituent cue and a lexical cue. The constituent cue was coded by an additional input unit that marked phrase boundaries by being activated following the constituents from

Box 5.3
Understanding Language Evolution through Computational Modeling

Computational modeling has been one of the key factors in bringing about the change in focus from biological to cultural evolutionary processes in theorizing about language evolution (for reviews, see Smith, 2014, and chapter 2). Although modeling of the biological evolution of language does exist (e.g., Nowak, Komarova, & Nyogi, 2001), it has been easier to derive empirically testable predictions from the modeling of cultural evolution. For example, Kirby, Dowman, and Griffiths (2007) used Bayesian modeling to show that cultural transmission across generations of learners could turn weak biases into strong constraints on linguistic patterning. In subsequent work, Kirby, Cornish, and Smith (2008) confirmed such amplification of weak biases when cultural evolution was implemented by having human learners receive as input what a previous learner had produced as output, thus simulating the cross-generational transmission of a linguistic system. Similarly, a shared sign system can emerge culturally from interactions between artificial (Steels, 2003) and human (Fay, Garrod, & Roberts, 2008) agents. Phylogenetic modeling of existing language patterns relating to word order also points to the cultural evolution of language (Dunn, Greenhill, Levinson, & Gray, 2011), in line with typological linguistic analyses (Evans & Levinson, 2009).

More generally, the evolutionary simulations discussed in the current chapter illustrate how computational modeling may inform theories of language evolution by providing the means for evaluating current theories, exploring new theoretical constructs, and/or offering existence proofs that specific hypotheses could work (Christiansen & Kirby, 2003). First, the simulations constitute an explicit *evaluation* of the degree to which subtle learning biases (see also chapter 7) can drive the cultural evolution of linguistic structure (e.g., toward word-order regularities). Second, the simulations *explore* how multiple cues may interact to facilitate the cultural evolution of language. Third, the simulation results provide an *existence proof* that a culturally evolving linguistic system can recruit cues and thereby become easier to learn and process by domain-general learners. Moreover, computational simulations of cultural evolution also make it possible to derive empirical predictions about extant language, which can subsequently be tested using human experimentation.

Table 5.2
The grammar skeleton used by Christiansen and Dale (2004)

S	→	{NP VP}	VP	→	{V (NP)}
NP	→	{N (PP)}	PP	→	{adp NP}
NP	→	{N PossP}	PossP	→	{poss NP}

Note. S = sentence, NP = noun phrase, VP = verb phrase, PP = adpositional phrase, PossP = possessive phrase, N = noun, V = verb, adp = adposition, poss = possessive marker. Curly brackets indicate that the order of constituents can be as is or the reverse. Parentheses indicate that the constituent is optional.

a particular phrase structure rule (e.g., N (PP) #, where # indicates the activation of the constituent cue after the NP and optional PP). The lexical cue was coded by another input unit that could be co-activated with any of the twenty-four words in the vocabulary. Thus, there were three potential sources of information for learning about the structure of a language in the form of distributional, constituent, and lexical cues.

The selection pressures of cultural evolution were simulated by having five different languages "compete" against one another, with fitness determined by how easily the SRNs learned each language. At the beginning of a simulation, the five languages were randomly generated based on the grammar skeleton with a random combination of constituent and lexical cues. Each language was then learned by five different SRNs, with a language's fitness being computed as the average across the five networks. The most easily learned language, along with four variations of it, then formed the basis for the next generation of languages, each being learned by five networks. Again, the most easily learned language was selected as the parent for the next generation, and the process repeated until the same language "won" for 50 consecutive generations. Language variation was implemented by randomly changing two of the three cues: 1) changing the head-order of a rule, 2) adding or deleting the constituent unit for a rule, or 3) adding or deleting the co-activation of the lexical unit for a word. Ten different simulations were run, each with different initial randomizations.

Of the ten simulations, one never settled, but the results of the remaining nine simulations followed a similar pattern. First, all languages ended up with a highly regular head ordering, with at least five of the six phrase structure rules being either all head-initial or all head-final. This fits the general tendency for word-order patterns in natural languages to be either head-initial or head-final (e.g., Dryer, 1992)[5]. Second, the constituent cue always separated NPs from other phrases, consistent with evidence from corpus analyses indicating that prosodic cues, such as pauses and pitch changes, are used to delineate phrases in both English and Japanese child-directed speech (Fisher & Tokura, 1996). Finally, the lexical cue reliably separated word classes, with six of the runs resulting in the lexical cue separating function words from content words. This is similar to the acoustic (Cutler, 1993) and phonological cue-based (Monaghan et al., 2005) differentiation of function and content words observed in English. To place these results in context, it is important to note that given the combination of the three different cues in these simulations, there were more than 68 billion[6] different possible linguistic systems that could have evolved through cultural evolution. However, only a tiny fraction of these 68 billion linguistic systems incorporate properties that closely resemble those observed in current languages. Thus, it is not a trivial result that the simulations culminated in natural language-like linguistic systems.

The simulations by Christiansen and Dale (2004) suggest that linguistic systems can recruit cues to facilitate learning when undergoing cultural evolution. The integration of these cues, in turn, allows language to become more complex while still being learnable by domain-general mechanisms. These results thus dovetail with the previous corpus analyses, SRN simulations, and human experimentation showing that both phonological and distributional cues are available in the input and are used in language acquisition and processing. A similar process of cue recruitment through cultural evolution has also been observed in the newly emerging Al-Sayyid Bedouin Sign Language (ABSL) that we discussed previously in chapter 2. Specifically, Sandler (2012) describes how ABSL has evolved across four generations of signers by gradually recruiting additional articulators (head, face, body, nondominant hand) to perform novel grammatical functions. The piecemeal addition of new articulators resulted in increasing linguistic complexity, paralleling the effects of cue recruitment observed in the Christiansen and Dale simulations. This lends support to our hypothesis that natural language is a culturally evolved multiple-cue integration system.

5. Recently, Dunn, Greenhill, Levinson, and Gray (2011) have shown that word order correlations tend to be lineage specific rather than universal as proposed by Dryer (1992). The current approach is consistent with these results under the assumption that the history of a particular language provides additional constraints on the specific path along which a language changes.

6. The number of possible linguistic systems was calculated as follows. There were six rules with four variations: two head-orderings, each with or without the constituent cue, and each language had a twenty-four-word vocabulary, each with two variants (each word may or may not be associated with the lexical cue). The total number of possibilities is therefore $4^6 \times 2^{24} = 68,719,476,736$. This number is many orders of magnitude larger than the number of possible "natural language-like" systems, which we estimate as 14 (five out of six rules consistent) $\times\ 2^5$ (five different constituent cue combinations, collapsing over NPs) $\times\ 2^4$ (four word classes with/without cues) = 7,168. Thus, there is less than a one in ten million chance of ending up with a natural system.

5.7 Further Implications of Multiple-Cue Integration

As discussed in chapter 3, our emphasis on the cultural evolution of language suggests that there may be few true language universals, i.e., properties common to all languages. If such universals exist, they are likely to result from the domain-general constraints that have shaped the cultural evolution of language, rather than from a genetically encoded language-specific endowment. Multiple-cue integration, as discussed here, may be a plausible candidate for a universal property common to all languages. Crucially, though, multiple-cue integration is not unique to language but also plays a key role in, for example, vision (e.g., Tsutsui, Taira, & Sakata, 2005) and sensori-motor control (e.g., Green & Angelaki, 2010). Indeed, it has been suggested that overlapping, redundant cues might be a precondition for perceptual, cognitive, and even social development (Bahrick, Lickliter, & Flom, 2004). Thus, the recruitment of multiple cues in language processing can be seen as part of a broader pressure to maximize the use of information to produce Right-First-Time responses in time-critical aspects of perception, action and cognition (as discussed in chapter 4).

5.7.1 A Trade-Off between Arbitrariness and Systematicity in Vocabulary Structure

The results of the corpus analyses and human experiments reviewed in this chapter demonstrate that language strikes a delicate balance between arbitrariness and systematicity in form-meaning mappings. The acquisition of the initial vocabulary is facilitated by the systematic relationship between sound and meaning in early-acquired words. At this stage, systematicity enables knowledge about known words to be extrapolated to constrain the meaning individuation of new words. However, as the vocabulary grows, such generalizations become less informative at the individual word level, as arbitrariness increases in later-acquired words (Monaghan et al., 2014). We hypothesize that arbitrariness then comes to facilitate Chunk-and-Pass language processing, by keeping the information sources from language input and context as independent as possible, and hence allowing better anticipation and recognition of chunks as they arrive. In light of Gasser's (2004) simulations, it may also be that as vocabulary size grows, arbitrariness facilitates vocabulary learning. Despite these pressures for an arbitrary relationship between form and meaning, systematicity across multiple cues and across *groups* of words is nonetheless significant, allowing learners to exploit systematic correspondences between phonological forms and lexical categories when acquiring new words.

The change in the role of phonological cues over development—from meaning individuation to lexical categories—may also signal a change in the relative usefulness of those cues for learning about nouns and verbs. Thus, whereas Monaghan et al. (2014) found no differences between nouns and verbs in terms of the impact of form-meaning systematicity on age of acquisition, phonological cues to lexical categories appear to work better for verbs than for nouns, as evidenced by both corpus analyses (Christiansen & Monaghan, 2006) and developmental experimentation (Fitneva et al., 2009). Because verbs, in comparison to nouns, appear to be conceptually harder to learn (e.g., Childers & Tomasello, 2006) and occur in fewer reliable distributional contexts (Monaghan et al., 2007), phonological cues may be particularly important for the acquisition of verbs (Christiansen & Monaghan, 2006). The structure of the vocabulary may in this way reflect a functional pressure in form-meaning mappings toward facilitating verb learning through phonological cues.

More generally, there appears to be a division of labor between arbitrariness and systematicity in word form-meaning mappings, deriving from opposing pressures from the task of learning the meanings of individual words, on the one hand, and the process of discovering how to use these words syntactically, on the other. Monaghan, Christiansen, and Fitneva (2011) present results from computational simulations, human experiments, and corpus analyses indicating that whereas one part of a word's phonological form may have a primarily arbitrary form-meaning mapping to facilitate meaning individuation, another part of the same word tends to incorporate systematicity to assist in the acquisition of lexical category information. In their corpus analyses of English and French, for instance, word beginnings were found to carry more information for individuation, whereas word endings supported grammatical-category level discovery (see also Hawkins & Gilligan, 1988; St. Clair, Monaghan, & Ramscar, 2009). Having category information at the ends of words may further be reinforced by a domain-general bias toward suffixation (Hupp, Sloutsky, & Culicover, 2009).

From this viewpoint, the structure of the vocabulary will be a product of multiple forces. As we have argued, the need to Chunk-and-Pass as quickly as possible, in order to overcome the Now-or-Never bottleneck, exert a strong pressure for sound-based input to be used as independently as possible from contextual information about meaning or syntactic class, thereby providing a selective pressure across cultural evolution in favor of the arbitrariness of the sign. But, as we have seen, selective pressures on lexical items to be learnable can push in the opposite direction, favoring form-meaning systematicity that allows the learner to gain a foothold in learning the language. As vocabulary size grows, systematic form-meaning mappings may even make acquisition more difficult, and, as we have just noted, the balance between these opposing selectional forces may operate differentially on different aspects of lexical items. In short, the patterns of arbitrariness and systematicity in the vocabulary are subtle and complex, but this is just as we would expect when construing the vocabulary as a culturally evolving system subject to a rich variety of selectional pressures imposed by endless processing episodes across generations of language users.

5.7.2 Multiple Cues to Vocabulary and Syntactic Structure

In this chapter, we have proposed that language has evolved through cultural evolution to exploit multiple cues in the context of selectional pressures to be easy to process rapidly, to be as expressive as possible, and to be readily learnable by domain-general mechanisms; the resulting patterns in the vocabulary yield a complex balance between arbitrariness and systematicity. We have seen how evolutionary simulations indicate how language may recruit multiple cues to facilitate learning. A prediction from this perspective on the cultural evolution of language is that each language should "evolve" its own constellation of statistical cues; we should not expect specific cues to be universal across languages. Cross-linguistic corpus analyses have confirmed this prediction with regard to phonological and distributional cues, additionally showing that the relationship between the sound of a word and how it is used is by no means arbitrary. Computational simulations have demonstrated that domain-general learners can take advantage of these cues in the context of multiple-cue integration—as can children when learning new words and how to use them in the proper syntactic context. Adult sentence processing experiments further indicate that the use of phonological cues becomes a crucial part of an emerging language processing system, built on domain-general mechanisms for multiple-cue integration. Hence, language has evolved to rely on multiple-cue integration in both acquisition and processing, making the rapid integration of multiple cues central to the computational architecture of our language system.

Going beyond learning the meaning of individual words, we have focused our discussion on phonological and distributional cues to syntactic structure, but these are, of course, not the only sources of information available for multiple-cue integration. Phonological and distributional information is likely also to be integrated with other cues to language acquisition, including prosody (e.g., Fisher & Tokura, 1996), semantics (e.g., Bowerman, 1973), and pragmatics (e.g., Tomasello, 2003) (see Monaghan & Christiansen, 2008, 2014; Morgan & Demuth, 1996; Weissenborn & Höhle, 2001, for reviews). Whereas semantic and pragmatic cues pose challenges for the kind of quantitative corpus analyses discussed here as well as for computational modeling in general (though see

McCauley & Christiansen, 2014b, for a review of initial progress), prosodic cues may be more easily amenable to corpus analyses and computational modeling. We expect future work to take advantage of prosodic information as well as to make inroads into applying semantic and pragmatic constraints as input to multiple-cue integration in language acquisition and use.

Importantly, though, because different natural languages employ different constellations of cues to signal syntactic distinctions, a key question for further research is exactly how the child's learning mechanisms discover which cues are relevant and for which aspects of syntax. This problem is compounded by the fact that the same cue may point in different directions across different languages. For example, nouns tend to contain more vowels and fewer consonants than verbs in English, whereas nouns and verbs in French show the opposite pattern (Monaghan et al., 2007). Moreover, most work on cue availability in the child's environment makes the simplifying assumption that all information is available to the child simultaneously. Relatedly, current computational models tend to capture the end-state of learning rather the developmental process itself. But this is, of course, an oversimplification. The gradual transitions over many years in what children can say and understand indicate that the whole of language is not acquired in one fell swoop (e.g., Tomasello, 2003), but rather is built up in overlapping phases of acquisition, where learning at any given point in time relies on the gradual accumulation of developmental progress that took place beforehand (see also chapter 4). The traces of this developmental trajectory are likely to be significant in shaping the adult language, just as the adult phenotype is profoundly influenced by its developmental biology. Fortunately, given that language acquisition is a matter of C-induction (learning to co-ordinate with other people as discussed in chapter 3), any consequent complexities in the variety of patterns in the language, while puzzling to scholars, may pose few learnability problems for children—because each child will automatically be following in the developmental footsteps of the past learners with whom they must co-ordinate. In any case, to reveal the true extent of multiple-cue integration and how it changes across development, researchers must capture the developmental trajectory of cue use across different phases of language acquisition.

5.8 Summary

The discussion in this chapter provides an illustration of the rich complexity that we should expect to arise if language is a culturally evolving system subject to a wide range of pressures from individual processing episodes and ease of acquisition; and, like any other evolving system, the present state of the vocabulary is not only a result of present selection pressures, but also the product of a long and idiosyncratic history. Observing such richness in isolation, it might seem inconceivable that such patterns can be learned afresh by each new generation of children without some innate biological endowment—indeed, in the context of syntax, rather than the vocabulary, such arguments have often been made. But we have seen that this is entirely the wrong conclusion to draw; the rich patterns of systematic structure in the vocabulary, and, we would argue throughout all aspects of language, are readily learned and processed precisely because they have evolved to fit with our cognitive machinery and processing constraints (such as the Now-or-Never bottleneck). We note, too, that the complex and shifting developmental constraints on language acquisition are shared between children (and, indeed, with those past learners who are now the adults providing much of the child's linguistic environment), thus ensuring that learners follow a common pathway of acquisition. As we stressed in chapter 2, language is readily learned and processed because it has been shaped by the brain through cultural evolution, rather than the other way around through some mysterious process of language-gene coevolution.

Yet the brain can shape language into a myriad of forms, just as the same ecological territory can shape a huge variety of anatomical forms in biological evolution. There are many specific ways in which linguistic forms can be readily processed and learned, just as there are many ways in which organisms can adapt and survive in a particular environment. We suggest that the search for "language universals," or the attempt to derive the structure of a putative UG, are likely to be less productive than the specific analysis of the systematic patterns, whether in the vocabulary, phonology, syntax, or semantics found in individual languages, and the exploration of their potential cognitive foundations.

6 Experience-Based Language Processing

Our native language is like a second skin, so much a part of us we resist the idea that it is constantly changing, constantly being renewed.
—Casey Miller and Kate Swift, 1980, *The Handbook of Nonsexist Writing*

Although we may not give it much thought, we are generally aware that the language we speak changes gradually over time. We regularly come across new words that quickly enter into our vocabulary. For example, the word *selfie* has rapidly become part of everyday English, referring to a photo of oneself (sometimes with other people and/or at notable locations or events), typically using a smartphone and uploaded to social networking services, such as Facebook, Snapchat, or Twitter. In 2013, *selfie* was selected as Word of the Year for both US and UK English by the Oxford English Dictionary.[1] The word (and the concept to which it refers) has been widely adopted across the world, with famous selfies including one taken by the Danish Prime Minister Helle Thorning Schmidt with US President Barack Obama and British Prime Minister David Cameron at Nelson Mandela's memorial service in December 2013, and comedian Ellen DeGeneres' selfie with nine Hollywood celebrities at the 2014 Oscar ceremony (which became the most retweeted photo ever with more than 3.3 million retweets[2]). The word *selfie* has even spawned activity-related variations such as *helfie* (a selfie of one's hair), *welfie* (a selfie during a workout), and *drelfie* (a selfie taken while being drunk). Of course, as new words like *selfie* enter into our vocabulary, others gradually disappear (e.g., the word *fax* discussed in chapter 3), resulting in an ever-changing vocabulary.

Whereas the shifting nature of our vocabulary is readily apparent, our intuition suggests that other aspects of language, including how we pronounce

1. http://blog.oxforddictionaries.com/2013/11/word-of-the-year-2013-winner.

2. DeGeneres, Ellen (2 March 2014). "If only Bradley's arm was longer. Best photo ever. #oscars" (http://pic.twitter.com/C9U5NOtGap). Twitter. Retrieved March 31, 2015.

words (phonology) and put them together to form sentences (syntax), are much more stable. However, as we already noted in chapter 3, our phonology is also subject to change over our lifetime, as evidenced by the drift in the vowels produced by the Queen of England in her yearly Christmas messages recorded between the 1950s and 1980s (Harrington et al., 2000). It is even possible to lose one's native language accent in circumstances where a second language becomes the primary means of communication (de Leeuw, Schmid, & Mennen, 2010). Such "first language attrition" can occur when adult immigrants live in linguistic environments with limited opportunity for utilizing their first language, resulting in a foreign accent when speaking their native language. But what about syntax? Does our grammatical knowledge change across the lifespan?

The standard generative perspective tends to see grammatical knowledge as being largely fixed after language acquisition is completed in childhood. Pinker (1994, p. 294) expressed this perspective succinctly: "... learning a language—as opposed to *using* a language—is perfectly useful as a one-shot skill. Once the details of the local language have been acquired from the surrounding adults, any further ability to learn (aside from vocabulary) is superfluous." Note, too, that Pinker highlights the separation of acquisition from processing (*learning* vs. *using* a language) that is characteristic of generative approaches to language from current versions of the Principles and Parameters Theory (e.g., Crain, Goro & Thornton, 2006; Crain & Pietroski, 2006), to the Simpler Syntax framework (Culicover & Jackendoff, 2005; Jackendoff, 2007; Pinker & Jackendoff, 2005) and the Minimalist Program (Boeckx, 2006; Chomsky, 1995), as discussed in chapter 1.

In contrast, we view acquisition and processing as fundamentally intertwined. As we argued in chapter 4, acquisition involves learning to carry out incremental Chunk-and-Pass processing to overcome the effects of the Now-or-Never bottleneck. This perspective assigns a fundamental role to linguistic experience in developing the appropriate processing skills to deal with the continual onslaught of the input, given various cognitive and communicative constraints. A key prediction from this account is that, just like the acquisition of any other skill, language learning never stops; we should continuously be influenced by new language input, and constantly be upgrading our language processing abilities, including our knowledge of syntactic regularities. Of course, this does not mean that the language of adults and children changes at the same rate. Consider how adding a teaspoon of red coloring to a glass of water will change the overall color considerably, whereas adding the same amount of color to a bathtub full of water has little effect. Likewise, linguistic input that has a substantial effect on a child's emergent processing skills may have relatively little impact on an adult language system, shaped by many years of accumulated linguistic experience. Still, with sufficient exposure adult language can change too, just as the bathtub water will turn red if we add enough coloring.

If linguistic experience plays an important role in shaping our native language ability, we would expect substantial individual differences in people's language skills as a function of variation in input. It is well established that differences in linguistic input can result in dramatic differences in vocabulary acquisition (e.g., Hoff, 2006). For example, a classic study by Hart and Risley (1995) found that middle-class children heard more than three times the number of words per hour compared to children in welfare-recipient families. At just three years of age, children from middle-class families had a cumulative vocabulary that was more than twice as large as the children from families receiving welfare. These childhood differences in vocabulary have been found to predict later language outcomes in large-scale studies (e.g., Burchinal et al., 2011; Farkas & Beron, 2004).

Effects of social background have also been found for basic language processing skills in infancy (Fernald, Marchman, & Weisleder, 2013). Longitudinal research indicates that these differences in language processing skill are associated with variations in the amount and richness of maternal speech (Hurtado, Marchman, & Fernald, 2008). Variations in linguistic input further predict the complexity of children's later syntactic development, both in terms of comprehension and production (Huttenlocher, Vasilyeva, Cymerman, & Levine, 2002). These differences continue into adulthood, as evidenced, for instance, by the considerable differences in syntactic abilities observed as a function of education (Dąbrowska, 1997). Importantly, such differences do not stem merely from failures to process obscure sentences with strange grammatical constructions. For example, Street and Dąbrowska (2010) found that adult native speakers of English with less than eleven years of education had problems comprehending passive sentences describing simple transitive events such as *The girl was photographed by the boy.* Thus, there are substantial individual differences in grammatical ability across adult native speakers (see Dąbrowska, 2012; Farmer, Misyak, & Christiansen, 2012, for reviews).

We see individual differences in language comprehension and production as a reflection of the importance of linguistic experience in shaping the Chunk-and-Pass processing skills necessary to deal with the fast pace of language input. In this chapter, we focus on the processing of relative clauses as a window into the role of experience in shaping language-processing skills more broadly. We consider evidence concerning how the differential processing difficulty of specific types of relative clause constructions reflects their distributional properties, when considered in combination with pragmatic constraints. Variation in exposure to relative clauses is further argued to be a determining factor in producing differences in the processing of these constructions across individuals. Of course, the role of linguistic experience is also shaped by the various cognitive and communicative constraints discussed in chapters 2 and 4. In this chapter, we focus on the cognitive constraints governing our ability to deal with sequentially presented material, whether linguistic or not. These sequence-processing mechanisms allow us to continuously update and maintain our knowledge of language, promoting not only long-term linguistic stability but also allowing for modification of our grammatical abilities, and hence our knowledge of the grammar, in response to changes in the input.

6.1 Processing Biases as a Reflection of Linguistic Input

Practice is generally considered to be key to successful skill learning. Similarly, to elaborate on Periander's famous saying,[3] practice makes perfect when it comes to language acquisition and processing. Through repeated experience with Chunk-and-Pass processing of various types of linguistic structure, children acquire their native language and adults hone their processing abilities. Thus, our general language processing biases relating to different types of syntactic constructions should, to a large extent, reflect the distribution of those very constructions in our linguistic input. This perspective is consistent with a number of usage-based approaches to language acquisition (e.g., Tomasello, 2003), processing (e.g., Arnon & Snider, 2010), and change (Bybee, 2006). A number of psycholinguistic studies have demonstrated the impact of distributional biases on the incremental processing of sentences involving syntactic ambiguities (e.g., Crocker & Corley, 2002; Desmet, De Baecke, Drieghe, Brysbaert, & Vonk, 2006; Jurafsky, 1996; MacDonald, Pearlmutter, & Seidenberg, 1994; Mitchell, Cuetos, Corley, & Brysbaert, 1995). Here, however, we focus on the influence of statistical information on the processing of unambiguous sentences involving embedded relative clauses.

A relative clause is a kind of subordinate clause that is typically used in English to modify a previously encountered noun or noun phrase. The previous referent can either be the subject or the object of the relative clause, as shown in (1) and (2), respectively (with the relative clauses underlined).

3. Periander—one of the Seven Sages of Greece—is attributed with the saying "practice does everything" (Laertius, 1853, p. 45), which is often reformulated, or perhaps plainly misquoted, as "practice makes perfect."

(1) The reporter that attacked the senator admitted the error.
(2) The reporter that the senator attacked admitted the error.

In both sentences, *the reporter* is the subject of the main clause (*the reporter... admitted the error*). The two sentences differ in the role that *the reporter* plays in the relative clause. In subject relative clauses as in (1), *the reporter* is also the subject of the relative clause (*the reporter attacked the senator*). This contrasts with the object relative clause in (2), where *the reporter* is the object of the relative clause (corresponding to *the senator attacked the reporter*).

Psycholinguistic experiments have shown that subject relative sentences such as (1) are easier to process than their object relative counterparts such as (2) (e.g., Ford, 1983; Holmes & O'Regan, 1981; King & Just, 1991; for a review, see Gibson, 1998). Theories differ in how they explain the observed difference in processing difficulty between subject and object relative sentences. Structure-based accounts inspired by generative grammar suggest that there is a universal preference for syntactic gaps (from movement) in the subject position (Miyamoto & Nakamura, 2003). This account predicts that subject relatives should always be easier to process than object relatives, irrespective of functional or discourse considerations. In contrast, functional accounts ascribe the differential processing difficulty to cognitive or communicative constraints. For example, working-memory-based approaches suggest that object relatives are harder to process because they require the language system temporarily to store more incomplete dependencies (because *the reporter* is the object of the subordinate verb *attacked*) than subject relatives (e.g., Gibson, 1998; Lewis, 1996). Some memory-based theories further highlight possible interference between constituents held in working memory (e.g., Bever, 1970; Gordon, Hendrick, & Johnson, 2001; Lewis & Vasishth, 2005): in object relative clauses, the two noun phrases (*the reporter* and *the senator*) may interfere with one another prior to the processing of the subordinate verb (*attacked*). Accounts focusing on communicative constraints suggest that object relative sentences, when presented in isolation, violate discourse-based expectations for object relatives to provide information about a previously mentioned *discourse-old* referent (e.g., Fox & Thompson, 1990; Mak, Vonk, & Schriefers, 2008; Roland, Mauner, O'Meara, & Yun, 2012). In contrast, subject relative sentences are less problematic when presented in isolation, as they tend to involve *discourse-new* referents. Finally, experience-based approaches suggest that a key factor in explaining the processing of subject versus object relative clauses is their relative distribution in the linguistic experience of individual language comprehenders (e.g., MacDonald & Christiansen, 2002; Reali & Christiansen, 2007a,b).

The Chunk-and-Pass processing perspective provides a possible experience-based framework within which to bring the cognitive and communicative factors together. As noted in chapter 4, experience is key to facilitate chunking at various levels of linguistic representations. In particular, repeated exposure to specific syntactic constructions, such as relative clauses, will make them easier to process, and allow for more chunks to be kept in memory (see also Jones, 2012). Consistent with this account, Roth (1984) found that extra experience with processing relative clauses improved three- to four-year-old children's comprehension of these constructions in comparison with a control group that received an equal amount of exposure to sentences involving different syntactic structures. Importantly, measures of the children's working memory indicated that the improvements in relative clause processing were not associated with increases in working memory capacity (as might be expected from pure working memory accounts; e.g., Just & Carpenter, 1992). Instead, we suggest that more efficient Chunk-and-Pass processing abilities explain such experience-based facilitation of relative clause comprehension. If so, then our ability to process different types of relative clause constructions should largely reflect their distributional properties in natural language (see box 6.1 on the widespread effects of frequency throughout cognition).

6.1.1 Pronominal Relative Clauses

Pronominal relative clauses provide a straightforward way to test this prediction. Previous work has shown that the processing difficulty associated with object relatives such as (2) is much diminished when the embedded clause involves a personal pronoun (such as *you*) in (3) rather than a full noun phrase (such as *the senator*) (e.g., Gordon et al., 2001; Warren & Gibson, 2002).

(3) The reporter <u>that you attacked</u> admitted the error.

To determine whether the comparative ease of processing pronominal object relative clauses might reflect the distributional properties of English, Reali, and Christiansen (2007a) conducted a large-scale corpus analysis, using the American National Corpus (ANC) (Ide & Suderman, 2004), which contains over eleven million words from both spoken and written sources. The corpus contains morpho-syntactic tags that allowed subject and object relative clause sentences to be extracted. When considering relative clause sentences with embedded full noun phrases (as in 1 and 2), subject relatives were twice as frequent as object relatives (68.3% vs. 31.7%). In contrast, when considering pronominal relative clause sentences, the picture was reversed: object relatives occurred almost twice as often as subject relatives (65.5% vs. 34.5%).

Box 6.1
Frequency in Processing

Frequency has ubiquitous effects throughout cognition, not just in language processing. Direct tests have studied how people estimate frequencies of items of all kinds: words, objects, pictures, or actions in daily life, and for repeated items presented in an experiment (Hasher & Zacks, 1984; Sedlmeier & Betsch, 2002). The mind is also indirectly sensitive to the frequency of stimuli of all kinds. For example, when people repeat a task many times, their performance speeds up, in a lawful way: the "power law" of practice was first documented in a study of Cuban cigar rollers with varying numbers of years of experience (Crossman, 1959). Similarly, when a perceptual stimulus, such as a face, Chinese character or painting, is presented for a second or third time, it is processed more fluently; and in consequence is often more preferred (the "mere exposure" effect, Zajonc, 1980). More generally, contemporary theories of perception and action have proposed that the cognitive system aims to build a probabilistic model, which captures the statistical structure of the external world (e.g., Chater, Tenenbaum, & Yuille, 2006). Thus, when confronted with an ambiguous stimulus, the perceptual system prefers the most probable interpretation, i.e., that which will be encountered most frequently. Moreover, many cognitive architectures naturally reflect frequency information during learning (e.g., Anderson, 1993; Logan, 2002; Rumelhart & McClelland, 1986), and frequency information may provide crucial clues that help people to perform tasks as diverse as categorization, prediction, causal learning, and choice of action (e.g., Cheng, 1997; Gallistel & Gibbon, 2000; Tversky & Kahneman, 1973).

Processing speed is strongly related to frequency across multiple levels in the language processing system, consistent with the Chunk-and-Pass perspective. For example, frequent words are read more rapidly (*elk* vs. *cat*) (Morton, 1969); familiar verb-argument combinations (*cut the cake* vs. *cut the pudding*; see Clifton, Frazier, & Connine, 1984), common grammatical constructions (e.g., a main clause vs. a reduced relative clause; Frazier & Fodor, 1978), and frequent multi-word chunks (*don't have to worry*; Arnon & Snider, 2010) are more easily and rapidly understood. Similarly, across linguistic levels, frequency speeds language production (Jaeger, 2006; Jescheniak & Levelt, 1994) and acquisition (Ellis, 2002).

Reali and Christiansen (2007a) then looked more closely at the pronominal relative clause results, conducting separate analyses for five different types of pronouns: first-person pronouns (*I, we, me, us*), second-person pronoun (*you*), third-person personal pronouns (*she, he, they, her, him, them*), third-person impersonal pronoun (*it*), and nominal pronouns (e.g., *someone, something*). These additional analyses revealed an intriguing pattern in which the pronominal relative clauses with a personal pronoun showed a distributional bias toward object relatives (first-person pronouns: 82% object relatives; second-person pronouns: 74% object relatives; third-person pronouns: 68% object relatives). However, this tendency was reversed when the relative clauses involved impersonal (34% object relatives) or nominal (22% object relatives) pronouns.

But are pronominal object relatives involving personal pronouns easier to process than their subject-relative counterparts, as would be expected if processing is facilitated by distributional frequency? To test this prediction, Reali and Christiansen (2007a) conducted three self-paced reading experiments involving first-person (*I/me*), second-person (*you*), and third-person (*they/them*) personal pronouns, as well as a fourth experiment involving a third-person impersonal pronoun (*it*). Examples of the subject and object relative clause stimuli used in the experiment with first-person personal pronouns can be seen in (4) and (5), respectively.

(4) The lady that visited me enjoyed the meal.
(5) The lady that I visited enjoyed the meal.

In the experiment with second-person personal pronouns, the subject and object relative materials are exemplified by (6) and (7).

(6) The lady that visited you enjoyed the pool in the back of the house.
(7) The lady that you visited enjoyed the pool in the back of the house.

Because the personal pronouns *they/them* are so-called referring pronouns, they need to be grounded in prior context when they are processed online—i.e., there needs to be something for them to refer back to. The examples of the subject and object relative clause stimuli in (8) and (9) therefore include a short preamble (*According to the students*), which provides the referent for *they/them*.

(8) According to the students, the teacher that praised them wrote excellent recommendation letters.
(9) According to the students, the teacher that they praised wrote excellent recommendation letters.

In the last experiment, the antecedent referent for the impersonal third-person pronoun *it* occurred in a brief sentence prior to the sentence containing the subject or object relative clauses, as in (10) and (11), respectively.

(10) The minivan was really fast. The car that chased it lost control suddenly.
(11) The minivan was really fast. The car that it chased lost control suddenly.

All four experiments measured online sentence processing using the so-called moving-window, self-paced reading task (Just, Carpenter, & Woolley, 1982). This experimental paradigm is a "work horse" of psycholinguistic research, in which participants read sentences on a computer screen, one word at a time, pressing a key to reveal the next word. The amount of time spent on each word provides a sensitive index of the difficulty people have in processing various parts of a sentence. Reali and Christiansen (2007a) reasoned that possible differences in processing difficulty associated with the subject and object relative versions of the sentences should show up in the two-word region that contrasted the two relative-clause types: VERB—PRONOUN (subject relatives) vs. PRONOUN—VERB (object relatives) (e.g., *chased it* vs. *it chased* in 10 and 11). The bottom of figure 6.1 shows the results from the four experiments, with reading times averaged across the critical two-word region. The top of the figure shows the predictions from distributional patterns of the same pronominal relative-clause types from the corpus analyses. As predicted by these analyses, object relatives were processed faster in sentences

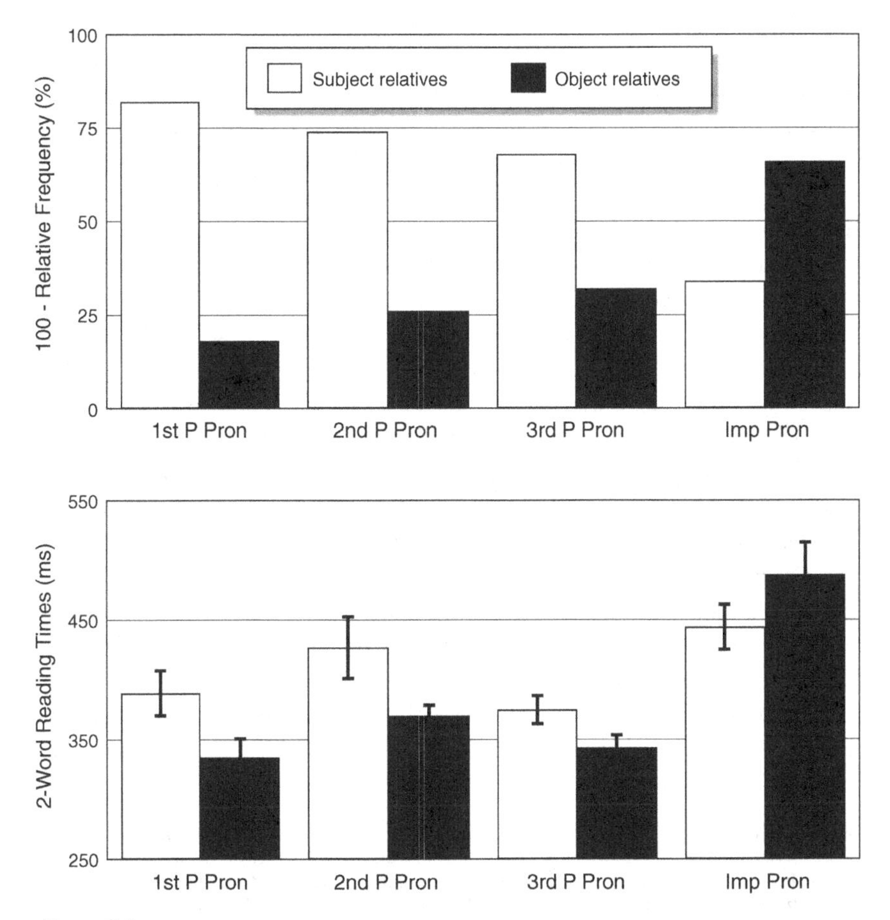

Figure 6.1
Results from the corpus analyses (top) and psycholinguistic experiments (bottom) for subject (white bars) and object relatives (black bars) involving personal and impersonal pronouns. The corpus-based predictions were generated by subtracting the observed relative frequency of a relative clause construction from 100% to yield qualitative predictions for reading times in the experiments. Error bars indicate the standard error of the mean. This figure is based on data from Reali and Christiansen (2007a).

with personal pronouns in the relative clauses (*I/me*, *you*, *they/them*). The opposite patterns were seen for impersonal pronouns (*it*), consistent with the hypothesis that Chunk-and-Pass processing of specific relative clause constructions is strongly influenced by their frequency of occurrence. A similar impact of distributional patterns is also observed cross-linguistically in children's acquisition of relative clauses, as we discuss further below.

The result of the experiment involving first-person personal pronouns (*I/me*) was replicated by Roland et al. (2012) with new items. They conducted additional analyses and experiments suggesting that differential discourse expectations for subject and object relatives provide an important source of constraints on relative clause constructions. Specifically, their results indicate that object relative clauses (whether involving pronouns or full noun phrases) tend to concern a discourse-old referent, whereas subject relative clauses tend to introduce discourse-new information. Extending this work, Heider, Dery, and Roland (2014) conducted corpus analyses of relative clause constructions involving the impersonal pronoun *it*. They found that if reduced relative clauses are included in the analysis (*The car that it chased … → The car it chased…*), then object relatives involving *it* are actually more frequent than the corresponding subject relatives, contrary to the original analyses of Reali and Christiansen (2007a). Additionally, Heider et al. were unable to replicate the results of Reali and Christiansen's sentence processing experiment with *it*, instead finding either no differences between subject and object relatives or faster processing of object relatives. They concluded that discourse-based expectations trump fine-grained frequency information at the level of specific relative clause patterns (*that* VERB *it* vs. *that it* VERB). However, although the Chunk-and-Pass processing perspective assigns a key role to top-down constraints from discourse anticipations in language comprehension, these are likely to be intertwined with fine-grained distributional expectations for specific word combinations.[4]

A study by Reali and Christiansen (2007b) provides initial support for this perspective. They compared the online processing of object relative clauses in

4. Indeed, the inconsistencies between the studies of *it* relative clause processing by Reali and Christiansen (2007a) and Heider et al. (2014) might be due to differences between the two studies in the fine-grained statistics of their stimuli. A preliminary analysis of these items using word chunk frequencies obtained from the Corpus of Contemporary American English (COCA; Davies, 2008) reveals subtle differences in the fine-grained statistics of the two stimuli sets, when considering the frequency of the three-word chunks *that* VERB *it* vs. *that it* VERB (where VERB refers to a specific verb such as *chased*). Moreover, the lack of difference between subject and object relatives that Heider et al. observed when replicating Reali and Christiansen's *it* experiment with the original stimuli may be attributed to experience-based differences between participant populations in their sensitivity to fine-grained statistics (students at an Ivy-league university vs. a state university—a possibility that Heider et al. highlight themselves).

which the combination of the first-person pronoun *I* and the subordinate verb either formed a high-frequency two-word chunk (*I loved* in 12) or low-frequency one (*I phoned* in 13):

(12) The actress who I loved phoned the comedian who presented his new show yesterday.
(13) The actress who I phoned loved the comedian who presented his new show yesterday.

They found that even though both sentence types were equally easy to understand offline, the sentences with the high-frequency *I*-VERB chunks were processed significantly faster. Additional analyses further demonstrated that not only did the frequency of the *I*-VERB chunk predict reading times for that critical region, but it was also associated with faster processing of the subsequent main verb. Crucially, these effects of fine-grained distributional information cannot be explained by discourse expectations or working memory constraints, as these are constant across the two sentence types.

Future work is needed to determine how fine-grained distributional cues interact with discourse-based expectations in relative clause processing. However, the Chunk-and-Pass perspective suggests that both types of constraints play a key role and are likely to be strongly affected by experience with language. We would therefore expect differences across development and between individuals in how distributional information and discourse factors will influence specific patterns of language comprehension and use, especially when combined with the variety of other cues discussed in chapter 5.

6.1.2 Cross-Linguistic and Developmental Patterns of Relative Clause Processing

The importance of fine-grained statistical information for relative clause processing becomes further evident when considering languages other than English. For example, Reali (2014) observed an English-like pattern in the distribution of relative clauses with pronouns and full noun phrases in Spanish, but also discovered a more fine-grained distributional pattern within object relative clauses. Spanish is a so-called pro-drop language, meaning that pronouns in object relative clauses may be dropped because verb inflection contains sufficient information about the pronominal subject of the subordinate clause. Thus, the pronoun *nosotros* (we) may be dropped in (14) without loss of information in Spanish because the inflection of the verb *perseguimos* unambiguously indicates that the subject is a first-person plural pronoun.

(14) El sapo que nosotros perseguimos.
[The toad that we chased]

Reali found that Spanish object relatives overwhelmingly tended to drop pronouns (whereas subject relatives always have overt pronouns), and incorporated changes to the order of subject (S) and verb (V), depending on whether the object relative involved a full noun phrase (VS) or an overt pronoun (SV). A subsequent self-paced reading-time study showed that Spanish speakers are sensitive to such fine-grained information. The results cannot be easily explained by discourse factors because these were held constant across the variations of the object relative clauses, using the same words in the alternative word orders. These and other cross-linguistic results (e.g., see contributions in Kidd, 2011) point to language-specific patterns of relative clause processing that are likely to result from the differential weighing of multiple factors, including fine-grained statistical information, working memory constraints and discourse-based expectations (see also Kidd & Bavin, 2002; O'Grady, 2011).

Similarly, cross-linguistic patterns of development provide further insight into the nature of relative clause processing. In a study with three-, four-, and five-year-old English speaking children, Kidd and Bavin (2002) found that the ability to deal with center-embedded relative clause sentences (such as 1–11 above) emerges gradually during development. In the study, children were asked to act out sentences in which relative clauses were either right-branching (as underlined in 15–16) or center-embedded (as underlined in 17–18).

(15) The kangaroo stands on the goat <u>that bumped the horse</u>.
(16) The horse jumps over the pig <u>that the kangaroo bumped</u>.
(17) The cow <u>that jumped over the pig</u> bumps the sheep.
(18) The sheep <u>that the goat bumped</u> pushes the pig.

Kidd and Bavin found that by four years of age, children had a good grasp of right-branching subject (15) and object (16) relative clauses, whereas their mastery of center-embedded subject (17) and especially object (18) relatives lagged behind. From a Chunk-and-Pass perspective, right-branching relative clause constructions are comparatively easy to process because they do not require "open" chunks (i.e., the main clause in 15 and 16 can be chunked into a single unit and passed up for further processing). In contrast, when processing center-embedded relative clauses (in 17 and 18), the chunking of the main clause cannot be completed until the relative clause has itself been chunked, creating a memory load that interferes with the processing of incoming input. Only once chunking becomes faster and more efficient—as a function of repeated experience—do children become better at processing center-embedded relative clauses (see O'Grady, 2011, for a related perspective). Thus, the Chunk-and-Pass perspective provides a potential processing account of children's gradual acquisition of complex sentence structure through *clause expansion*, as advocated by Bloom (1991), Diessel (2004), Kidd and Bavin (2002), and Tomasello (2003) (we discuss this issue further in chapter 7).

Interestingly, experience with specific types of relative clause constructions appears to play a key role in the development of children's processing abilities. Indeed, several studies have shown cross-linguistically that children's processing improves markedly when they are presented with relative clause constructions that are representative of the input they receive (e.g., Brandt, Kidd, Lieven, & Tomasello, 2009; Arnon, 2010; Kidd, Brandt, Lieven, & Tomasello, 2007). For example, Arnon (2010) reports results from an analysis of Hebrew child-directed speech as well as child productions, finding in both cases that object relative clauses tended to involve pronouns (or have no overt subject), whereas subject relatives primarily contained full lexical noun phrases. When testing three- to five-year-old Israeli children on Hebrew relative clause constructions, she found that they performed better on pronominal relative clauses, especially when it came to object relatives. However, in contrast to the results with adults (Reali & Christiansen, 2007a; Roland et al., 2012), the presence of pronouns did not make subject relatives easier to comprehend and produce than object relatives. As Arnon notes, this might be because the experimental items did not closely reflect the fine-grained distributional information regarding object relatives. In the children's input, the subject of the main clause, which is being modified by the object relative clause, is inanimate most of the time (as underlined in 19), whereas the experimental items all involved an animate main-clause subject (as underlined in 20).

(19) The ball that I like to play with.
(20) The girl that I am drawing.

Indeed, Kidd et al. (2007) found that children are sensitive to such animacy constraints in English and German and that, when presented with sentences in line with such constraints, they process object relatives as easily as subject relatives. In a related study, Brandt et al. (2009) tested English- and German-speaking three-year-olds' comprehension of relative clauses using a referential task in which children were asked to pick a toy in response to questions such as (21) and (22):

(21) Can you give me the ball that he just threw?
(22) Can you give me the donkey that he just fed?

As predicted by corpus data, children were much better at understanding pronominal object relatives modifying an inanimate noun (as in 21) rather than an animate noun (as in 22). Moreover, when animacy was taken into account, pronominal object relatives were better understood than pronominal subject relatives.

From the viewpoint of Chunk-and-Pass processing, the picture that emerges from these studies is that repeated exposure to specific constructions over development (and in adulthood) may result in the representation of chunk-related information at different levels of granularity, from individual word combinations (e.g., involving specific pronouns, such as *I*, as shown by Reali & Christiansen, 2007b) to abstract constructions (subject relatives: *that—VERB—NP* – object relatives: *that—NP—VERB*). Top-down information from discourse expectations may further lead to the formation of prototypical schemas, such as INANIMATE NP—PRONOUN—VERB for object relatives (Brandt et al., 2009). Importantly, though, such schemas are themselves probabilistic in nature, reflecting the distribution of the relevant patterns in the language in general (e.g., as suggested cross-linguistically by the results of Kidd et al., 2007, and Reali, 2014). The distribution of these patterns changes across time as children receive an increasing proportion of their input from reading: children's production of object relatives actually drops compared to passive relative clauses (e.g., *The toy that was carried by the girl*) with increased reading experience, reflecting the higher proportion of the latter in written language (Montag & MacDonald, 2015). More generally, variations in the properties of the language being learned (e.g., regarding word-order flexibility, availability of cues from gender and case markings, etc.) will result in different processing biases (e.g., O'Grady, 2011) and different rates of acquisition (e.g., Kidd et al., 2007), as children learn to integrate the multiple cues relevant for Chunk-and-Pass processing of relative clauses in their native language.

6.2 Individual Differences in Chunk-and-Pass Processing

Cross-linguistic data on the acquisition and processing of relative clauses have revealed a strong relationship between the patterns of different types of subject and object relatives as they occur in everyday language and the difficulty with which they are processed by children and adults (e.g., Arnon, 2010; Kidd et al., 2007; Reali & Christiansen, 2007a). This highlights the role of linguistic experience in shaping Chunk-and-Pass processing relative to a specific language, in some cases down to the level of particular word combinations (e.g., Arnon & Snider, 2010; Bannard & Matthews, 2008; Reali & Christiansen, 2007a). But given that the breadth and variety of linguistic experience can vary quite substantially (e.g., as a product of social, economic, and educational factors), does this then result in substantial differences in people's language abilities?

There are, as we have noted, indeed substantial differences in language ability—even within the otherwise fairly homogenous group of undergraduate students often used in psycholinguistic experiments (see Dąbrowska, 2012; Farmer, Misyak, & Christiansen, 2012, for reviews—see also box 6.2). These differences have been attributed to several factors, including variations in working memory capacity (e.g., Just & Carpenter, 1992), cognitive control (e.g., Novick, Trueswell, & Thompson-Schill, 2005), and perceptual processing (e.g., Leech, Aydelott, Symons, Carnevale, & Dick, 2007). Here we focus on the potential role of language experience in explaining individual differences in the processing of relative clauses.

Box 6.2
The Standard View of Individual Differences in Language Processing

There are considerable variations in syntactic processing abilities across individuals, not only across people with different educational backgrounds (e.g., Dąbrowska, 1997) but also within the narrow group of young college-aged individuals used in most psycholinguistic studies (e.g., Farmer, Misyak, & Christiansen, 2012). In addition to relative clause processing, which is the focus of the current chapter, individual differences have also been demonstrated in other aspects of sentence processing, including in the ability to resolve syntactic ambiguities (MacDonald, Just & Carpenter, 1992), the use of context in such ambiguity resolution (Just & Carpenter, 1992), and the interpretation of pronouns (Daneman & Carpenter, 1980). But how are such individual differences explained?

Taking a universal grammar perspective implies that there are no significant individual differences in linguistic competence: each speaker is endowed with the very same innate grammatical knowledge. Differences in language ability therefore are assumed to derive only from variations in vocabulary and performance factors, such as variations in working memory capacity across individuals. From this viewpoint, the role of linguistic experience is seen as largely irrelevant, restricted to affecting the acquisition of vocabulary and idioms.

The role of experience is generally also underplayed in theories that explain individual differences in terms of cognitive constraints. Syntactic knowledge is typically considered to be relatively static and thus assumed to contribute little to explanations of variation in language processing across individuals. Instead, the focus is on biological differences in the relevant cognitive factors. For example, Just and Carpenter (1992) proposed an account in which variations in working memory capacity determined the ability of an individual to process syntactic structure. On this account, syntactic knowledge is stored in a grammar and words are stored in a lexicon. The grammar and lexicon are assumed to be separate both from one another, and from a working memory where syntactic information is brought together with relevant lexical material during the processing of a sentence. The size of the working memory capacity determines how much information can be maintained and used for processing. A smaller capacity allows less information to be maintained, creating processing problems when dealing with complex syntactic structures such as the long-distance dependencies in relative clauses. This working memory capacity perspective on individual differences leaves little room for experience. In contrast, we argue that experience is crucial for determining individual variations in the Chunk-and-Pass processing, which in turn determines the capacity of an individual for language processing.

6.2.1 Individual Differences in Relative Clause Processing

King and Just (1991) provided some of the first systematic evidence of individual differences in relative clause processing. In a self-paced reading task, they presented participants with sentences containing subject and object relative clauses with full noun phrases (as in 1 and 2, repeated here as 23 and 24).

(23) The reporter that attacked the senator admitted the error.
(24) The reporter that the senator attacked admitted the error.

As a measure of individual differences in verbal working memory for language, they further administered a reading span test (Daneman & Carpenter, 1980) in which participants read aloud progressively larger sets of sentences, one at a time, while retaining the sentence-final words for later recall. The number of sentence-final words that could be correctly recalled by participants—their reading span—was then taken as a measure of their working memory capacity for language processing. King and Just found a main effect of working memory capacity at the main verb (*admitted*), where processing complexity is high: high-span individuals read sentences faster than low-span participants. They additionally obtained a main effect of sentence type: subject relatives were read faster than object relatives. Finally, they observed an interaction between working memory capacity and sentence type: while both high- and low-span participants read the subject relatives at about the same pace, the low-span participants read the object relatives significantly more slowly than the high-span individuals. King and Just interpreted these results as suggesting that only high-span individuals had sufficient working memory capacity to process object relatives without too much difficulty (see also Just & Carpenter, 1992).

MacDonald and Christiansen (2002) provided an alternative interpretation of the King and Just (1991) results, suggesting that the individual differences observed in this study derive from variations in linguistic experience rather than putative differences in working memory capacity. Specifically, they noted that when processing subject relatives (*that attacked the senator*), it is possible to piggyback on the processing of simple transitive sentences (*the reporter attacked the senator*) because both involve the canonical English (S)VO word order. In contrast, this is not possible when processing object relatives (*that the senator attacked*) because the object of the embedded clause (*reporter*) occurs before the subject, yielding a non-canonical (O)SV word order that does not map onto the structure of the corresponding simple transitive sentence (*the senator attacked the reporter*). This means that experience with simple transitive sentences will not facilitate the processing of object relatives. Instead, proficiency in processing object relatives therefore requires direct experience with such constructions, whereas the processing of subject relatives can rely to a large degree on structural overlap with the frequently occurring simple transitive sentences. Brandt et al. (2009) put forward a similar argument with regard to the acquisition of relative clauses in English and German, while pointing to the opposite pattern in Chinese: object relatives follow the canonical word order, apparently making them easier to process than subject relatives (Hsiao & Gibson, 2003; O'Grady, 2011; but see Vasishth, Chen, Li, & Guo, 2013, for an alternative experience-related perspective).

6.2.2 The Frequency × Regularity Interaction in Language Processing

This experienced-based explanation of the King and Just (1991) results can be seen as exemplifying the broader relevance of linguistic experience in fine-tuning Chunk-and-Pass processing skills across development and into adulthood. The relationship between linguistic exposure and structural overlap inherent in this interpretation is, moreover, an instance of the Frequency × Regularity interaction that we mentioned in chapter 2. This interaction has been observed across many levels of linguistic processing, from auditory (Lively, Pisoni, & Goldinger, 1994) and visual (Seidenberg, 1985) word recognition, to English past tense acquisition (Hare & Elman, 1995) and aspects of sentence processing (Juliano & Tanenhaus, 1994; Pearlmutter & MacDonald, 1995). The Frequency × Regularity interaction suggests that direct exposure is needed for irregular patterns, whereas the overlap between regular patterns allows experience to be accumulated across different instances.

As an example, consider word recognition. Regularly spelled words, which have a fairly straightforward mapping from letters to sound, are easy to recognize, independent of frequency, because they can be read by virtue of the reader's experience with similar words. Thus, even though *pave* is relatively rare, its recognition is facilitated by pattern overlap with similarly spelled words, such *gave, save, rave, cave, shave*, etc. In contrast, the recognition of irregular words is highly frequency sensitive: both *pint* and *have* are irregularly spelled words (cf. *lint, mint, tint* and *gave, save, cave,* respectively) but because *pint* has a low frequency of occurrence, it is much harder to recognize than *have*, which occurs very frequently. Crucially, Seidenberg (1985) found that amount of reading experience modulates the effects of the Frequency × Regularity interaction: skilled readers more easily recognized irregularly spelled words than poor readers. MacDonald and Christiansen (2002) suggested that differential experience similarly gives rise to the individual differences in relative clause processing observed by King and Just (1991).

Subject relatives have the standard VO word order characteristic of English; they are "regular" in the sense of the Frequency × Regularity interaction.

Chunk-and-Pass processing of subject relatives is therefore facilitated by extensive experience with the same VO structure in simple transitive declarative sentences. However, object relatives have an "irregular" OV order, so that Chunk-and-Pass processing of object relatives will be sensitive to the frequency with which those constructions occur. The amount of linguistic experience therefore should affect object relatives more than subject relatives, providing an alternative explanation of the King and Just (1991) results. From this viewpoint, participants described as having "high working memory capacity" perform better on object relatives because (a) they read more, (b) reading increases exposure to both subject and object relatives, and (c) increased exposure is more important for object relatives than subject relatives. This experience-based explanation challenges standard working memory accounts (e.g., Gibson, 1998; Just & Carpenter, 1992) but is consistent with the role of experience in reading proficiency (e.g., Stanovich & Cunningham, 1992, 1993) as well as in skill acquisition more generally (e.g., Ericsson & Kintsch, 1995).

6.2.3 Simulating the Role of Experience in Relative Clause Processing

To demonstrate the differential role of experience for subject and object relative clauses predicted by the Frequency × Regularity interaction, MacDonald and Christiansen (2002) conducted a set of computational simulations involving Simple Recurrent Networks (SRNs; Elman, 1990; as introduced in chapter 5). These networks have a set of recurrent connections that allow them to learn to process sentences, one word at a time. MacDonald and Christiansen created 10 different SRNs, each randomized with a different set of initial weights, and each exposed to a different corpus consisting of 10,000 sentences. The aim was, very roughly, to capture the fact that language learners approach acquisition with different initial conditions and are exposed to different samples of their native language. To test the potential role of the Frequency × Regularity interaction in explaining individual differences in relative clause processing, the corpora were designed so that they contained 95% simple (transitive/intransitive) sentences and 5% sentences with relative clause constructions (equally divided between subject and object relatives[5]). Experience with language was manipulated by allowing the networks one, two, or three exposures to the corpus. After training, each network was tested on a separate test set,

5. As discussed above, naturally occurring English tends to have a ratio of 2/3 subject relatives to 1/3 object relatives involving full noun phrases (as in the King & Just, 1991, experiment). However, in order not to bias their results, MacDonald and Christiansen (2002) adopted an equal distribution of subject and object relatives in their simulations.

involving ten novel subject relatives and ten novel object relatives, not seen during training.

MacDonald and Christiansen (2002) trained their SRNs on a prediction task in which they had to predict the next word in the sentence being processed. This means that the networks learned a probability distribution of possible next items given previous context. To assess performance, MacDonald and Christiansen calculated the Grammatical Prediction Error (GPE; Christiansen, & Chater, 1999), which measures the SRN's ability to make grammatically correct predictions for a specific point in a sentence. GPE maps onto human reading times, with low GPE values reflecting a prediction of fast reading times, and high GPE values indicating slow predicted reading times. The simulation results are shown in figure 6.2 (lower panels) along with the original King & Just (1991) results (upper panels). As predicted by the Frequency × Regularity interaction, additional experience benefitted the processing of object relatives more than subject relatives. Direct experience is needed to build up efficient Chunk-and-Pass processing of object relatives, whereas the processing of subject relatives can piggyback on the processing of simple declarative sentences. Notably, the simulations revealed a pattern similar to the King and Just study[6], in which less trained networks resemble "low-capacity" readers and more experienced networks look like "high-capacity" readers. This suggests that individual differences, previously attributed to variations in working memory capacity, may be better explained in terms of different amounts of linguistic experience. Thus, from a Chunk-and-Pass processing perspective, the reading span task is simply another measure of language processing skill[7], rather than a measure of a dedicated working memory capacity (see also Ericsson & Kintsch, 1995; MacDonald & Christiansen, 2002; Martin, 1995).

6.2.4 Inducing Individual Differences in Relative Clause Processing

To further underscore the importance of linguistic experience for explaining individual differences in relative clause processing, Wells, Christiansen, Race,

6. There may appear to be a discrepancy between MacDonald and Christiansen's (2002) simulations of the subject relative constructions and the results from King and Just (1991): the SRNs have less difficulty with processing the final word (*senator*) in the subordinate clauses compared to human readers. This is probably due to variations in the length of the subject relative clauses in the King and Just materials. When the length is uniform (as in the simulations), readers tend to experience less slow-down, as can be seen in figure 6.3 below.

7. Note, however, that because an individual's language processing skill reflects a number of interacting factors—including language exposure as well as a host of domain-general components—there will not be a perfect one-to-one mapping between reading span scores and linguistic experience (see Farmer, Fine, Misyak, & Christiansen, in press, for discussion).

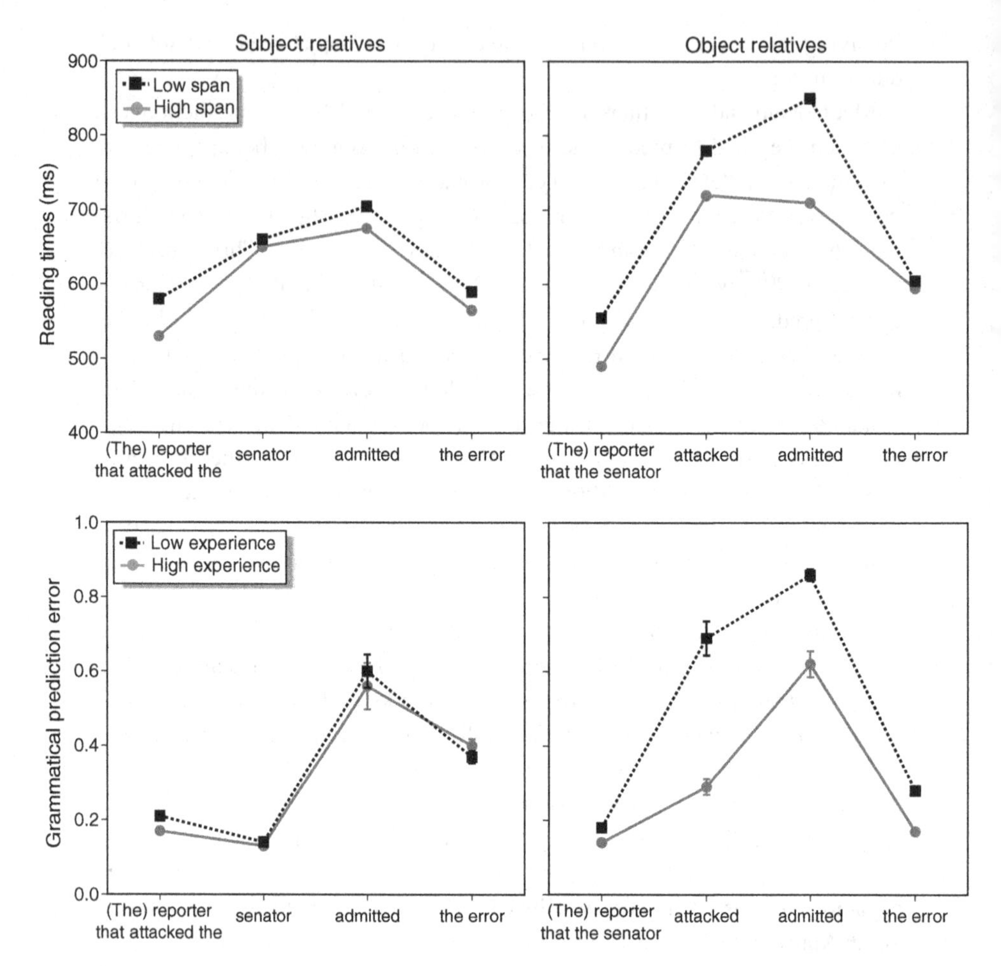

Figure 6.2
Processing patterns for subject (left) and object (right) relative clauses: (upper panels) human reading times for low- and high-span individuals from King and Just (1991) and (lower panels) mean Grammatical Prediction Error for low- and high-experience SRNs from MacDonald and Christiansen (2002) (error bars indicate standard error of the mean).

Acheson, and MacDonald (2009) directly manipulated people's exposure to language. Specifically, by analogy to the SRN simulations, they wanted to determine whether individual differences in relative clause processing could be induced simply by providing people with more opportunity to process relative clauses. Wells et al. first used the standard self-paced reading task to assess their participants' baseline processing of subject and object relative sentences (similar to the sentences used by King & Just, 1991). Over three separate exposure sessions, four to eight days apart, participants were then asked to read a total of eighty subject and eighty object relative sentences, such as (25) and (26), respectively (along with a number of filler sentences).

(25) The police officers that searched for the missing child discovered several homeless children in an abandoned house.
(26) The former policeman that the store manager hired caught a thief red-handed.

The exposure sentences were presented one by one, with each sentence appearing all at once on the computer screen (in contrast to the word-by-word presentation in the self-paced reading task). After the participants had read a sentence, it would disappear, and they were asked to choose between one of two statements, only one of which was compatible with the meaning of the original sentence. This comprehension probe ensured that the participants read the exposure sentences for their meaning. Participants were then tested again using the self-paced reading task (with novel stimuli) on their ability to process subject and object relative clauses four to eight days after their final exposure session. To ensure that any potential improvements in relative clause processing were due to experience with these specific syntactic constructions, a control group went through the same testing and training regime as the experimental participants but with exposure to complex sentences involving either sentential complements (as in 27) or conjoined sentences (as in 28) instead of relative clauses constructions.

(27) The angry prosecutor denied that the police had tainted the evidence from the crime scene.
(28) The police officers searched for a missing child and discovered several homeless children in an abandoned house.

These control-group training sentences were also chosen to cover similar semantic themes, use similar words, and be of similar average length. Finally, the two groups were matched on both reading span and basic reading skill prior to the exposure manipulation.

As expected, Wells et al. (2009) found that the difference in processing difficulty between the subject and object relative clauses became smaller for the participants in the experimental condition as compared to the control group, demonstrating the expected effect of relative clause experience. Figure 6.3 (upper panels) shows the improvement from before and after training for the experimental participants. Consistent with predictions from MacDonald and Christiansen's (2002) simulations, experience with relative clause processing facilitated object relatives more than subject relatives (Hutton & Kidd, 2011, observed a similar differential experience-related effect on structural

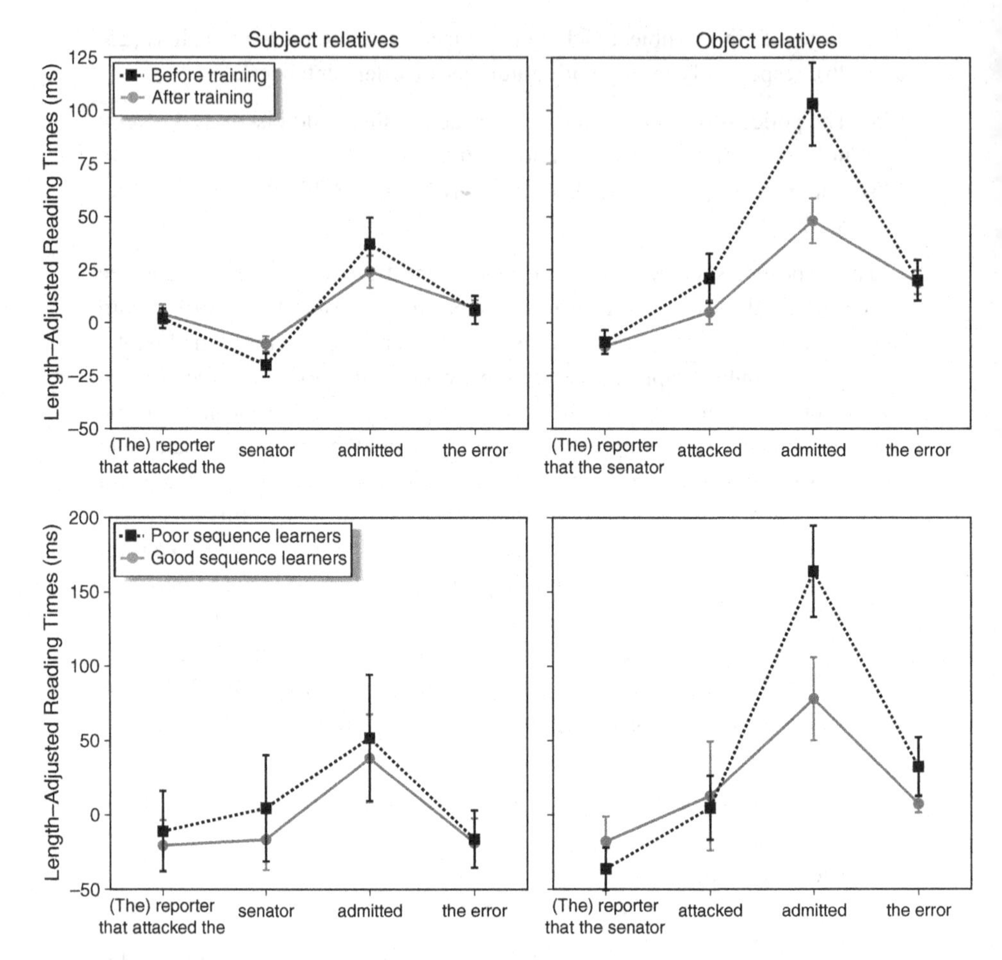

Figure 6.3
Processing patterns for subject (left) and object (right) relative clauses: (upper panels) reading times for experimental participants before and after training in Wells et al. (2009) and (lower panels) reading times for individuals with poor and good sequence learning skill from Misyak et al. (2010). In both studies, reading times are length-adjusted to control for word length (error bars indicate standard error of the mean).

priming of subject and object relatives). After training, the experimental group performance closely resembled the participant group labeled as "high-capacity" readers in the King and Just (1991) study, whereas before training the very same participants looked like "low-capacity" readers. Thus, as predicted by the Frequency × Regularity interaction, variations in adults' experience with relative clauses can explain processing differences previously attributed to working memory capacity.

6.2.5 The Role of Sequence Learning and Chunking in Relative Clause Processing

So far, we have discussed evidence that highlights the role of language experience in Chunk-and-Pass processing—in particular as related to relative clauses. Even a short amount of exposure within a single experimental session can change processing patterns, as shown in studies of so-called *syntactic adaptation* (e.g., Farmer, Monaghan, Misyak, & Christiansen, 2011; Fine, Jaeger, Farmer, & Qian, 2013). But what cognitive mechanisms could mediate such effects of linguistic experience? In chapter 2, we pointed to sequence learning as one of the cognitive mechanisms subserving language (see also Calvin, 1994; Christiansen & Ellefson, 2002; Conway & Christiansen, 2001; Greenfield, 1991). There is a close connection between the general problem of sequence learning and Chunk-and-Pass language processing: both require the detection and encoding of elements occurring in temporal sequences (or spatio-temporal sequences in the case of sign language). Perhaps, then, variations in people's abilities to pick up statistical regularities among sequence elements might explain some of the experience-based differences across individuals in language processing skills.

Misyak, Christiansen, and Tomblin (2010) sought to address this question by investigating the connection between the learning of nonadjacent relationships among elements in a sequence and online processing of long-distance dependencies produced by relative clauses. To quantify individual differences in sequence learning, they developed a novel experimental paradigm: the AGL-SRT task. This task integrates two previous implicit learning paradigms, combining the structured, probabilistic input of *artificial grammar learning* (AGL; e.g., Reber, 1967) with the online learning of a *serial reaction-time task* (SRT; Nissen, & Bullemer, 1987). Following Gómez (2002), the sequences all had the form, a_iXb_i, where $a_i\_b_i$ consisted of three nonadjacent pairs, in which a_i was always followed by b_i (i.e., $a_1\_b_1$, $a_2\_b_2$, $a_3\_b_3$), and the middle element X was drawn randomly from a set of 24 other items. Each element in a sequence was represented by nonsense words (e.g., *pel, wadim, tood*), which would be presented visually on a computer screen in all caps (e.g., PEL, WADIM, TOOD). Participants heard spoken forms of the nonsense words and used the computer mouse to click on the corresponding written word on the computer screen as quickly as possible. For each of the three elements in a sequence, the participant has to choose between two nonsense words on the screen: a *target* and a *foil.* After multiple blocks of exposure to these sequences, participants showed evidence of having picked up on the nonadjacent relationship between $a_i\_b_i$, slowing down their responses when presented with sequences that violated this pattern (e.g., $*a_1\_b_2$).

Given the importance of prediction for language processing (e.g., see chapter 4), Misyak et al. (2010) gave their participants a prediction task at the end of the AGL-SRT task. Specifically, participants would hear and respond to the first two words of a sequence as before, but were then asked to predict which of two written nonsense words would come next (e.g., a_1X_{17}_ where the two response options would be the target, b_1, and a foil, b_2). Misyak et al. observed considerable individual differences among the participants on this task (from 25 to 100% correct), which they correlated with performance on the standard self-paced reading task incorporating the same relative clause stimuli as in the Wells et al. (2009) study. Individual variation in performance on the AGL-SRT prediction task was negatively correlated with reading times at the main verb in object relatives (e.g., *admitted* in 24) where the long-distance dependency with the subject noun (e.g., *the reporter* in 24) needs to be resolved: better sequence-learning ability was associated with shorter reading times. Strikingly, when Misyak et al. divided the participants into high- and low-performing sequence learners based on whether they scored above or below chance (50%) on the prediction task, a familiar pattern of reading times for subject and object relative clauses emerged. As illustrated in figure 6.3, the pattern of reading times for high- and low-performing sequence learners (bottom panel) closely resembles that of the individuals before and after training in the Wells et al. study (top panel). Thus, individual differences in sequence learning contribute to variations in the ability to use linguistic experience to fine-tune Chunk-and-Pass processing of language (see box 6.3 for an application of this perspective to so-called *specific language impairment*).

Because of the processing pressures from the Now-or-Never bottleneck described in chapter 4, we might expect that a basic ability for chunking sequential input would underlie the close connection between sequence learning and language. That is, individual differences in chunking, as a fundamental memory skill, should predict variations in language processing (see also Jones, 2012; Jones, Gobet, Freudenthal, Watson, & Pine, 2014). To explore this possibility, McCauley and Christiansen (2015) devised a novel twist on a classic psychological memory paradigm—the serial recall task—to pinpoint chunking ability. They used a corpus analysis to extract sublexical units of two or three consonants, differing in their frequency of occurrence as a chunk. These consonant bigrams (pairs) and trigrams (triples) were then concatenated according to frequency into strings consisting of eight or nine letters, respectively, to be used in the recall task. An example of a high-frequency trigram string is *x p l n c r n g l*, whereas *v s k f n r s d* is a low-frequency bigram string. In order to factor out potential effects of basic short-term memory, attention, and motivation, control items were created by pseudo-randomizing the experimental

Box 6.3
How Specific Is Specific Language Impairment?

The existence of children with *specific language impairment* (or SLI) would seem to pose a challenge to our focus on experience in language processing. After all, these children apparently receive the same linguistic input as everyone else but do not develop normal language. Children with SLI show delayed language development and typically end up with long-term limitations on their spoken and receptive language, but do not have hearing loss or other neurodevelopmental conditions, such as autism and mental retardation (Tomblin, Records, & Zhang, 1996). They also have problems dealing with grammatical morphology (e.g., Bedore & Leonard, 1998) and long-distance syntactic dependencies in relative clauses (Novogrodsky & Friedmann, 2006). A standard interpretation of SLI has been to postulate that some aspect of language acquisition relating to grammatical processing is impaired (e.g., van der Lely & Pinker, 2014). But an alternative perspective is that children with SLI may have deficits in some of the multitude of general cognitive mechanisms that subserve language processing. For example, one explanation for phonological language deficits focuses on difficulties with rapid processing of complex acoustic material (e.g., Tallal et al., 1996). Limitations on working memory capacity have also been proposed as an explanation of SLI (e.g., Gathercole & Baddeley, 1990). Importantly, given the emphasis on sequence learning in the current chapter and the next, individuals with SLI have been shown not only to have difficulties in learning the patterns in a serial reaction-time task (Tomblin, Mainela-Arnold, & Zhang, 2007) but are also to be unable to learn nonadjacent dependencies of the type, a_iXb_i, (used in by Misyak et al., 2010) when tested in a statistical learning task (Hsu, Tomblin, & Christiansen, 2014). Indeed, it has been suggested that a procedural learning deficit, not specific to language, may lie at the center of SLI (Ullman & Pierpont, 2005). Together, these results suggest that the term SLI is a misnomer: SLI does not appear to be specific to language after all. Instead, children with SLI represent the tail end of the distribution of language skills (Dale & Cole, 1991; Leonard, 1987; Tomblin & Zhang, 1999): the variety of linguistic deficits that fall under this category result from a range of cognitive limitations in processes that underpin language, and mediate the role of experience in normally developing children.

items to minimize bigram/trigram information (e.g., *l g l c n p x n r* is the matching control string to the above high-frequency item). When participants were asked to recall these stimuli, McCauley and Christiansen found a strong effect of chunk frequency. Because there were no vowels in the strings, they could not be easily pronounced and thereby rehearsed. Successful recall therefore required generalizing past experience with the relevant sublexical consonant combinations during reading to the new non-linguistic context of the memory task.

McCauley and Christiansen (2015) further observed considerable individual differences in participants' performance on the chunking task. To assess variation across individuals in language processing, the participants were also administered a standard self-paced reading task incorporating the same relative clause materials used by Wells et al. (2009). To derive an individual difference measure of chunking ability that controls for basic short-term memory span, McCauley and Christiansen calculated the mean difference in recall rate between experimental and control items (i.e., measuring how much performance was facilitated by the presence of chunks in the experimental items). As expected, individual differences in chunking ability predicted variation in online processing of relative clauses—better chunking was associated with faster processing—with a stronger effect for object relatives than subject relatives. When participants were divided into "good chunkers" and "poor chunkers" (using a median split), the familiar differential pattern (from figures 6.2 and 6.3) of subject and object relative processing emerged. The pattern of processing for the good chunkers resembled that of the good sequence learners and more experienced readers, with relatively little difference between subject and object relatives. In contrast, the poor chunkers looked like poor sequence learners and less experienced readers, experiencing particular processing difficulty with object relatives compared to subject relatives. These results provide initial, tantalizing support for the hypothesis that basic chunking abilities might arise from a single learning and memory mechanism that deals with both linguistic and non-linguistic sequential input, through their integration into Chunk-and-Pass language processing. More research is needed, though, to fully substantiate this hypothesis, as well as to examine individual differences in chunking ability across developmental time, tracing the impact of chunking on specific aspects of language acquisition, including the early development of complex sentence processing.

6.3 Summary

In this chapter, we have explored the role of experience with language in developing and fine-tuning Chunk-and-Pass processing skills. Focusing on relative clause processing, we discussed how our ability to process specific types of subject and object relatives is strongly affected by the frequency with which they occur in natural language. Of course, other factors, such as discourse expectations, cognitive control, and memory limitations also play a role, though in the current framework these factors themselves are likely to be adapted to linguistic experience (as discussed in chapter 4). The key role of exposure to specific patterns of language was further underscored by evidence from neural network simulations and human experimentation, which suggested that individual differences previously attributed to variations in working memory capacity may emerge from variations in experience and processing architecture. In line with our evolutionary arguments in chapter 2, we considered evidence suggesting that at least part of our ability to carry out Chunk-and-Pass processing of relative clauses appears to rely on cognitive mechanisms for sequence learning, which mediate the role of linguistic experience, likely through basic memory skills for chunking.

When evaluating the impact of linguistic experience on language processing, it is important to consider potential interactions between different types of grammatical regularities, as exemplified by the role of the Frequency × Regularity interaction in creating individual differences in relative clause processing. This means that Chunk-and-Pass processing skills are not only affected by the distribution of individual construction types but also by complex interactions between multiple, partially overlapping, syntactic patterns (of which the overlap between simple transitive sentences and subject relatives is one example). Indeed, Fitz, Chang, and Christiansen (2011) manipulated the relative frequency of different types of relative clause constructions in a neural network simulation. The results showed that the difficulty in processing a particular type of relative clause depended not only on its structural complexity, but also on its frequency in comparison with other types of relative clauses. That is, for three constructions of the type, A, B, and C, the ease with which A is processed may depend on the frequency and the nature of the overlap with B and C. Thus, the language to which we are exposed forms a complex, integrated system of constructions (as noted in chapter 2; see also Beckner et al. 2009). As the language changes across time, so, too, do our Chunk-and-Pass processing skills—though in most cases such changes will be almost imperceptible to us, like a second skin.

7 Recursion as a Usage-Based Skill

This is the farmer sowing his corn, that kept the cock that crow'd in the morn, that waked the priest all shaven and shorn, that married the man all tatter'd and torn, that kissed the maiden all forlorn, that milk'd the cow with the crumpled horn, that tossed the dog, that worried the cat, that killed the rat, that ate the malt, that lay in the house that Jack built.
—British nursery rhyme

One of the most celebrated properties of human language is that it allows us to produce and understand an indeterminate number of different sentences. We have already discussed in earlier chapters how the adoption of new words into the language provides a seemingly unending source of new ways of expressing ourselves. However, the most powerful component of our linguistic productivity can be found within the structure of language itself: the reuse of the same grammatical construction multiple times within a given sentence. This recursive ability[1] adds tremendous expressive power to the language, as exemplified by the traditional British nursery rhyme "The house that Jack built." Indeed, this nursery rhyme can be seen as an exercise in recursion. It starts with the sentence *This is the house that Jack built*, which contains a single object relative clause (of the sort discussed in chapter 6), *that Jack built*. The nursery rhyme continues by adding a subject relative construction, *that lay in the house*, in front of the original object relative clause, *This is the malt that lay in the house that Jack built*. A further subject relative clause, *that ate the malt*, is then added before the previous one, producing the next line, *This is the rat that ate the malt that lay in the house that Jack built*. This process is repeated

1. Strictly speaking, repeating the same construction multiple times does not necessarily involve recursion but could stem from a simpler process of *iteration* (e.g., a loop that repeats a particular structure). For a structure, rule, process, or function to be recursive, it has to contain a self-reference or somehow "call" (or apply to) itself. Recursive structure may also arise from the interaction of multiple rules rather than a single rule calling itself. So, for example, Rule A might call Rule B, and Rule B might call Rule A (see box 7.1; Lobina 2014, for discussion).

until the complete cumulative tale is told through the gradual addition of subject relative clauses before the first sentence, as seen in the quotation above. Thus, by reusing the same subject relative construction nine times, we are able to produce the rather long, 70-word sentence that corresponds to the final form of the nursery rhyme.

The house that Jack built illustrates the simplest kind of recursion (see box 7.1), often referred to as *tail recursion* (or right-branching recursion), where new structure is added recursively at the right edge of the previous material (even though the subject relatives are added in front of previous constituents in the nursery rhyme, the resulting grammatical structure is tail recursive). This kind of recursive structure is rather easy for us to process because we can chunk each of the relative clauses separately and pass it up for further processing, without imposing significant demands on memory. Similarly, left-branching recursive structure, as exemplified by possessive phrases such as *Jack's friend's uncle's cat* (which has two levels of recursion), also tends to cause relatively few processing problems. Note that incremental interpretation is

Box 7.1
Varieties of Recursion and Recursive Structure

Traditionally, recursion has been viewed as a property of linguistic grammars. Chomsky (1956) proposed a hierarchy of languages, couched in terms of grammars comprised by so-called rewrite rules of the general form $\alpha \rightarrow \beta$, meaning that α can be rewritten as β, and with α and β corresponding to some combination of terminals (i.e., words) and symbols (variables like α and β). Recursion occurs when a left-hand side symbol appears on the right-hand side of the same rule (e.g., $A \rightarrow wA$; assuming that A is a symbol and w a string of terminals, possibly empty). Recursive structure may further occur as a consequence of a set of rules, none of which are recursive on their own (e.g., $A \rightarrow w_1B$, $B \rightarrow w_2A$). Every class of languages within the Chomsky hierarchy is defined in terms of the restrictions imposed on its rewrite rules. Restricting α to be a single symbol and β either a single symbol followed by a (possibly empty) string of terminals, or vice versa, yields the narrowest class: the *regular languages* (also known as finite-state languages). For example, a right-branching rewrite rule takes the form $A \rightarrow wB$, whereas a left-branching rule has the format of $A \rightarrow Bw$. The less restrictive class of *context-free languages* allows β to be any (non-empty) string of symbols and terminals, making center-embedded recursion possible, e.g., $A \rightarrow w_1Aw_2$. The *context-sensitive languages* constitute an even broader class, loosening the restriction on α to permit more than one symbol to occur on the left-hand side (though still ensuring that β is at least of the same length as α). As an example, consider the rewrite rule $A_1B_1 \rightarrow A_1A_2B_1B_2$ which takes A_1B_1 and expands it such that the dependency between A_1 and B_1 crosses that of A_2 and B_2. Finally, the broadest class of *unrestricted languages* does not have any restrictions on α and β. Situating human language within the Chomsky hierarchy is a matter of much debate (e.g., Christiansen & Chater, 1999; Jäger & Rogers, 2012; Martins, 2012; Petersson, 2005; Stabler, 2009). Here, we distinguish between recursive *mechanisms* and recursive *structures*. In linguistics and computer science, there has often been a focus on recursive formalisms that far exceed human language processing abilities. By contrast, we aim to provide a processing account of the limited human ability to deal with recursive linguistic constructions.

fairly straightforward: on hearing *Jack*, we can figure out which person this is; and Jack's friend can then be identified, now that we know who Jack is; and Jack's friend's uncle is easy to pick out, now that we know who Jack's friend is; and finally this person's cat is the referent of the whole phrase. The same is true when we interpret the right-branching recursive structure encountered in "The house that Jack built": we can build up our interpretation of the entire nursery rhyme phrase by phrase. We will refer to left- and right-branching (tail) recursive constructions as *iterative recursive structure*, because such structures can always be generated in a non-recursive manner by an iterative process (e.g., Aho, Sethi, & Ullman, 1986; Pulman, 1986).

There are, however, other kinds of recursive structure that quickly result in great processing difficulty. Consider the simple transitive sentence *The rat ate the malt.* As in the original nursery rhyme, we can expand on this sentence using tail recursion, turning it into a subject relative clause in (1). Alternatively, we can express roughly the same sentiment by center-embedding the new material as an object relative clause within the original sentence, as illustrated by (2).

(1) The cat killed the rat that ate the malt.
(2) The rat that the cat killed ate the malt.

As discussed in chapter 6, (1) is somewhat easier to process than (2), though both sentences should be perfectly understandable. We can repeat both of these processes, creating the right-branching sentence in (3) and the center-embedded sentence in (4).

(3) The dog worried the cat that killed the rat that ate the malt.
(4) The rat that the cat that the dog worried killed ate the malt.

Whereas the right-branching sentence in (3) with two levels of recursion is reasonably easy to read, the doubly center-embedded sentence in (4) is close to impossible for most people to understand. Of course, this differential effect of increased levels of recursion only becomes more pronounced if we repeat the recursive processes once more, producing (5) and (6).

(5) The cow tossed the dog that worried the cat that killed the rat that ate the malt.
(6) The rat that the cat that the dog that the cow tossed worried killed ate the malt.

Again, the right-branching sentence in (5) is still comprehensible, though perhaps awkward, and now begins to resemble parts of the old nursery rhyme that inspired it. In contrast, the triply center-embedded sentence in (6) is completely incomprehensible, even though it is expressing more or less the same meaning as (5), and with precisely the same words. The center-embedded recursive constructions exemplified by (2), (4), and (6) cannot be generated in an unbounded manner by an iterative process; we will refer to such recursive structures as *complex recursive structures.*

The processing difficulty associated with sentences containing multiple center-embeddings is well known in psycholinguistics. English doubly center-embedded sentences (such as 4) are read with the same intonation as a list of random words (Miller, 1962), cannot easily be memorized (Foss & Cairns, 1970; Miller & Isard, 1964), are difficult to paraphrase (Hakes & Foss, 1970; Larkin & Burns, 1977) as well as to comprehend (Blaubergs & Braine, 1974; Hakes, Evans, & Brannon, 1976; Hamilton & Deese, 1971; Wang, 1970), and are judged to be ungrammatical (Marks, 1968). Even when facilitating the processing of center-embeddings by adding semantic biases or providing training, only modest improvement is seen in performance (Blaubergs & Braine, 1974; Powell & Peters, 1973; Stolz, 1967). Notably, the limitations on processing center-embeddings are not confined to English. Similar patterns have been found in a variety of languages, ranging from French (Peterfalvi & Locatelli, 1971), German (Bach, Brown, & Marslen-Wilson, 1986), and Spanish (Hoover, 1992) to Hebrew (Schlesinger, 1975), Japanese (Uehara & Bradley, 1996) and Korean (Hagstrom & Rhee, 1997). It would seem reasonable to assume, then, that humans would tend to avoid multiple center-embeddings. Indeed, corpus analyses of Danish, English, Finnish, French, German, Latin, and Swedish (Karlsson, 2007) indicate that doubly center-embedded sentences are almost entirely absent from spoken language and only exist to a very limited extent even in written text.

Given the difficulty experienced with multiple center-embedded object relative clauses (as in 4 and 6), it may seem surprising that such complex recursive structures have been at center stage of debates over the nature of the representational machinery needed for language (see box 7.2). Since the mid-1950s, it has been suggested that the existence of center-embedded structures requires language to be represented by grammars that are at least *context-free* (e.g., Chomsky, 1956; Jäger & Rogers, 2012; Stabler, 2009; see box 7.1). Indeed, because of the existence of cross-dependency structures in Swiss-German and Dutch (in which grammatical dependencies cross over one another, as discussed further below), natural language grammars have been argued to be more powerful still—perhaps to be mildly *context-sensitive* (e.g., Shieber, 1985; see box 7.1). Whether context-free or context-sensitive, the resulting recursive grammar formalisms are very powerful, sanctioning sentences with multiple center-embeddings as grammatical and thus making them part of the language.

Box 7.2
Debates over Natural Language Recursion

The discussion of recursion in natural language is often marred by a failure to distinguish between recursion as a generative mechanism (typically a property of grammar) and recursive linguistic structure (repeated use of the same construction). A recursive grammar generates sentences with recursive structure, but the latter does not automatically imply the former (see also Heine & Kuteva, 2007). Just because a sentence can be analyzed in terms of recursive structure, that does not imply that it was generated by recursive mechanisms. Tail-recursive structure can be captured by iteration (e.g., a loop) and bounded complex recursive structures can be captured by finite-state systems (e.g., Christiansen & Chater, 1999; Karlsson, 2010; Petersson, 2005)

Discussions are further muddled by the selective use of different types of recursive structures to drive intuitions about recursion. For example, when questioning the adaptive value of recursion in language evolution, Premack (1985) challenged readers to devise an evolutionary scenario in which being able to express the following sentence would confer any selective advantage: *"Beware of the short beast whose front hoof Bob cracked when, having forgotten his own spear back at camp, he got in a glancing blow with the dull spear he borrowed from Jack"* (p. 282). Notice that this sentence involves a mix of different types of recursive structures, including several embedded relative clauses, making it cumbersome to read. Pinker (1994) defended the adaptationist perspective on language evolution by noting the potential usefulness of right-branching recursive sentences, such as *"He knows that she thinks that he is flirting with Mary,"* (p. 368) to express complex internal belief states. This kind of tail-recursive structure is relatively easy to process; but, as we have noted, it can be captured by iteration and therefore does not provide definitive evidence in favor of recursion as a property of the grammar. It might seem obvious, then, that the debate should focus on complex recursive structures that cannot readily be accounted for by iterative processing. Notably, however, doubly center-embedded sentences, such as *The cat the dog the mouse bit chased ran away*, are rarely, if ever, used to convince readers of the unbounded nature of natural language recursion. Of course, such sentences are close to impossible to understand, revealing the problems with imposing a mathematical structure on language which does not fit our actual linguistic abilities, and which can only be made to work by forcing a square peg into a round hole.

By making complex recursion a built-in property of grammar, the proponents of such linguistic representations are faced with a fundamental problem: the grammars generate sentences that can never be understood and that would never be produced. The standard solution is to propose a distinction between an infinite linguistic *competence* and a limited observable psycholinguistic *performance* (e.g., Chomsky, 1965). The latter is assumed to be bounded by memory limitations, attention span, lack of concentration, and other processing constraints, whereas the former is construed as being essentially infinite by virtue of the recursive nature of grammar. There are a number of methodological and theoretical issues with the competence/performance distinction (e.g., Christiansen, 1992; Petersson, 2005; Pylyshyn, 1973; Reich, 1969; see also our discussion in chapter 8). Here, however, we focus on a substantial challenge to the standard solution, deriving from the considerable variation across languages and between individual speakers in the use of recursive structures—differences that cannot readily be ascribed to performance factors.

In a recent review of the pervasive differences that can be observed throughout all levels of linguistic representations across the world's approximately 7,000 languages, Evans and Levinson (2009) remarked that

"... recursion is not a necessary or defining feature of every language. It is a well-developed feature of some languages, like English or Japanese, rare but allowed in others (like Bininj Gun-wok), capped at a single level of nesting in others (Kayardild), and in others, like Pirahã, it is completely absent." (p. 443)

Using examples from Central Alaskan Yup'ik Eskimo, Khalkha Mongolian, and Mohawk, Mithun (2010) further notes that recursive structures are far from uniform across languages, nor are they static within individual languages. Hawkins (1994) observed substantial offline differences in perceived processing difficulty of the same type of recursive constructions across English, German, Japanese, and Persian. Moreover, a self-paced reading study involving center-embedded sentences found differential processing difficulties in Spanish and English (even when morphological cues were removed in Spanish; Hoover, 1992).

Considerable variations in the ability to deal with recursive constructions have also been observed developmentally. Dickinson (1987) showed that the production of recursive structures emerge gradually, in a piecemeal fashion. Regarding comprehension, we have previously noted that training improves comprehension of singly embedded relative clause constructions both in three- to four-year old children (Roth 1984) and in adults (Wells, Christiansen, Race, Acheson, & MacDonald, 2009), independent of other cognitive factors. Level of education further correlates with the ability to comprehend complex recursive sentences (Dąbrowska, 1997). More generally, these developmental differences are likely to reflect individual variations in experience with language—as discussed in the previous chapter—differences that may further be amplified by variations in the structural and distributional characteristics of the languages being spoken.

Together, these individual, developmental, and cross-linguistic differences in dealing with recursive linguistic structure are hard to explain in terms of performance factors, such as memory, processing, or attentional constraints, limiting an otherwise infinite recursive grammar. Invoking such limitations would appear to require different biological constraints on working memory, processing, or attention for speakers of different languages, which seems highly unlikely. Similarly, such constraints cannot easily accommodate the impact of specific linguistic experience, resulting in differential processing difficulty for the same syntactic construction across development and individuals, as discussed in chapter 6. To resolve these issues, we need to separate claims about *recursive mechanisms* from claims about *recursive structure*. Specifically, the ability to deal with a limited amount of recursive structure in language does not necessitate the postulation of recursive processing mechanisms. Thus, we suggest that instead of treating recursion as an a priori property of whatever formalism is used to represent language, we need to provide a mechanistic account of the actual amount of recursive structure that humans are able to process—no more, no less.

In this chapter, we offer an alternative account of the processing of recursive structure that builds upon construction grammar and usage-based approaches to language. The essential idea is that the ability to process recursive structure does not depend on a built-in property of a competence grammar but, rather, is an acquired skill, learned through experience with specific instances of recursive constructions and limited generalizations over these (Christiansen & MacDonald, 2009). Performance limitations emerge naturally through interactions between linguistic experience and cognitive constraints on learning and processing, so that recursive abilities degrade in line with human performance across languages and individuals. We show how our usage-based account of recursion can accommodate human data on the differential processing of different types of complex recursive structures in German (center-embedding) and Dutch (cross-dependencies). Further predictions are derived and experimentally corroborated, suggesting that even our ability to process simple recursive constructions, such as multiple prepositional phrases, is bounded. Building on the arguments of chapter 6, we discuss genetic, cognitive neuroscientific, and comparative evidence suggesting that our ability to process recursive structures may have evolved on top of our broader abilities for complex sequence learning. Hence, we argue that Chunk-and-Pass processing, implemented by domain-general mechanisms—not recursive grammars—is what endows language with its hallmark productivity, allowing it to "... make infinite employment of finite means," as the celebrated German linguist, Wilhelm von Humboldt (1836/1999: p. 91), noted more than a century and a half ago.

7.1 Sequence Learning as the Basis for Recursive Structure

Although recursion has often figured in discussions of the evolution of language (e.g., Chomsky, 1988; Christiansen, 1994; Corballis, 1992; Pinker & Bloom, 1990; Premack, 1985), the new millennium saw a resurgence of interest in the topic following the publication of a paper by Hauser, Chomsky, and

Fitch (2002), controversially raising the possibility that recursion may be the one and only core component of the language faculty (i.e., a putative innately-specified, language-specific mechanism, unique to humans). Subsequent papers have covered a wide range of topics, from criticisms of the Hauser et al. claim (e.g., Parker, 2006; Pinker & Jackendoff, 2005) and how to characterize recursion appropriately (e.g., Lobina, 2014; Tomalin, 2011) to its potential presence (e.g., Gentner, Fenn, Margoliash, & Nusbaum, 2006) or absence in animals (e.g., Corballis, 2007), and its purported universality in human language (e.g., Everett, 2005; Evans & Levinson, 2009; Mithun, 2010) and cognition (e.g., Corballis, 2011; Vicari & Adenzato, 2014) (see box 7.3 for discussion). Our focus here, however, is to advocate a usage-based perspective on the processing of recursive structure, suggesting that it relies on evolutionarily older abilities for dealing with temporally presented sequences of input.

Box 7.3
Recursion across Cognition and Species

It has recently been suggested that recursion may be a crucial, and perhaps even the only, piece of cognitive machinery that is specific to language (Hauser et al., 2002). If so, then we should expect recursion to be absent from other aspects of human cognition; and not to be observed in nonhuman animals. Yet both these claims have been widely challenged. Indeed, standard psychological theories of action, planning and problem solving typically assume a degree of recursion: e.g., goals are broken into subgoals, which may be further recursively divided into sub-sub-goals, and so on (e.g., Miller, Galanter, & Pribram, 1960). Similarly, our ability to think through complicated scenarios, such as "she thinks that I think that she thinks that ..." has also been highlighted as evidence of recursion in the human theory of mind (e.g., Corballis, 2011). Even the species-specificity of recursion has been questioned. In the wild, the complex sequences of actions to prepare certain plant foods by mountain gorillas may be described as being recursively structured (Byrne & Russon, 1998). However, controlled lab experiments with non-human primates have largely failed to demonstrate the learning of complex recursive structure—but research on songbirds has been more successful. For example, European starlings appear able to learn A^nB^n recursive structures, involving *n* A elements followed by *n* B elements (Gentner et al., 2006). This study has been criticized on methodological grounds (e.g., Corballis, 2007); moreover, although such "counting recursion" requires a context-free grammar, this type of complex recursive structure is not obviously relevant to recursion in natural language (Christiansen & Chater, 1999). Another study involving Bengalese finches demonstrated both sensitivity to predictability in an artificial context-free grammar and the ability to learn singly center-embedded recursive structure (Abe & Watanabe, 2011). Again, this work has been subject to methodological criticisms (e.g., Beckers, Bolhuis, Okanoya, & Berwick, 2012). Nonetheless, it is important to note that if similar criticisms were applied to artificial grammar learning in human children and adults, then many of these studies would not count as evidence of learning recursive structure (e.g., Saffran, 2002; Saffran et al., 2008; Gómez, 2002). More generally, we want to stress that the presence of recursive structure in *behavior* does not necessarily require explanations in terms of recursive *mechanisms*—whether or not the behavior in question concerns human cognition (including language) or the behavioral repertoire of another species.

7.1.1 Comparative, Genetic, and Neural Connections between Sequence Learning and Language

Language processing involves extracting regularities from highly complex sequentially organized input, pointing to a connection between general sequence learning (e.g., planning, motor control, etc.; Lashley, 1951) and language: after all, both involve the extraction and further processing of discrete elements occurring in temporal sequences (see also e.g., Bybee, 2002; Conway & Christiansen, 2001; de Vries, Christiansen, & Petersson, 2011; Greenfield, 1991, for similar perspectives). For sequence learning, there is comparative, genetic, and neural evidence suggesting that humans may have evolved specific abilities for dealing with complex sequences. Experiments have shown that non-human primates are able both to learn fixed sequences, akin to a phone number (e.g., Heimbauer, Conway, Christiansen, Beran, & Owren, 2012), as well as probabilistic sequences, similar to "statistical learning" in human studies (e.g., Heimbauer, Conway, Christiansen, Beran, & Owren, 2010, submitted; Wilson et al., 2013). However, when it comes to dealing with recursive sequence structure, non-human primates appear to have significant limitations in comparison with human children (e.g., in recursively sequencing actions to nest cups within one another; Greenfield, Nelson, & Saltzman, 1972; Johnson-Pynn, Fragaszy, Hirsch, Brakke, & Greenfield, 1999). Although more carefully controlled comparisons between the sequence learning abilities of human and non-human primates is needed (see Conway & Christiansen, 2001, for a review), the currently available data suggest that humans may have evolved a superior ability to deal with more complex sequences involving recursive structures.

The current knowledge regarding the *FOXP2* gene is consistent with the suggestion of a human adaptation for sequence learning (for a review, see Fisher & Scharff, 2009). *FOXP2* is highly conserved across species, but two amino acid changes have occurred after the split between humans and chimps, and these became fixed in the human population about 200,000 years ago (Enard et al., 2002). In humans, mutations to *FOXP2* result in severe speech and orofacial motor impairments (Lai, Fisher, Hurst, Vargha-Khadem, & Monaco, 2001; MacDermot et al., 2005). Studies of *FOXP2* expression in mice and imaging studies of an extended family pedigree with *FOXP2* mutations have provided evidence that this gene is important to neural development and function, including the cortico-striatal system (Lai, Gerrelli, Monaco, Fisher, & Copp, 2003). When a humanized version of *Foxp2*[2] was inserted into mice,

2. The upper case, italicized *FOXP2* is used to denote the gene; *Foxp2* is used to refer to the homologous non-human gene.

it was found to specifically affect cortico-basal ganglia circuits (including the striatum), increasing dendrite length and synaptic plasticity (Reimers-Kipping, Hevers, Pääbo, & Enard, 2011). Indeed, the humanized *Foxp2* mice were subsequently shown to be better at learning action sequences than their wild-type littermates (Schreiweis et al., 2014). More generally, the cortico-basal ganglia system has been shown to be important for sequence (and other types of procedural) learning in humans (Packard & Knowlton, 2002). Remarkably, preliminary findings from a mother and daughter pair with a translocation involving *FOXP2* indicate that they have problems with both language and sequence learning (Tomblin, Shriberg, Murray, Patil, & Williams, 2004). Finally, we note that sequencing deficits also appear to be associated with specific language impairment (SLI) more generally (e.g., Hsu, Tomblin, & Christiansen, 2014; Lum, Conti-Ramsden, Page, & Ullman, 2012; Tomblin, Mainela-Arnold, & Zhang, 2007; see Lum, Conti-Ramsden, Morgan, & Ullman, 2014, for a review, and box 6.3).

Hence, both comparative and genetic evidence suggest that humans have evolved complex sequence learning abilities, which, in turn, appear to have been pressed into service to support the emergence of our linguistic skills. This evolutionary scenario would predict that language and sequence learning should have considerable overlap in their neural underpinnings. This prediction is substantiated by a growing body of research in the cognitive neurosciences highlighting the close relationship between sequence learning and language (see Conway & Pisoni, 2008; Ullman, 2004, for reviews). For example, violations of learned sequences elicit the same characteristic event-related potential (ERP) brainwave response as ungrammatical sentences, and with the same topographical scalp distribution (Christiansen, Conway, & Onnis, 2012). Similar ERP results have been observed for musical sequences (Patel, Gibson, Ratner, Besson, & Holcomb, 1998). Additional evidence for a common domain-general neural substrate for sequence learning and language comes from functional imaging (fMRI) studies showing that sequence violations activate Broca's area (Lieberman et al. 2004; Petersson, Forkstam, & Ingvar, 2004; Forkstam, Hagoort, Fernández, Ingvar, & Petersson, 2006; Petersson, Folia, & Hagoort, 2012), a region in the left inferior frontal gyrus forming a key part of the cortico-basal ganglia network involved in language. Results from a magnetoencephalography (MEG) experiment further suggest that Broca's area plays a crucial role in the processing of musical sequences (Maess, Koelsch, Gunter, & Friederici, 2001).

If language is subserved by the same neural mechanisms as used for non-linguistic sequence processing, then we would expect a breakdown of syntactic processing to be associated with impaired sequencing abilities. Christiansen, Kelly, Shillcock, and Greenfield (2010) tested this prediction in a population of agrammatic aphasics, who have severe problems with natural language syntax in both comprehension and production due to lesions involving Broca's area (e.g., Goodglass, 1993; Goodglass & Kaplan, 1983—see Martin, 2006; Novick, Trueswell, & Thompson-Schill, 2005, for reviews). They confirmed that agrammatism was associated with a deficit in sequence learning in the absence of other cognitive impairments. Similar impairments to the processing of musical sequences by the same population were observed in a study by Patel, Iversen, Wassenaar, and Hagoort (2008). Moreover, diffusion tensor magnetic resonance imaging (DTI) data suggest that white matter integrity in Broca's areas predicts success in sequence learning, with higher degrees of integrity resulting in better learning (Flöel, de Vries, Scholz, Breitenstein, & Johansen-Berg, 2009). Further underscoring the functional role of Broca's area in sequence learning, studies applying transcranial direct current stimulation (tDCS) during training (de Vries, Barth, Knecht, Zwitserlood, & Flöel, 2010) or repetitive transcranial magnetic stimulation (rTMS) during testing (Uddén et al., 2008) have found that sequencing performance is positively related to stimulation of this area (specifically, Brodmann Areas 44/45). Together, these cognitive neuroscientific studies point to considerable overlap in the neural mechanisms involved in language and sequence learning,[3] as predicted by our evolutionary account (see also Christiansen, Dale, Ellefson, & Conway, 2002; Conway & Pisoni, 2008; Hoen et al., 2003; Ullman, 2004; Wilkins & Wakefield, 1995, for similar perspectives).

7.1.2 Cultural Evolution of Recursive Structures Based on Sequence Learning

Comparative and genetic evidence is consistent with the hypothesis that humans have evolved more complex sequence-learning mechanisms, whose neural substrates subsequently were recruited for language. But how might recursive structure arise on top of such complex sequence learning abilities? Reali and Christiansen (2009) explored this question using simple recurrent networks (SRNs; Elman, 1990). We have previously seen that such networks are able to accommodate both multiple-cue integration (chapter 5) and experience-based individual differences in relative clause processing (chapter 6). Other work has established that SRNs are also competent sequence learners (e.g., Botvinick & Plaut, 2004; Servan-Schreiber, Cleeremans, & McClelland,

3. Some studies purportedly indicate that the mechanisms involved in syntactic language processing are not the same as those involved in most sequence learning tasks (e.g., Friederici, Bahlmann, Heim, Schibotz, & Anwander, 2006; Musso et al., 2003; Peña, Bonnatti, Nespor, & Mehler, 2002). However, the methods and arguments used in these studies have subsequently been severely challenged (de Vries, Monaghan, Knecht, & Zwitserlood, 2008, Marcus, Vouloumanos, & Sag, 2003, and Onnis, Monaghan, Richmond, & Chater, 2005, respectively). Overall, the preponderance of the evidence suggests that sequence-learning tasks tap into the mechanisms involved in language acquisition and processing (see Petersson et al., 2012, for discussion).

1991). To capture the difference in sequence-learning skills between humans and non-human primates, Reali and Christiansen first "evolved" a group of networks to improve their performance on a sequence-learning task in which they had to predict the next digit in a five-digit sequence generated by randomizing the order of five digits, 1–5 (based on a human task developed by Lee, 1997). At each generation, the best performing network was selected, and its initial weights (prior to any training)—i.e., their "genome"—were slightly permuted to produce a new generation of networks. After 500 generations of this simulated biological evolution, the resulting networks performed significantly better than the first generation SRNs.

Reali and Christiansen (2009) then introduced language into the simulations. Building on the simulations by Christiansen and Dale (2004) discussed in chapter 5, each miniature language was generated by a context-free grammar derived from the grammar skeleton in table 7.1. In this case, the grammar skeleton allowed more flexibility in word order insofar as the material on the right-hand side of each rule could be ordered as shown in the table (right-branching), in the reverse order (left-branching), or have a flexible order (i.e., with half the sentences generated by the right-branching grammar in table 7.1 and the other half by the reverse-ordered left-branching grammar). The grammar skeleton produces 3^6 (= 729) distinct grammars, with differing amounts of consistency in the ordering of sentence constituents. Reali and Christiansen implemented both biological and cultural evolution in their simulations: As with the evolution of better sequence learners, the initial weights of the network that best acquired a language in a given generation were slightly altered to produce the next generation of language learners—with the additional constraint that performance on the sequence learning task had to be maintained at the level reached at the end of the first part of the simulation (to capture the fact that humans have evolved to be better sequence learners than non-human primates). As in Christiansen and Dale (2004), cultural evolution

Table 7.1
The grammar skeleton used by Reali and Christiansen (2009)

S	→	{NP VP}
NP	→	{N (PP)}
PP	→	{adp NP}
VP	→	{V (NP) (PP)}
NP	→	{N PossP}
PossP	→	{NP poss}

Note. S = sentence, NP = noun phrase, VP = verb phrase, PP = adpositional phrase, PossP = possessive phrase, N = noun, V = verb, adp = adposition, poss = possessive marker. Curly brackets indicate that the order of constituents can be as shown, can be reversed, or either order occurs with equal probability (i.e., flexible word order). Parentheses indicate an optional constituent.

of language was simulated by having the networks learn several different languages at each generation and selecting the best learned language as the basis for the next generation. The best learned language was then modified slightly by changing the directions of a rule to produce a set of related "off-spring" languages for each generation.

Although the simulations started with a language that was completely flexible, and thus had no reliable word-order constraints, after less than one hundred generations of cultural evolution, the resulting language had adopted consistent word-order constraints in all but one of the six grammatical rules. When comparing the networks from the first generation at which language was introduced to the final generation, Reali and Christiansen (2009) observed no difference in the abilities of the networks to learn language. By contrast, when comparing the final language to the initial (completely flexible) language, the SRNs had a substantially easier time learning the former. Together, these two analyses suggest that it was the cultural evolution of language, rather than biological evolution of better learners, that allowed for the language to become more easily learned and more structurally consistent across these simulations. This is, of course, as we would expect if language has been shaped by the brain, as argued in chapter 2.

7.1.3 Sequence Learning and Recursive Consistency

A key result of the simulations by Reali and Christiansen (2009) was that the sequence learning constraints embedded in the SRNs tend to favor what we will refer to as *recursive consistency* (Christiansen & Devlin, 1997). Consider rewrite rules 2 and 3 from table 7.1 (NP → {N (PP)}, PP → {adp NP}). Together, these two skeleton rules form a recursive rule set, because each refers to (or "calls") the other. Ignoring the flexible version of these two rules, we get the four possible recursive rule sets shown in table 7.2. Using these rule sets, we can generate the complex noun phrases shown in (7)–(10):

(7) $[_{NP}$ buildings $[_{PP}$ from $[_{NP}$ cities $[_{PP}$ with $[_{NP}$ smog]]]]]

(8) $[_{NP}$ $[_{PP}$ $[_{NP}$ $[_{PP}$ $[_{NP}$ smog] with] cities] from] buildings]

Table 7.2
Recursive rule sets

Right-Branching	Left-Branching	Mixed	Mixed
NP → N (PP)	NP → (PP) N	NP → N (PP)	NP → (PP) N
PP → prep NP	PP → NP post	PP → NP post	PP → prep NP

Note. NP = noun phrase, PP = adpositional phrase, prep = preposition, post = postposition, N = noun. Parentheses indicate an optional constituent

(9) [$_{NP}$ buildings [$_{PP}$ [$_{NP}$ cities [$_{PP}$ [$_{NP}$ smog] with]] from]]
(10) [$_{NP}$ [$_{PP}$ from [$_{NP}$ [$_{PP}$ with [$_{NP}$ smog]] cities]] buildings]

The first two rule sets from table 7.2 generate recursively consistent structures that are either right-branching (as in 7) or left-branching (as in 8). The prepositions and postpositions, respectively, are always in close proximity to their noun complements, making it easier for a Chunk-and-Pass learner to discover their relationship. In contrast, the final two rule sets generate recursively inconsistent structures, involving center-embeddings: all nouns are either stacked up before all the postpositions (9) or after all the prepositions (10). In both cases, the learner has to work out that *from* and *cities* together form a prepositional phrase, despite being separated from each other by another prepositional phrase involving *with* and *smog*. This process is further complicated by an increase in memory load caused by the latter intervening prepositional phrase. From a Chunk-and-Pass processing perspective, it should therefore be easier to acquire the recursively consistent structures found in (7) and (8) compared to the recursively inconsistent structures in (9) and (10) (see also Hawkins, 1994). Indeed, all the simulation runs in Christiansen and Reali (2009) resulted in languages in which both recursive rule sets were consistent (rules 2–3 and 5–6 in table 7.1). It turns out that consistent rule sets are, indeed, naturally created by cultural evolution over SRNs: the SRNs impose a selectional pressure in favor of languages that they can readily learn and process.

Christiansen and Devlin (1997) had previously shown that SRNs perform better on recursively consistent structures (such as those in 7 and 8). However, if human language has adapted through cultural evolution to avoid recursive inconsistencies (such as 9 and 10), then we should expect people to be better at learning recursively consistent artificial languages than recursively inconsistent ones. Reeder (2004), following initial work by Christiansen (2000), tested this prediction by exposing participants to one of two artificial languages, generated by the artificial grammars shown in table 7.3. Notice that the consistent grammar instantiates a left-branching grammar from the grammar skeleton used by Reali and Christiansen (2009), involving two recursively consistent rule sets (2–3 and 5–6). The inconsistent grammar differs only in the direction of two rules (3 and 5), which are right-branching, whereas the other three rules are left-branching. The languages were instantiated using 10 spoken nonwords to generate sentences to which the participants were exposed. Participants in the two language conditions would thus see sequences with exactly the same lexical items, only differing in their order of occurrence as dictated by the respective grammar (e.g., consistent: *jux vot hep vot meep nib* vs. inconsistent: *jux meep hep vot vot nib*). Note that as in child language

Table 7.3
The grammar used by Christiansen (2000) and Reeder (2004)

Consistent grammar			Inconsistent grammar		
S	→	NP VP	S	→	NP VP
NP	→	(PP) N	NP	→	(PP) N
PP	→	NP post	PP	→	prep NP
VP	→	(PP) (NP) V	VP	→	(PP) (NP) V
NP	→	(PossP) N	NP	→	(PossP) N
PossP	→	NP poss	PossP	→	poss NP

Note. S = sentence, NP = noun phrase, VP = verb phrase, PP = adpositional phrase, PossP = possessive phrase, N = noun, V = verb, post = postposition, prep = preposition, poss = possessive marker. Parentheses indicate an optional constituent.

acquisition, the structural relationships between words are not immediately obvious but have to be learned from distributional information. After training, the participants were presented with a new set of items, one by one, and asked to judge whether or not these new items were generated by the same rules as those they had seen previously. Half of the new items incorporated subtle violations of the sequence ordering (e.g., grammatical: *cav hep vot lum meep nib* vs. ungrammatical: *cav hep vot <u>rud</u> meep nib*, where *rud* is ungrammatical in this position).

The results of this artificial language learning experiment showed that the consistent language was learned significantly better (61.0% correct classification) than the inconsistent one (52.7%). It is important to note that, because the consistent grammar was left-branching (and thus more like languages such as Japanese and Hindi), knowledge of English cannot explain the results. Indeed, if anything, the two right-branching rules in the inconsistent grammar bring that language closer to English. To further demonstrate that the preferences for consistently recursive sequences is a domain-general bias, Reeder (2004) conducted a second experiment, in which the sequences were instantiated by black abstract shapes that cannot be easily verbalized. The results of the second study closely replicated those of the first, suggesting that there may be general sequence-learning biases that favor recursively consistent structures, as predicted by Reali and Christiansen's (2009) evolutionary simulations.

The question remains, though, whether such sequence-learning biases can drive cultural evolution of language in humans. That is, can sequence-learning constraints promote the emergence of language-like structure when amplified by processes of cultural evolution? To answer this question, Cornish, Dale, Kirby, and Christiansen (submitted) conducted an iterated sequence learning experiment, modeled on the human iterated learning studies that we discussed in chapters 2 and 4. Participants were asked to take part in a memory experiment, in which they were presented with 15 consonant strings. Each string was presented briefly on a computer screen after which the participants attempted to type out the string. After multiple repetitions of the 15 strings, the participants were asked to recall all of them: that is, they were asked to keep generating strings until they had provided 15 unique strings. The recalled 15 strings were then recoded by permuting their specific letters (e.g., *X* might be replaced throughout by *T*, *T* by *M*, etc.) to control for trivial biases such as the location of specific letters on the computer keyboard and the presence of potential acronyms. The resulting set of 15 strings, with the same underlying structure as prior to recoding, was then used as the training strings for the next participant. A total of 10 participants were run within each "evolutionary" chain.

The initial set of strings used for the first participant in each chain was deliberately designed to have minimal distributional structure (all consonant pairs, or bigrams, had a frequency of one or two). Because recalling fifteen arbitrary strings is close to impossible given normal memory constraints, it was expected that many of the recalled items would be strongly affected by sequence-learning biases. The results showed that, as these chunk-based biases became amplified across generations of learners, the sequences gained more and more distributional structure (as measured by the relative frequency of repeated two- and three-letter chunks). Importantly, the emerging system of sequences became increasingly easy to learn. Initially, participants could only recall about four of the fifteen strings correctly but by the final generation this had doubled, so that participants could recall more than half the strings. Importantly, this increase in learnability did not evolve at the expense of string length, which did not decrease across generations. Instead, the sequences became easy to learn and recall because they formed a *system*, allowing subsequence chunks to be reused productively. Using network analyses (see Baronchelli, Ferrer-i-Cancho, Pastor-Satorras, Chater, & Christiansen, 2013, for a review), Cornish et al. demonstrated that the way in which this chunk-based productivity was implemented strongly mirrored that observed for child-directed speech.

The results from Cornish et al. (submitted) suggest that the kind of sequence learning constraints explored in the simulations by Reali and Christiansen (2009) and demonstrated by Reeder (2004) can promote the cultural evolution of language-like distributional regularities that facilitate learning. This supports our hypothesis that chunk-based constraints on sequence learning, amplified by cultural transmission, have shaped natural language, including its limited use of embedded recursive structure.

This kind of evolutionary account is consistent with evidence for a similar kind of "cultural evolution" in zebra finch song (Fehér, Wang, Saar, Mitra, & Tchernichovski, 2009). Male zebra finches normally learn their species-typical song from other male birds. To model cultural evolution, an initial bird was raised in isolation and developed a raspy, arrhythmic song. This bird then became the tutor for another young male, who had to learn from this unusual input. Afterwards, the second bird became the tutor for a third bird, and so on. After three to four generations, the resulting song had evolved into a decent approximation of the wild type song, indicating how general constraints on song learning can shape the evolution of zebra finch songs over time, thus paralleling what we have hypothesized for human language.

7.2 A Usage-Based Model of Complex Recursive Structure

So far, we have discussed converging evidence supporting the theory that language in important ways builds upon evolutionarily prior neural mechanisms for chunk-based sequence learning. But can a domain-general sequence-learning device capture our ability to process the kinds of complex recursive structures that have been argued to require powerful grammar formalisms (e.g., Chomsky, 1956; Jäger & Rogers, 2012; Shieber, 1985; Stabler, 2009)? From our *usage-based* perspective, the answer does not necessarily require the postulation of recursive mechanisms as long as the proposed mechanisms can deal with the same level of complex recursive structure observed in humans language processing. In other words, what needs to be accounted for is the *empirical evidence* regarding human processing of complex recursive structures, and not *theoretical presuppositions* about recursion as a stipulated property of our language system.

Christiansen and MacDonald (2009) conducted a set of computational simulations to determine whether a sequence-learning device such as the SRN would be able to capture human processing performance on complex recursive structures. Building on prior work by Christiansen and Chater (1999), they focused on the processing of sentences with center-embedded and cross-dependency structures. These two types of recursive constructions produce multiple overlapping nonadjacent dependencies, as illustrated in figure 7.1, resulting in rapidly increasing processing difficulty as the number of embeddings grows. We have already discussed how performance on center-embedded constructions breaks down at two levels of embeddings (e.g., Blaubergs & Braine, 1974; Hakes, Evans, & Brannon, 1976; Hamilton & Deese, 1971; Wang, 1970). The processing of cross-dependencies, which exist in Swiss-German and Dutch, has received less attention, but the avail-

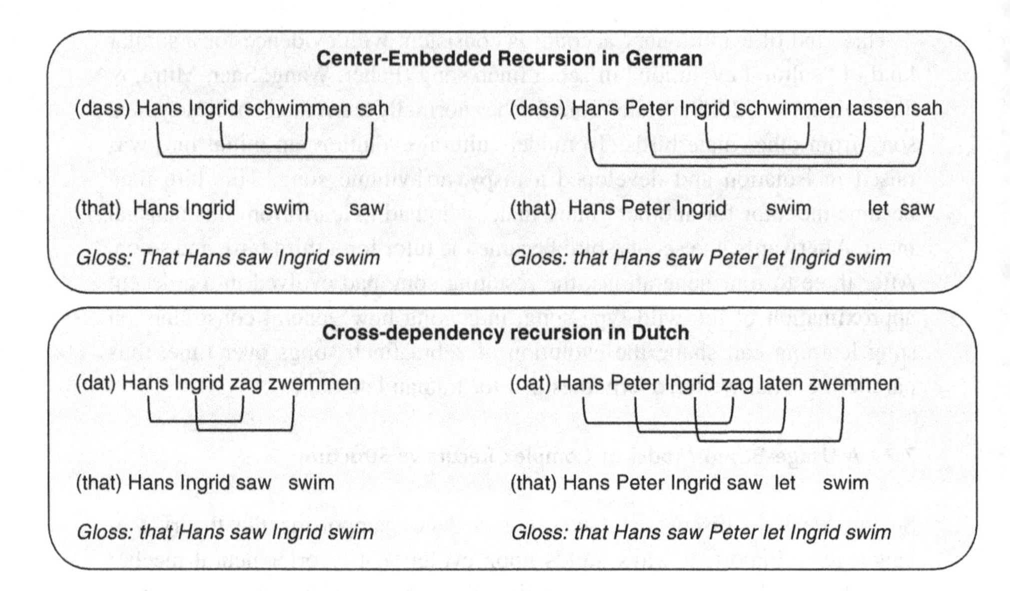

Figure 7.1
Examples of complex recursive structures with one and two levels of embedding: center-embeddings in German (top panel) and cross-dependencies in Dutch (bottom panel). The lines indicate noun-verb dependencies.

able data also point to a decline in performance with increased levels of embedding (Bach et al., 1986; Dickey & Vonk, 1997). Christiansen and MacDonald trained networks on sentences derived from one of the two grammars show in table 7.4. Both grammars incorporated left-branching recursive structure in the form of prenominal possessive genitives and right-branching recursive structure in the form of subject relative clauses, sentential complements, prepositional modifications of noun phrases, and noun phrase conjunctions. The grammars also incorporated subject noun/verb agreement and three additional verb argument structures (transitive, optionally transitive, and intransitive). The only difference between the two grammars was the type of complex recursive structure they contained: center-embedding versus cross-dependency.

The grammars could generate a variety of sentences, with varying degrees of syntactic complexity, from simple transitive sentences (such as 11) to more complex sentences, involving different kinds of recursive structure (such as 12 and 13).

(11) Mary kisses John.
(12) John knows that Mary's girls' cats see mice.
(13) Mary who loves John thinks that men say that girls chase boys.

Table 7.4
The grammars used by Christiansen and MacDonald (2009)

Rules common to both grammars		
S	→	NP VP
NP	→	N I NP PP I PossP N I N rel I N *and* NP
PP	→	prep N (PP)
PossP	→	(PossP) N poss
rel_{sub}	→	*who* VP
VP	→	V_i I V_t NP I V_o (NP) I V_c *that* S

Center-embedding grammar			Cross-dependency grammar		
rel_{obj}	→	*who* NP $V_{t\|o}$	S_{cd}	→	N_1 N_2 $V_{1(t\|o)}$ $V_{2(i)}$
			S_{cd}	→	N_1 N_2 N $V_{1(t\|o)}$ $V_{2(t\|o)}$
			S_{cd}	→	N_1 N_2 N_3 $V_{1(t\|o)}$ $V_{2(t\|o)}$ $V_{3(i)}$
			S_{cd}	→	N_1 N_2 N_3 N $V_{1(t\|o)}$ $V_{2(t\|o)}$ $V_{3(t\|o)}$

Note: S = sentence, NP = noun phrase, VP = verb phrase, PP = prepositional phrase, PossP = possessive phrase, rel = relative clauses (subscripts, $_{sub}$ and $_{obj}$, indicate subject/object relative clause), N = noun, V = verb, prep = preposition, poss = possessive marker. For brevity, NP rules have been compressed into a single rule, using 'I' to indicate exclusive options. The subscripts $_i$, $_t$, $_o$, and $_c$ denote intransitive, transitive, optionally transitive, and clausal verbs, respectively. Subscript numbers indicate noun-verb dependency relations. Parentheses indicate an optional constituent.

The generation of sentences was further restricted by probabilistic constraints on the complexity and depth of recursion. Following training on either grammar, the networks performed well on a variety of recursive sentence structures, demonstrating that the SRNs were able to acquire complex grammatical regularities (including multiple instances of right-branching subject relatives as exemplified by the nursery rhyme with which we began this chapter; see also Christiansen, 1994).[4] The networks acquired sophisticated abilities for generalizing across constituents in line with usage-based approaches to constituent structure (e.g., Beckner & Bybee, 2009; see also Christiansen & Chater, 1994). Differences between networks were observed, though, regarding how they processed the complex recursive structures permitted by the two grammars.

To model human data on the processing of center-embedding and cross-dependency structures, Christiansen and MacDonald (2009) relied on a study conducted by Bach et al. (1986), in which sentences with two center-embeddings in German were found to be significantly harder to process than comparable sentences with two cross-dependencies in Dutch. Bach et al. asked native German speakers to provide comprehensibility ratings for German

4. All simulations were replicated multiple times (including with variations in network architecture and corpus composition), yielding qualitatively similar results.

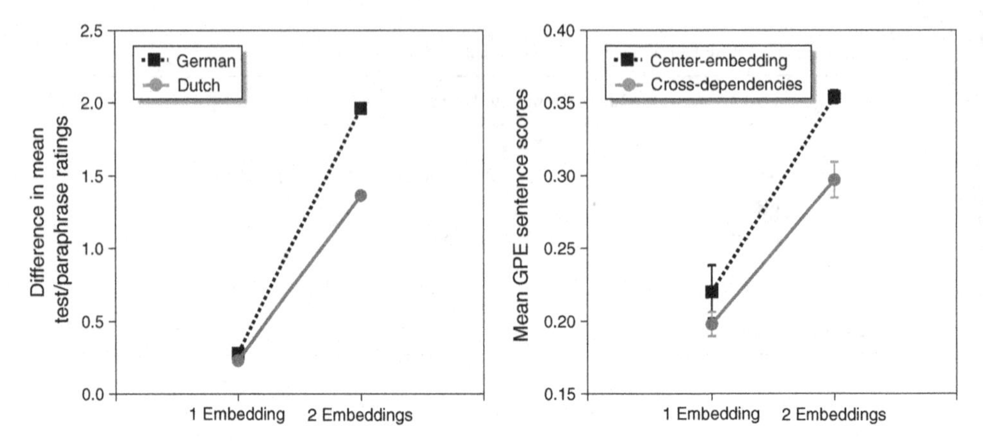

Figure 7.2
Human performance (from Bach et al., 1986) on center-embedded constructions in German and cross-dependency constructions in Dutch with one or two levels of embedding (left panel). SRN performance (from Christiansen & MacDonald, 2009) on similar complex recursive structures (right panel).

sentences with varying depths of recursive structure in the form of center-embedded constructions and corresponding right-branching paraphrases with the same meaning. Native Dutch speakers were tested using similar Dutch materials but with the center-embedded structures replaced by cross-dependency constructions. The left-hand side of figure 7.2 shows the Bach et al. results, with the ratings for the right-branching paraphrase sentences subtracted from the matching complex recursive test sentences, to remove effects of processing difficulty due to length. The SRN results—the mean sentence Grammatical Prediction Error (GPE) scores averaged over 10 novel sentences (similar to chapter 6)—are displayed on the right-hand side of figure 7.2. For both humans and SRNs, there is no difference in processing difficulty for the two types of complex recursive structure at one level of embedding. However, for doubly embedded constructions, center-embedded structures (in German) are harder to process than comparable cross-dependencies (in Dutch). These simulation results thus demonstrate that the SRNs exhibit the same kind of qualitative processing difficulties as people on the two types of complex recursive constructions (see also Christiansen & Chater, 1999).

7.2.1 Bounded Recursive Structure

Christiansen and MacDonald (2009) demonstrated that a sequence learner such as the SRN is able to mirror the differential human performance on center-embedded and cross-dependency recursive structures. Notably, the networks were able to match human performance without needing the complex external memory devices (such as a stack of stacks; Joshi, 1990) or external memory constraints (Gibson, 1998) of previous accounts. The human-like performance of the SRN can be attributed to an interaction between *intrinsic* architectural constraints (Christiansen & Chater, 1999) and the statistical properties of its input experience (MacDonald & Christiansen, 2002; see also chapter 6). By analyzing the internal states of SRNs before and after training with right-branching and center-embedded materials, Christiansen and Chater found that this type of network has a built-in architectural bias toward locally bounded dependencies, similar to those typically found in right-branching recursion. However, in order for the SRN to process multiple instances of such recursive structure, exposure to specific recursive constructions is required. Such exposure is even more crucial for the processing of center-embeddings because the network in this case also has to overcome its architectural bias toward local dependencies. Hence, the SRN does not have a built-in ability for recursion, but instead it develops its pattern of human-like processing of different recursive constructions through exposure to repeated instances of such constructions in the input.

As noted earlier, this usage-based approach to recursion differs from many previous processing accounts, in which unbounded recursion is implemented as part of the representation of linguistic knowledge (typically in the form of a rule-based grammar). This means that traditional systems can process complex recursive constructions, such as center-embeddings, to a degree that is far beyond human capabilities. Since Miller and Chomsky (1963), the standard solution to this mismatch has been to impose extrinsic memory limitations exclusively aimed at capturing human performance limitations on doubly center-embedded constructions (see Lewis, Vasishth, & Van Dyke, 2006, for a review). Examples include limits on stack depth (Church, 1982; Marcus, 1980), limits on the number of allowed sentence nodes (Kimball, 1973) or partially complete sentence nodes in a given sentence (Stabler, 1994), limits on the amount of activation available for storing intermediate processing products, as well as for executing production rules (Just & Carpenter, 1992), the "self-embedding interference constraint" (Gibson & Thomas, 1996), and an upper limit on sentential memory cost (Gibson, 1998).

In contrast, no comparable extrinsic limitations are typically imposed on the processing of right- and left-branching recursive constructions. This may be due to the fact that even finite-state devices with bounded memory are able to process such recursive structures of infinite length (Chomsky, 1956). It has been widely assumed that depth of recursion does not affect the acceptability (or processability) of right- and left-branching recursive structures in any interesting way (e.g., Chomsky, 1965; Church, 1982; Foss & Cairns, 1970;

Gibson, 1998; Reich, 1969; Stabler, 1994). Indeed, many studies of center-embedding in English have used right-branching relative clauses as baseline comparisons and found that performance was better relative to the center-embedded stimuli (e.g., Foss & Cairns, 1970; Marks, 1968; Miller & Isard, 1964). A few experimental studies have reported more detailed data on the effect of depth of recursion in right-branching constructions and found that comprehension also decreases as depth of recursion increases in these structures, although not to the same degree as with center-embedded stimuli (e.g., Bach et al., 1986; Blaubergs & Braine, 1974). However, it is not clear from these results whether the decrease in performance is caused by recursion per se or is merely a byproduct of increased sentence length. To go beyond fitting existing human data, Christiansen and MacDonald (2009) therefore derived a series of novel predictions regarding the processing of right-, left-, and center-embedded recursive structures in English.

When exposed to multiple instances of right-branching recursion, the SRN learns to represent each level of recursive structure slightly differently from the previous one (Elman, 1991). As a consequence, the SRN faces increased processing difficulty as the level of recursion grows because it has to keep track of each level of recursion separately, suggesting that depth of recursion in right-branching constructions should affect processing difficulty beyond a mere length effect. This prediction contrasts with expectations from most traditional models of sentence processing in which multiple levels of right-branching recursion are represented by exactly the same structure occurring several times. Thus, Christiansen and MacDonald (2009) derived specific predictions for sentences of similar length involving zero, one, or two levels of right-branching recursion in the form of prepositional phrase modifications of a noun phrase as shown in (14)-(16) (with the prepositional phrases indicated by square brackets):

(14) The nurse with the vase says that the flowers [by the window] resemble roses.

(15) The nurse says that the flowers [in the vase] [by the window] resemble roses.

(16) The blooming flowers [in the vase] [on the table] [by the window] resemble roses.

The mean GPE score across the whole sentence was used to predict human goodness ratings for sentences involving the same recursive constructions.

Christiansen and MacDonald (2009) elicited human ratings by first having participants read a sentence, word-by-word, while at each step they decided whether the sentence was grammatical or not using a variation of the "stop making sense" sentence-judgment paradigm (Boland, 1997; Boland, Tanenhaus, & Garnsey, 1990; Boland, Tanenhaus, Garnsey, & Carlson, 1995). Following the presentation of each sentence, participants rated it on a 7-point scale according to how good it seemed to them as a grammatical sentence of English (with 1 indicating that the sentence was "perfectly good English" and 7 indicating that it was "really bad English"). The predictions from the SRN revealed an effect of level of recursion that was confirmed by the human ratings: as the level of recursion increased, the worse the sentences were judged to be. Thus, increasing the depth of right-branching recursion has a negative effect on processing difficulty that cannot be attributed to a mere length effect. This result is not predicted by most other current models of sentence processing, in which right-branching recursion does not cause processing difficulties beyond potential length effects (although see Lewis & Vasishth, 2005).

Relevant to the local nature of Chunk-and-Pass processing, Christiansen and MacDonald (2009) also tested an observation made by Christiansen (1994) suggesting that the depth of recursion effect in left-branching structures varied in its severity, depending on the sentential position in which such recursion occurs. When processing left-branching recursive structures involving multiple prenominal genitives, the SRN learns that it is not crucial to keep track of what occurs before the final noun. This tendency is efficient early in the sentence, but creates a problem with recursion toward the end of the sentence because the network becomes somewhat uncertain about where it is in the sentence. Christiansen and MacDonald tested this observation in the context of multiple possessive genitives occurring in either subject (17) or object (18) positions in transitive constructions:

(17) Jane's dad's colleague's parrot followed the baby all afternoon.
(18) The baby followed Jane's dad's colleague's parrot all afternoon.

The SRN predicted that having two levels of recursion in a noun phrase involving left-branching prenominal genitives should be less acceptable in an object position than in a subject position, which was confirmed by the human ratings for similar sentence types.

To further investigate the nature of the SRN's intrinsic constraints on the processing of multiple center-embedded constructions, Christiansen and MacDonald (2009) explored an intriguing previous result from Christiansen and Chater (1999): that SRNs found ungrammatical versions of doubly center-embedded sentences with a missing verb *more* acceptable than their grammatical counterparts[5] (for similar SRN results, see Engelmann & Vasishth, 2009). A previous offline (paper-and-pencil) rating study by Gibson and Thomas (1999) had found that when the middle verb phrase (*was cleaning every week*) was removed from (19), the resulting ungrammatical sentence in (20) was rated no worse than the original grammatical version in (19).

(19) The apartment that the maid who the service had sent over was cleaning every week was well decorated.
(20) The apartment that the maid who the service had sent over was well decorated.

However, when Christiansen and MacDonald tested the SRN on similar doubly center-embedded constructions, they obtained predictions for (20) to be rated better than (19). This prediction was confirmed when they asked participants to rate such constructions *online* (i.e., to read each word one by one, before making a judgment at the end of the sentence), using the same stimuli as the Gibson and Thomas study.

The original stimuli from Gibson and Thomas (1999) had certain properties that could have affected the outcome of the online rating experiment. Firstly, there were substantial length differences between the ungrammatical and grammatical versions of a given sentence (e.g., 20 is four words shorter than 19). Secondly, the sentences incorporated semantic biases making it easier to line up a subject noun with its respective verb (e.g., *apartment—decorated, service—sent over* in 20). To control for these potential confounds, Christiansen and MacDonald (2009) replicated their experiment using semantically-neutral stimuli controlled for length (adapted from Stolz, 1967), as illustrated by (21) and (22).

(21) The chef who the waiter who the busboy offended appreciated admired the musicians.
(22) The chef who the waiter who the busboy offended frequently admired the musicians.

The results from this second online rating experiment yielded the same results as the first, thus replicating the "missing verb" effect. These results have subsequently been confirmed by online ratings in French (Gimenes, Rigalleau, & Gaonac'h, 2009) and a combination of self-paced reading and eye-tracking experiments in English (Vasishth, Suckow, Lewis, & Kern, 2010).

5. Importantly, Christiansen and Chater (1999) demonstrated that this prediction is primarily due to intrinsic architectural limitations on the processing of doubly center-embedded material rather than insufficient experience with these constructions. Moreover, they further demonstrated that the intrinsic constraints on center-embedding are independent of the size of the hidden unit layer.

However, evidence from German (Vasishth et al., 2010) and Dutch (Frank, Trompenaars, & Vasishth, in press) indicates that speakers of these languages do not show the missing verb effect but instead find the grammatical versions easier to process. Because verb-final constructions are common in German and Dutch, requiring the listener to track dependency relations over a relatively long distance, substantial prior experience with these constructions likely has resulted in language-specific processing improvements, as discussed in chapter 6 (see also Engelmann & Vasishth, 2009; Frank et al., in press, for similar perspectives). Nonetheless, in some cases the missing verb effect may appear even in German, under conditions of high processing load (Trotzke, Bader, & Frazier, 2013).

Figure 7.3 summarizes the SRN's novel predictions for the right-, left-, and center-embedded recursive structures along with the human online ratings for these constructions. Importantly, this model was not developed for the purpose of fitting these data but was, nevertheless, able to predict the patterns of human ratings across these different kinds of recursive structure. Indeed, the figure shows that the SRN predictions not only provide a close fit with the human ratings *within* each experiment, but also capture the increased complexity evident *across* the experiments. Strikingly, the remarkably good fit between the model and the human data both within and across the experiments was obtained without changing any parameters across the simulations. In contrast, the present pattern of results provides a challenge for most other accounts of human sentence processing that rely on arbitrary, externally specified, limitations on memory or processing to explain patterns of human performance.

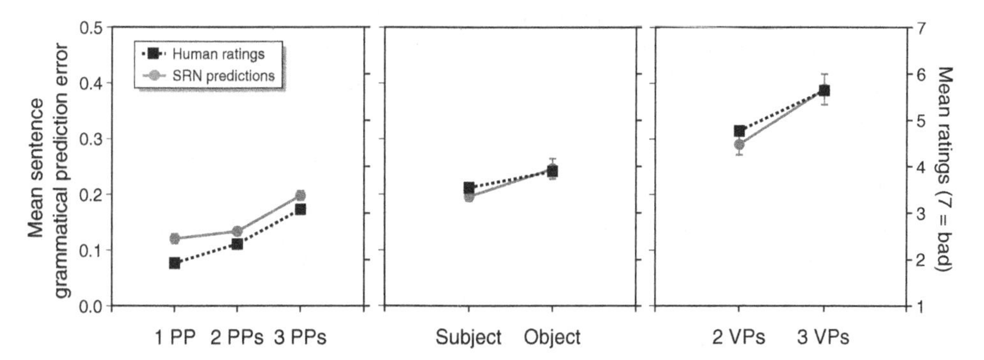

Figure 7.3
Comparison between SRN predictions of sentence complexity (left *y*-axes: mean Grammatical Prediction Error) and human ratings (right *y*-axes: mean grammaticality ratings) for three recursive sentence types from Christiansen and MacDonald (2009): (a) multiple prepositional (PP) modifications of noun phrases (left panel), (b) multiple possessive genitives in subject or object positions (center panel), (c) doubly center-embedded sentences with two or three verb phrases (VPs) (left panel).

7.2.2 Sequence Learning Limitations Mirror Constraints on Complex Recursive Structure

In chapter 6, we discussed evidence suggesting that Chunk-and-Pass processing of singly embedded relative clauses was affected by experience, mediated by sequence learning skills. Can our limited ability to process multiple complex recursive embeddings similarly be shown to reflect constraints on sequence learning? Importantly, the embedding of multiple complex recursive structures—whether in the form of center-embeddings or cross-dependencies—results in several pairs of overlapping nonadjacent dependencies (as illustrated by figure 7.1). In contrast, the work on learning nonadjacent dependencies discussed previously only involved a single pair of nonlocal dependencies (e.g., Misyak, Christiansen, & Tomblin, 2010). Nonetheless, the SRN simulation results reported above suggest that a sequence learner might also be able to deal with the increased difficulty associated with multiple, overlapping nonadjacent dependencies.

Multiple nonadjacent dependencies have often been seen as one of the key defining characteristics of human language. Indeed, when a group of generativists and cognitive linguists recently met to determine what is special about human language (Tallerman et al., 2009), one of the few things they could agree about was that long-distance dependencies constitute one of the hallmarks of human language, and not recursion (contra Hauser et al., 2002). Using a variation of the AGL-SRT task (Misyak et al., 2010; see chapter 6), de Vries, Geukes, Zwitserlood, Petersson, and Christiansen (2012) investigated whether the limitations on the processing of multiple nonadjacent dependencies might depend on general constraints on human sequence learning, instead of being unique to language. Participants were asked to use the computer mouse to select one of two written words (a target and a foil) presented on the screen as quickly as possible, given auditory input. Stimuli consisted of sequences with two or three nonadjacent dependencies, with either center-embeddings or cross-dependencies. The dependencies were instantiated using a set of dependency pairs that were matched for vowel sounds: *ba-la, yo-no, mi-di,* and *wu-tu*. Examples of each of the four types of stimuli are presented in (23)-(26), where the subscript numbering indicates dependency relationships.

(23) ba_1 wu_2 tu_2 la_1

(24) ba_1 wu_2 la_1 tu_2

(25) ba_1 wu_2 yo_3 no_3 tu_2 la_1

(26) ba_1 wu_2 yo_3 la_1 tu_2 no_3

Thus, (23) and (25) implement center-embedded recursive structure and (24) and (26) involve cross-dependencies. To determine the potential effect of linguistic experience on the processing of complex recursive sequence structure, participants were either native speakers of German (which has center-embedding but not cross-dependencies) or Dutch (which has cross-dependencies). Participants were only exposed to one of the four kinds of stimuli, e.g., doubly center-embedded sequences as in (25) (in a fully crossed length × embedding × native language design).

De Vries et al. (2012) first evaluated learning by administering a block of ungrammatical sequences in which the learned dependencies were violated. As expected, the ungrammatical block produced a similar pattern of response slow-down for both center-embedded and cross-dependency items involving two nonadjacent dependencies (similar to the results Bach et al. [1986] obtained with natural language). However, an analogue of the missing verb effect was observed for the center-embedded sequences with three nonadjacencies but not for the comparable cross-dependency items. Indeed, an incorrect middle element in the center-embedded sequences (e.g., where *tu* is replaced by *la* in 26) did not elicit any slow-down at all, indicating that participants were not sensitive to violations at this position. This result suggests that basic limitations on memory for sequential information may underlie the missing verb effect, rather than constraints specific to syntactic processing (e.g., Gibson & Thomas, 1999).

Sequence learning was further assessed using a prediction task at the end of the experiment (after a recovery block of grammatical sequences). In this task, participants heard a beep replacing one of the elements in the second half of the sequence and were asked simply to click on the written word that they thought had been replaced. Participants exposed to the sequences incorporating two dependencies performed reasonably well on this task, with no difference between center-embedded and cross-dependency stimuli. However, as for the response times, a missing verb effect was observed for the center-embedded sequences with three nonadjacencies. When the middle dependent element was replaced by a beep in center-embedded sequences (e.g., ba_1 wu_2 yo_3 no_3 <*beep*> la_1), participants were more likely to click on the foil (e.g., *la*) than the target (*tu*). This was not observed for the corresponding cross-dependency stimuli, once more mirroring the Bach et al. (1986) psycholinguistic results.

Contrary to psycholinguistic studies of German (Vasishth et al., 2010) and Dutch (Frank et al., in press), de Vries et al. (2012) found an analogue of the missing verb effect in speakers of both languages. Because the sequence-learning task involved nonsense syllables, rather than real words, it may not have tapped into the statistical regularities that play a key role in real-world language processing[6]. Instead, the results reveal fundamental limitations on the learning and processing of complex recursive structured sequences. However, these limitations may be mitigated to some degree, given sufficient exposure to the "right" patterns of linguistic structure—including statistical regularities over morphological and semantic cues—and thus minimizing the impact of sequence processing constraints that would otherwise result in the missing verb effect for doubly center-embedded constructions. Whereas the statistics of German and Dutch appear to support such amelioration of Chunk-and-Pass processing, the statistical make-up of linguistic patterning in English and French apparently does not. This is consistent with the findings of Frank et al. (in press), demonstrating that native Dutch and German speakers show a missing verb effect when processing English (as a second language), even though they do not show this effect in their native language (except under extreme processing load; Trotzke et al., 2013). Together, the results suggest that the constraints on human processing of multiple long-distance dependencies in recursive constructions derive from limitations on sequential learning interacting with linguistic experience.

7.3 Summary

In this chapter, we argued that our ability to process recursive structure does not rely on recursion as a fundamental property of the grammar, but instead emerges gradually by piggybacking on top of domain-general sequence-learning abilities. Evidence from genetics, comparative work on non-human primates, and cognitive neuroscience suggests that humans have evolved complex sequence-learning skills, which were subsequently pressed into service to accommodate language. Constraints on sequence learning therefore have played an important role in shaping the cultural evolution of linguistic structure, including our limited abilities for processing recursive constructions. We have shown how this perspective not only can account for the degree to which humans are able to process complex recursive structure, whether center-embeddings or cross-dependencies, but also suggests that even the processing of right- and left-recursive structure is bounded—albeit to a lesser degree than

6. De Vries et al. (2012) did observe a nontrivial effect of language exposure: German speakers were faster at responding to center-embedded sequences with two nonadjacencies than to the corresponding cross-dependency stimuli. No such difference was found for the Germans learning the sequences with three nonadjacent dependencies, nor did the Dutch participants show any response-time differences across any of the sequence types. Given that center-embedded constructions with two dependencies are much more frequent than with three dependencies (see Karlsson, 2007, for a review), this pattern of differences may reflect the German participants' prior linguistic experience with center-embedded, verb-final constructions.

complex recursive constructions. These processing limitations on recursive structure derive from chunk-based constraints on sequence learning, modulated by our individual native language experience.

We have taken the first steps toward a usage-based account of recursion, where our recursive abilities are acquired piecemeal, construction by construction, in line with developmental evidence. As in chapter 6, we have highlighted the key role of language experience in explaining cross-linguistic similarities and dissimilarities in the ability to process different types of recursive structure. And although we have focused on the important role of sequence learning in explaining the limited human abilities, we want to stress that Chunk-and-Pass language processing includes other domain-general factors. Whereas distributional information clearly provides important input to language acquisition and processing, it is not sufficient, but must be complemented by numerous other sources of information, from phonological and prosodic cues (discussed in chapter 5) to semantic and discourse information. Thus, while our account is, of course, incomplete, it does offer the promise of a usage-based perspective on recursion. Just as being able to recite the old nursery rhyme "The house that Jack built" requires considerable practice, so is our ability to process other kinds of recursive structure likewise a usage-based skill.

8 From Fragmentation to Integration

The terms "splitters" and "lumpers" come from taxonomy, where the classifiers were separated into those who liked to create new taxa because of small differences and those who preferred to coalesce categories because of similarities. The concept has found wider applicability as knowledge in all fields expands. Specialists are confined to ever-narrowing domains while generalists survey the immensity of information in an effort, one hopes, to find higher orders of structure. It is clear that in the university and intellectual community ... the splitters are in command and the lumpers are in serious disarray, unable to keep up with the output of printouts that are generated in such a variety of ways. It is saddening to witness the loss of status of those engaged in integrative thought, for one sees in it the fragmentation of scientific and humanistic disciplines.
—Harold Morowitz, 1979, *Splitters and Lumpers*

The argument of this book is that languages emerge from myriad individual communicative exchanges. Each communicative exchange may be measured in seconds, and key processing operations involved in producing and understanding such exchanges may be measured in tens of milliseconds. And each use of language leaves traces in the minds of its participants, shaping future uses. Language is a skill that, like other skills, we acquire by *doing*. From this usage- or item-based perspective on language acquisition (e.g., Tomasello, 1992), the cumulative impact of many years of these myriad interactions will be the piecemeal accumulation, construction-by-construction, of a mastery of the language. And we have argued that the traces left in the minds of each participant in a communicative exchange have another, and potentially even longer-term impact: shaping the cultural evolution of the language itself. Linguistic patterns that are easily acquired and repeated across communities will tend to become established; patterns that are difficult to acquire or unstable when acquired will be eliminated. We have suggested that the processes of language change observed over decades and centuries are continuous with the evolution of language itself, presumably over many tens of thousands of years.

From this perspective, then, language processing, acquisition, and evolution are closely intertwined over a wide range of timescales. Constraints operating on how the cognitive system can process language, such as the Now-or-Never bottleneck that we discussed in chapter 4, inevitably shape what can or cannot be acquired; and what is readily acquired will, conversely, shape the linguistic environment with which the language system must deal. Constraints on processing and acquisition will be powerful selectional forces operating on the cultural evolution of language, both within populations and across successive generations. The result of such cultural evolutionary pressures will shape the patterns exhibited by the world's natural languages.

Given these considerations, a natural research strategy is to attempt to build a unified framework for understanding the structure of language, and the multiple and closely interrelated forces in processing, acquisition, and cultural evolution that have led to its piecemeal creation over an unimaginably vast number of momentary communicative episodes. Setting out a sketch of such a framework, and outlining how it might be applied (e.g., in case studies described in chapters 5, 6, and 7), has been the goal of this book.

8.1 Integration or Conflation?

Yet this attempt to sketch an integrated framework for the language sciences threatens to ride roughshod over a variety of widely-used theoretical distinctions in mainstream generative grammar: between core and periphery (Chomsky, 1981), competence and performance (Chomsky, 1965), acquiring a language and learning to process (e.g., Crain & Lillo-Martin, 1999), language evolution and language change (e.g., Hauser, Chomsky, & Fitch, 2002), and, as we shall see later, between grammatical structure and processing history (e.g., Fodor, Bever, & Garrett, 1974) (see box 8.1). We have argued that breaking down such divisions provides a richer understanding of language, and allows theories regarding different aspects of language to constrain each other in powerful ways. But might we instead be breaking down divisions that are central to organizing the language sciences? Might we be conflating distinct issues, rather than helping to create a constructive synthesis?

In this final chapter, we respond to this concern directly. We will reconsider some of the traditional distinctions that conflict, whether directly or indirectly, with the approach developed here. Such traditional distinctions aim to split apparently related issues into two: that is, to suggest that, superficial indications apart, Topic A and Topic B require different kinds of answers (where the topics might concern core and periphery, or competence and performance, and

Box 8.1
The Fracturing of the Language Sciences

> The advent of generative grammar more than a half-century ago raised the possibility of a new and rigorous approach to the science of language that might elucidate, among other things, the structure of language, the computational mechanisms by which it can be processed and acquired, the biological basis of language, the patterns of variation seen across languages, and the processes of language change and evolution. Yet the prospect of an integrated biological and computational theory of language has become ever more distant. Early research on the *derivational theory of complexity* (Miller & McKean, 1964) explored whether the number of "transformations" required to generate a sentence might be related to processing difficulty. Difficulties arose in connecting generative linguistic theory with psycholinguistic data—but this did not lead to revised proposals. Instead, the connection between the two was downplayed. Furthermore, it became increasingly clear that transformational grammar was ill-suited to computational language processing, both regarding efficient online parsing and building up compositional semantic representations. Computationally-oriented researchers were driven to create new syntactic formalisms, severing links with mainstream generative grammar (e.g., Gazdar, Klein, Pullum, & Sag, 1985; Joshi & Schabes, 1997; Steedman, 2000). Similarly, mainstream generative grammar is difficult to reconcile with the piecemeal character of language acquisition, or indeed the very possibility of learning an abstract grammar from an apparent "impoverished" linguistic input (Chomsky, 1980b). Attempting to clarify how language is learnable is one motivation for a variety of syntactic proposals that depart from the mainstream generative tradition (e.g., Clark, 2011; Culicover & Jackendoff, 2005; Goldberg, 2006; Jackendoff, 2002). The objective of this book is to reinforce attempts to build stronger links between different aspects of the language sciences: to provide a framework in which it may be possible to construct a more unified science of language.

so on), and hence that the study of Topic A and Topic B should proceed in relative, or even complete, isolation from one other. In terms of the quotation from the biologist Harold Morowitz at the beginning of the chapter, the concern we address here is that the approach we have been advocating involves an overenthusiastic lumping together of topics that would better remain cleanly split apart.

Our argument for integration rather than fragmentation follows a common pattern, borrowed and extended from an insightful discussion by Culicover (1999). We shall see that addressing one of the putatively distinct topics, say Topic A, turns out to extend naturally to provide a viable account of Topic B. So, by maintaining that a separate theory of Topic B is required (which is utterly independent of Topic A), we are stuck with two distinct and potentially rivalrous theories of Topic B. Parsimony strongly favors a unified account of both topics, and concludes that the distinction is not required after all. We shall refer to this style of reasoning as the *Culicover argument.*

Let us begin by seeing how this style of argument applies to the distinction between core and periphery in natural language syntax, before extending the discussion to a variety of further distinctions.

8.1.1 The Structure of Language: Core and Periphery

Following construction grammar in linguistics (e.g., Croft, 2001; Fillmore, Kay & O'Connor, 1988; Goldberg, 1995, 2006), and closely related item- and exemplar-based accounts of language acquisition and processing (e.g., Bod, 2009; Daelemans & Van den Bosch, 2005; Tomasello, 1992), we have presupposed that the structure of language can be understood in terms of many fairly independent constructions.

Culicover (1999) notes that this approach appears to conflict with two distinctions important in traditional generative grammar: between the grammar and the lexicon, and between two parts of grammar: *core* and *periphery* (Chomsky, 1981). The grammar-lexicon distinction is blurred because lexical items are just another type of construction, and it is assumed that there are no grammatical rules, over and above the collection of constructions. That is, all the interesting structure in the language is presumed to be inherent in the constructions (including individual words), rather than in some set of potentially very abstract, complex, and subtle grammatical rules or constraints, which might, for example, be presumed to capture universal features of all possible human languages (Chomsky, 1980b, 1981).

In a construction- or item-based approach to language, without a distinct set of grammatical principles, the distinction between core and peripheral aspects of the grammar cannot even be easily stated. In Chomsky's (1981) formulation, the core of the grammar captures deep underlying regularities in language, and is the primary target of generative grammar. The periphery, by contrast, consists of the myriad capricious grammatical idiosyncrasies that individual languages exhibit. Here, we find irregular morphology, loan constructions from other languages, and the plethora of peculiar constructions that seem to violate "normal" syntactic rules. Chomsky introduces the split between core and periphery to allow the development of syntactic theory to focus on fundamental regularities in language, rather than attempting to capture endless linguistic quirks.

As Culicover (1999) points out, this strategy has the immediate danger that any awkward cases for a particular syntactic theory may all too readily be brushed aside, as belonging to the periphery. He notes, moreover, that this danger is especially serious, given the vast richness and variety of such cases, apparently at the heart of everyday language. Among a blizzard of fascinating examples, Culicover points out that we say *so big*, *too big*, *as big*, *sufficiently big*, but that we say *big enough* rather than **enough big*; we can speak of *the responsible parties*, *the guilty parties*, *the parties responsible*, but, oddly, we can't speak of **the parties guilty*.

Culicover's argument against the core-periphery distinction is not, though, merely that the periphery turns out to be alarmingly large. It is that whatever grammatical machinery is required to capture the periphery, with all its spectacular variation and intricacy, will automatically be powerful enough to deal with the much more regular core. Indeed, Culicover argues that the difference between constructions traditionally assigned to the periphery, rather than the core, is a matter of degree of generality, rather than any fundamental difference in kind.

In short, the language system needs to deal with the complex and idiosyncratic periphery one way or another; and if it can do so, surely it can use the very same mechanisms to deal with the much less challenging core. Referring back to the general pattern of the Culicover argument above, why distinguish Topic A and Topic B, if any solution to Topic A (here, the so-called periphery) is likely to be powerful enough to deal with Topic B (here, the so-called core) as a straightforward corollary?

Now, in principle, of course, there might be reasons to maintain the core-periphery distinction. For example, if some types of regularity were completely fossilized, while others were completely productive, then there might be independent grounds for splitting aspects of the language that must merely be listed, on the one hand, from aspects which follow general rules, on the other. Culicover argues convincingly, however, that the degree of productivity of linguistic forms also lies on a continuum, further undermining any motivation for maintaining a theoretical split between core and periphery. And, if there is no theoretical need for a deep grammatical core, and the structure of the language can be captured by layers of constructions of different levels of generality, then the distinction between lexical items and abstract patterns (e.g., concerning agreement or verb argument structure), is also more naturally modeled, not as a binary split (between lexicon and grammar), but as a continuum.

8.1.2 Competence and Performance

Mainstream generative grammar takes the primary data of linguistics to be native speaker intuitions (e.g., Chomsky, 1965). Yet, creating an elegant syntactic theory may appear to be plagued not merely by the capriciousness of language (from the so-called periphery), but by the limitations of our ability to process language (see box 8.2).

At this point, the competence-performance distinction appears to come to the rescue. Chomsky (1965, p. 3–4) explains that "Linguistic theory is concerned primarily with an ideal speaker-listener … who … is unaffected by

Box 8.2
Meta-linguistic Judgments, Grammaticality, and Center-Embedding

Fodor (1975) notes that the chant *bulldogs bulldogs bulldogs fight fight fight* can, according to standard generative grammar, be assigned a center-embedded syntactic structure, and indeed a meaning (see also chapter 7). This initially baffling claim can be justified by something like the following line of argument. Clearly, *bulldogs fight* is an unproblematic English sentence with a clear interpretation; *bulldogs that spaniels love fight* is a little strange, but unproblematic enough. But what if we wish to talk about bulldogs that are fought by (other) bulldogs, rather than loved by spaniels? Then we can replace that *that spaniels love* with *that bulldogs fight* to yield: *bulldogs that bulldogs fight fight*. The complementizer *that* is optional, by analogy with similar constructions. So we have: *bulldogs bulldogs fight fight*.

Now, the final step. To get from *[bulldogs]$_{NP}$ fight* to *[bulldogs bulldogs fight]$_{NP}$ fight* we have simply replaced one noun phrase with another. That is, *[bulldogs]$_{NP}$* is switched with *[bulldogs bulldogs fight]$_{NP}$* . So surely we can play the same trick again, but this time with the second occurrence of bulldogs. Thus, *[bulldogs [bulldogs]$_{NP}$ fight]$_{NP}$ fight* can surely be mapped to *[bulldogs [bulldogs bulldogs fight]$_{NP}$ fight]$_{NP}$ fight*, which, if we remove the brackets is, of course, just *bulldogs bulldogs bulldogs fight fight fight*. And, by following a parallel line of argument, we should, in principle, surely be able to figure out what the sentence means (although, according to the unaided linguistic intuition, it does seem utterly impenetrable!).

What does this type of example illustrate? One possibility is that it shows the extraordinarily severe limitations of human language processing: that while the language itself may allow all sorts of complex structures, which should be captured by any suitably elegant grammatical theory, the language processing system is rapidly overloaded, and cannot deal with such sentences. So a gulf appears to open up between unaided syntactic intuitions, which roundly rejects the "bulldog" sentence, and the results of certain kinds of linguistic reasoning, which accept it as both grammatical and meaningful. Another possibility, of course, is that such sentences are not part of the language at all—but only an *extension* of the language created by adopting particular formal assumptions, and following complex meta-linguistic reasoning, along the lines described above, which stem from our general thinking and problem solving abilities, and have no relation to the language system proper.

such grammatically irrelevant conditions as memory limitations, distractions, shifts of attention and interest, and errors (random or characteristic) in applying his knowledge of this language in actual performance." This ideal knowledge of language is linguistic *competence*; the actual use of the language, and creation of grammatical judgments, is a matter of linguistic *performance*. For Chomsky, linguistic theory should abstract away from performance issues, and focus on capturing linguistic competence.

It would be possible, in principle, to hold on to a competence-performance distinction, while still forging a close link between linguistic and psycholinguistic theory. This would be true, for example, if the theory of language processing consisted of a specification of linguistic competence, and a set of specific processing rules and memory restrictions that would explain how that linguistic competence maps on to performance (see chapters 6 and 7 for discussions of problems with this approach). In practice, though, the mainstream generative tradition has taken a very different tack: syntactic theory has been developed in a way that has almost no connection either to psychological data on language processing, or to computational models of language processing and production that might potentially indicate how linguistic performance might be explained. For many decades, computational linguists faced with practical natural language processing challenges have had to create very different linguistic formalisms to support computational models of language processing (e.g., Generalized Phrase Structure Grammar, Gazdar, Klein, Pullum, & Sag, 1985; Combinatorial Categorial Grammar, Steedman, 2000; Tree Adjoining Grammar, Joshi & Schabes, 1997; and wide variety of statistical models of language, Manning and Schütze, 1999).

In light of the apparent divergence between theories of competence in the mainstream generative grammar tradition and theories of linguistic performance, it might be tempting to suggest that while processing proposals (such as the Chunk-and-Pass account described in chapter 4) may help explain incremental language production and comprehension (e.g., given performance constraints, such as the Now-or-Never bottleneck), it will never explain, and must be psychologically separate from, the idealized linguistic competence, which will include all manner of complex sentences that people are entirely unable to process straightforwardly. More broadly, there has been a tendency in the mainstream generative tradition to distance claims about linguistic structure from claims about the operation of the language processing system as far as possible.

Yet this use of the competence-performance distinction falls victim to the Culicover argument that we described above. Suppose, indeed, that performance is entirely separate from competence. Then, whatever computational machinery is required to understand and produce sentences (and, presumably, indirectly to help evaluate whether they are grammatical) must embody enough language structure to allow us to deal with the language as it is used by native speakers. Thus, if we could characterize this knowledge, then surely we would have spelled out the structure of language sufficiently to deal with anything that would ever be said or heard. But then the job of linguistic theory would be done! Or, at least, if we were to insist that there is, in addition, an entirely separate theory of linguistic competence, then we would seem to have *two* theories of language structure, which overlap on all interesting cases, and differ only on sentences that people don't spontaneously produce, and wouldn't understand, even if they were produced.

What, then, are we to make of peculiar sentences such as the "bulldog" sentence in box 8.2? Rather than assuming that they are part of the language in good standing, yet lying tantalizingly beyond human processing limitations, a more straightforward approach is to deny that they are part of the domain of linguistic theory in the first place. Now, of course, it is entirely possible to create a string of meta-linguistic arguments, such as that described in box 8.2, for the "bulldog" sentence, in which we make various assumptions about substitutability; and such arguments could lead us to give a syntactic analysis, and even a meaning, to apparently impenetrable sentences. But such extrapolations are surely not part of the knowledge of the language, but are instead theoretically-guided extensions of the language, which draw not on fundamental knowledge of the language itself, but on the application of our general reasoning abilities.

We suggest, moreover, that the specific mathematical analysis of language in early generative grammar (e.g., aiming to locate natural language in the Chomsky hierarchy; Chomsky, 1956; see box 7.1) may have led to an unfortunate break between the study of grammar and theories of processing. A conventional generative grammar creates a huge space of possible sentences, of which only an infinitesimal fraction could actually be said or understood, and this dissonance led to linguists embracing this vast space of unsayable sentences as linguistically bona fide. Instead, we suggest, theories of language structure should focus on sentences people actually use, and whatever knowledge allows us to process these sentences will be sufficient to characterize their structure.

Note, finally, that to reject the competence-performance distinction does not require concluding that grammatical theory should blindly incorporate speech errors and half-finished sentences. Language, like any other empirical phenomenon, is the product of many factors, and explaining language requires separating the influence of these different factors. When considering astronomical data, we allow the possibility that our measuring instruments are inaccurate or malfunctioning, that the atmosphere may have a distorting effect, that our data analysis methods may be imperfect, and so on. Understanding and taking account of these factors as much as possible will lead to the best possible inferences about the properties of the solar system or outer space, which are of primary interest. The same is true in the study of language. In interpreting either observational or experimental data about human language processing, we should be aware of the wide range of possible factors, in addition to the operation of the language processing mechanism itself, which may be needed to explain a particular piece of data. To take a particularly simple example, if a person stops speaking mid-flow due to the sound of a gunshot, we clearly do not want the resulting half-sentence to be classified as a grammatical sentence. But the competence-performance distinction, as it has been used in the generative tradition, has a much wider role than this: driving a wedge between the study of language processing operations, and the knowledge of the structure of language itself. In light of the perspective outlined in chapters 1 and 4–7, this division may have been especially unhelpful. We suggested that the Now-or-Never bottleneck encountered in online language processing (a performance factor par excellence) is crucial to explain many aspects of linguistic structure (such as duality of patterning, the local nature of linguistic dependencies, and so on). Attempting to understand the structure of language by abstracting away from the computational machinery that produces and understands it may turn out to have been a methodological wrong-turn.

8.1.3 Identifying a Grammar versus Learning to Speak a Language

As discussed in chapter 4, the apparently miraculous process by which children spontaneously learn their native language appears, at first blush, to involve acquiring a *skill*. To understand language, the child has to learn to map streams of sounds (or signs) onto meanings; and to produce language, the child must map meanings back onto streams of sounds (or signs). The complexity of the perceptual, cognitive, and motor processing involved in human language use is, of course, remarkable, but one would imagine that the process of acquisition would be continuous with the acquisition of other skills, such as reading, arithmetic, drawing, chess, playing music, and so on.

Yet many theorists of language development, primarily working within the generative framework, have taken a very different viewpoint. Holding fast to the competence-performance distinction, they have sharply distinguished the problem of identifying the grammar of the language (viewed as an issue of competence) from any challenges the child faces in actually processing language (as matters of performance). And questions of competence have been center stage.

The difficulties that we have identified with the competence-performance distinction carry over directly to cast doubt on the usefulness of treating learning to process language (attaining skilled linguistic "performance") and acquiring an abstract representation of the grammar (the putative linguistic "competence") as distinct topics. Following the now-familiar line of the Culicover argument, note that any account that can explain how a child learns to process language (i.e., to develop the relevant language processing *skill*) has to account for how these processes can handle the full range of linguistic input that it typically receives. So explaining how the child acquires the skill of processing language seems automatically to require explaining how the child learns the structure of the language. Thus, to propose that the structure of language that is embodied by the processing operations is utterly distinct from a more abstract linguistic competence, also somehow represented in the child's "language faculty," implies an apparently unnecessary duplication.

The focus on language acquisition as the identification of an idealized grammatical competence, independent of processing constraints, may have had a number of deleterious consequences. First, to the extent that the structure of language is adapted to processing constraints, such as the Now-or-Never bottleneck, then the child does not need to *learn* to impose these constraints at all: they are built into the child's basic cognitive machinery. But, abstracting from processing considerations, the patterns of constraints on language are likely to appear arbitrary, and hence will appear to pose a substantial learnability challenge for the child. Such learnability challenges constitute one motivation for the postulation of an innate universal grammar (UG). If, by contrast, language has been shaped by the general processing, learning, and communicative biases of the human brain, then the appropriate constraints will be embodied, "for free," by the cognitive mechanisms by which the child learns to process, as we argued in chapters 2 and 3.

A second deleterious consequence of splitting apart the study of identifying a grammar from learning to process is that the problem of acquiring natural language is often then assimilated to the problem of learning one language from the entire class of formal languages (e.g., as in the famous theorems by Gold, 1967). So, for example, if we choose to idealize English by a stochastic context-free phrase structure grammar, then it is all too easy to assume that the child's learning machinery must be able to identify an *arbitrary* stochastic context-free phrase structure grammar, from exposure to the output from such a grammar (see Alexander Clark & Lappin, 2011, for extensive discussion). This leads to a focus on the question of which formal class of languages includes the natural languages: the broader the class of languages, the more difficult the learning problem appears to be (see also box 7.1 in chapter 7). In the face of this apparently daunting challenge of learning, it is a short step to the conclusion that, perhaps, language acquisition does not involve a substantial degree of learning at all—that language acquisition is more like the maturational development of an organ such as the liver or the heart, than learning a skill (e.g., Chomsky, 1980b; Pinker, 1999). Applying this perspective to the study of child language leads to a range of apparently very counterintuitive proposals, including that early language production, while appearing to violate adult natural language syntax, corresponds, in fact, to an adult-level linguistic competence, subject to severe performance constraints (Crain, 1991); and apparent grammatical errors by young children should be construed as indicating that they are speaking a *different* natural language from that spoken in the linguistic environment (because one or more parameters is set inappropriately; e.g., Yang, 2006).

Note, too, that a great deal of attention has been paid to the question of how language can be acquired from positive evidence alone (Brown & Hanlon, 1970; Chater & Vitányi, 2007; Pinker, 1979, 1984; Hsu, Chater, & Vitányi, 2013)—i.e., without explicit correction. Referring to puzzling idiosyncrasies of language, such as those noted by Culicover, Baker (1979) wondered how the "gaps" in such patterns are learnable without explicit correction. For example, how is the child to learn that *the rabbit hid*, *the rabbit disappeared*, and *I hid the rabbit* are acceptable, but that **I disappeared the rabbit* is not, without the child ever receiving explicit negative feedback? As Baker pointed out, it is clear that such idiosyncratic and unpredictable irregularities cannot be part of UG. But whatever learning mechanisms underlie the child's ability to learn the spectacularly intricate patterns in the linguistic periphery, which cannot possibly be part of UG, can surely successfully learn the more regular patterns typically attributed to the linguistic core. So the textbook assumption that lack of negative evidence implies innate structure clearly needs to be revised (e.g., Crain & Lillo-Martin, 1999). The same line of argument applies to "Poverty of the Stimulus" arguments for innate linguistic structure more broadly (Chomsky, 1980b; see Clark and Lappin, 2011, for discussion).

A further challenge for the language-acquisition-as-grammar-identification perspective arises when considering the learning of a second language through explicit instruction. The learner deliberately and effortfully tries to master vocabulary, idioms, and grammatical rules, and initially experiences great difficulties with processing language in real-time. Learning a second language seems, indeed, to be a paradigm example of skill learning; and the skill is learned step-by-step, through extensive practice, and with continual errors. In short, second language learning through instruction does not appear to fit well with the idea that the learner is developing an idealized linguistic competence, perhaps by fine-tuning a pre-existing UG. The response to this observation, within much of mainstream linguistics and language acquisition research, has been to argue that language acquisition mechanisms involved in first language acquisition are utterly distinct from those involved in second language acquisition (e.g., Chomsky, 1965). When acquiring the first language, the child is assumed to have access to UG, but this is assumed not to be available for second language learning (e.g., Pinker, 1994). Indeed, from this point of view, language acquisition is sometimes seen as showing a "critical period" (e.g., Lenneberg, 1967; Pinker, 1994), analogous to that observed in some songbirds (Nottebohm, 1969) and for other aspects of brain development (e.g., Hubel & Wiesel, 1970; see box 8.3). But if this is right, then the Culicover argument applies yet again: if it is possible to learn to speak and understand language *without* access to UG, then surely those same learning mechanisms could just as well be applied to first language acquisition. If so, is UG really essential to language acquisition after all?

The standard response to this line of thinking is to stress that the problems of first and second language acquisition are very different. For example, second language acquisition typically involves instruction, rather than mere immersion in the language. But, of course, this difference is a matter of degree. For most second language learners, *using* the language seems a critical part of learning, and many adults can learn a second language entirely from immersion in a new linguistic environment without direct instruction. Moreover, explicit instruction is inevitably extremely limited, given the full complexity of the grammar of any language. Defenders of a fundamental distinction

Box 8.3
Critical Periods for Language Acquisition?

Can the claim that language acquisition involves skill learning, applying general cognitive mechanisms to linguistic experience, be reconciled with the claim that there are "critical periods" for language acquisition (Lenneberg, 1967)? The latter claim is generally assumed to include (1) unless the first language is learned before puberty, language acquisition is at best partial; and (2) second language acquisition differs profoundly, and is inferior to, first language acquisition. It is difficult to draw firm conclusions from cases of first language deprivation (Curtiss, 1977), as they typically correlate with minimal social contact, and impoverished non-linguistic environments. Another possible source of evidence is second language acquisition: investigating how well people can learn a second language after puberty. Studies have shown that many late second language learners achieve high levels of fluency, overlapping considerably with the surprisingly wide variation in linguistic ability in native speakers (Dąbrowska, 2012). Some aspects of the second language, especially phonology, appear especially difficult to learn. But this is consistent with the experience-based viewpoint, because the representations developed during first language acquisition will be re-used to encode the second language. Thus, for example, if the second language involves phonological distinctions absent in the first language, the second language learner may build Chunk-and-Pass representations of the second language input that omit such distinctions. Age of immigration to a new country, and hence exposure to the new language, provides a test of the critical period hypothesis. The key question is whether there is a discontinuity in ultimate linguistic competence for immigrants who arrived before or after the end of the critical period. Hakuta, Bialystok, and Wiley (2003) conducted a large-scale study using U.S. Census data, and found no evidence of such a discontinuity, although later arrivals had lower competence in English. This general age-of-arrival effect had been observed by Johnson and Newport (1989) studying the English abilities of Asian immigrants to the United States. A critical period account predicts that later learners of English should be particularly affected by aspects of the language that are part of UG, rather than for the linguistic periphery, which should be relatively untouched. This prediction has not been confirmed (Johnson & Newport, 1991). An experience-based perspective predicts that the degree to which second language learners succeed will depend crucially on the amount of linguistic experience they receive, and the relationship between the first and second languages. Both these predictions have been observed empirically in a study of immigrants to Spain (Birdsong & Molis, 2001).

between first and second language acquisition may also stress the apparent difference in the "final state" of such learning; late second language learners tend to retain aspects of the phonology, and sometimes syntax, of their first language. Nonetheless, many late second language learners acquire the grammar of the new language well enough to function in it at a very high level, including, in some cases, producing great literature. For example, Polish-born Joseph Conrad—author of *Heart of Darkness, Nostromo,* and many other celebrated works—only attained a good command of English in his twenties. He had a strong accent, yet was an acclaimed English writer. If Conrad was able to achieve this impressive level of linguistic command without access to UG, and, indeed, without properly *identifying* the grammar of English, then it is natural to suggest that whatever skill-learning mechanisms he used in his second language acquisition might also be deployed successfully in first language acquisition to good effect.

8.1.4 Language Evolution versus Language Change

We have argued in this book that language is a product of cultural evolution over myriads of processing episodes by many generations of speakers, and is shaped by the learning, processing and communicative biases of the human brain. So, from this viewpoint, the evolution of language is simply language change writ large. Thus, one might expect the insights gained from historical linguistics to provide insight into language change over tens of thousands of years; and processes, such as grammaticalization, that have been postulated to explain aspects of language change, would be expected to help explain the very long-term evolution of language (e.g., Heine & Kuteva, 2002a, 2007, 2012; Tomasello, 2003).

Yet, by contrast, language evolution and language change have generally been treated as entirely separate domains within the mainstream linguistic tradition (e.g., Berwick, Friederici, Chomsky, & Bolhuis, 2013). The assumption has been that the core grammar, which is assumed to be universal across languages, is innately specified. The innate UG is assumed to be a product of genetic change, either through adaptation (e.g., Pinker & Bloom, 1990) or through some other, possibly sudden, genetic change (e.g., Chomsky, 2010), and/or a cognitive reorganization generated by some threshold being reached in any underlying continuous quantity, such as brain size (e.g., Bickerton, 1995; Lightfoot, 2000). Language change over historical time is assumed to be confined to the non-core aspects of language, such as the grammatical periphery and the lexicon. So language evolution and language change are assumed to be driven by different mechanisms—genetic vs. cultural—and to concern distinct and non-overlapping aspects of the language. For adaptationists (e.g., Pinker & Bloom, 1990), the processes of biological and cultural evolution are entwined, of course; nonetheless, it is still assumed that there are two distinct processes at work.

As we outlined in chapter 2, there is no biologically viable account of how a genetically encoded UG could become established. Instead, language has been shaped by the pre-existing processing, learning, and communicative biases of the brain. Hence, as we argued in chapter 3, the ability of children, and for that matter, adults, to acquire language so rapidly, arises because the most easily learnable variants of language are those that are positively selected during the cultural evolution of language. And, as we described in chapter 4, what is easy to learn will be, among other things, easy to *process* under severe cognitive constraints, such as the Now-or-Never bottleneck. From this point of view, the apparent divide between language change and language evolution is entirely illusory: it is a side-effect of a theoretical position that is no longer tenable. It remains, of course, of great interest to understand the biological evolutionary history that led to the cognitive pre-requisites for the cultural evolution of language. Candidate mechanisms include joint attention, large long-term memory, sequence processing ability, appropriate articulatory machinery, auditory processing systems, and so on. But this is the study not of language evolution per se, but of the evolution of the biological precursors of language.

8.1.5 Grammar, Trees, and Processing

Language structure and language processing have, as we have seen, been deliberately disconnected in much recent discussion in mainstream generative linguistics (although a great deal of linguistics, computational linguistics, and psycholinguistics have proceeded outside the generative mainstream). Yet the Chunk-and-Pass perspective outlined in chapter 4 implies that there is a direct link between processing and traditional linguistic notions (also cf. Hawkins, 1994, 2004). In both production and comprehension, the language system creates a sequence of chunking operations, which links different linguistic units together across multiple levels of structure—that is, the syntactic structure of a given utterance is reflected in its processing history (Christiansen & Chater, in press). This conception is reminiscent of previous proposals in computational linguistics, in which syntax is viewed as a control structure for guiding semantic interpretation (e.g., Kempson Meyer-Viol, & Gabbay, 2001). For example, in describing his incremental parser-interpreter, Pulman (1985, p. 132) notes, "Syntactic information is used to build up the interpretation and to guide the parse, but does not result in the construction of an independent level of representation." Steedman (2000) adopts a closely related perspective when introducing his Combinatory Categorial Grammar, which aims to map surface structure directly onto logic-based semantic interpretations, given rich lexical representations of words that include information about phonological structure, syntactic category, and meaning: "… syntactic structure is merely the characterization of the process of constructing a logical form, rather than a representational level of structure that actually needs to be built …" (Steedman, 2000, p. xi). Thus, in these accounts, the syntactic structure of a sentence is not explicitly represented by the language system, but plays the role of a processing "trace" of the operations used to create or interpret the sentence (see also O'Grady, 2005).

To take an analogy from constructing objects rather than sentences, the process by which components of an IKEA-style flat-pack cabinet are combined provides a "history" (combine a board, handle, and screws to construct the doors; combine frame and shelf to construct the body; combine doors, body, and legs to create the finished cabinet). The history by which the cabinet was constructed may thus reveal the intricate structure of the finished item, but this structure need not be explicitly represented during the construction process. Similarly, we can "read off" the syntactic structure of a sentence from its processing history, revealing the syntactic relations between various constituents (perhaps with a rather "flat" structure; Frank, Bod, & Christiansen, 2012; Sanford & Sturt, 2002). Syntactic representations are neither computed nor in any way represented during comprehension or production; instead, syntax is reflected in the history of processing operations. That is, we view *linguistic structure as processing history.* Importantly, this perspective implies that syntax is not privileged but is only one part of the language processing system—and it is not independent of the other components.

From this standpoint, a rather minimal notion of grammar specifies how the chunks from which a sentence is built can be composed. There may be several ways in which such combinations can occur, just as operations for furniture assembly may be somewhat flexibly carried out (but not completely without constraints—it might turn out that the body must be screwed together before a shelf can be attached). In the context of producing and understanding language, the process of construction is likely to be much more constrained: each new component is presented in turn, and must be used immediately or it will be lost due to the Now-or-Never bottleneck. Moreover, viewing Chunk-and-Pass processing as an aspect of skill acquisition, we might expect that the precise nature of chunks may change with expertise: highly overlearned material might, for example, gradually come to be treated as a single chunk (see Arnon & Christiansen, submitted, for a review).

As with other skills, the cognitive system will tend to be a *cognitive miser* (Fiske & Taylor, 1984), generally following a *Principle of Least Effort* (Zipf, 1949). As processing proceeds, there is a complex interplay of top-down and bottom-up processing to alight on the message as rapidly as possible. The language system needs only to construct enough chunking structure so that, when combined with prior discourse and background knowledge, the intended message can be inferred incrementally. This observation relates to some interesting contemporary linguistic proposals. For example, from a generative perspective, Culicover (2013) highlights the importance of incremental processing, arguing that the interpretation of a pronoun depends on which discourse elements are available when it is encountered. This implies that the linear order of words in a sentence (rather than hierarchical structure) plays an important role in many apparently grammatical phenomena, including weak cross-over effects in referential binding. From an emergentist perspective, O'Grady (2015) similarly emphasizes the importance of real-time processing constraints for explaining differences in the interpretation of reflexive pronouns (*himself, themselves*) and plain pronouns (*him, them*). The former are resolved locally, and thus almost instantly, whereas the antecedents for the latter must be searched for over a broader domain (causing problems in acquisition because of a bias toward local information).

More generally, our view of grammatical knowledge as processing history offers a way of integrating the formal linguistic contributions of construction grammar (e.g., Croft, 2001; Goldberg, 2006) with the psychological insights from usage-based approaches to language acquisition and processing (e.g., Bybee & McClelland, 2005; Tomasello, 2003). Specifically, we propose to view constructions as *computational procedures*[1]—specifying how to process and produce a particular chunk—where we take a broad view of constructions as involving chunks at different levels of linguistic representation from morphemes to multiword sequences. A procedure may integrate several different aspects of language processing or production, including chunking acoustic input into sound-based units (phonemes, syllables), mapping a chunk onto meaning (or vice versa), incorporating pragmatic or discourse information, and associating a chunk with specific arguments (see also, O'Grady, 2005, 2013).

1. The term "computational procedure" is also used by Sagarra and Herschensohn (2010) but they see these as developing "in tandem with the growing grammatical competence" (p. 2022). Likewise, Townsend and Bever (2001) discuss "frequency-based perceptual templates that assign the initial meaning." (p. 6). However, this only results in "pseudosyntactic" structure, which is later checked against a complete derivational structure. In contrast, we argue that the computational procedures are all there is to grammar; a proposal that dovetails with O'Grady's (2005, 2013) notion of "computational routines," but with a focus on chunking in our case.

As with other skills (e.g., Heathcote, Brown, & Mewhort, 2000; Newell & Rosenbloom, 1981), there will be practice effects, where the repeated use of a given chunk results in faster processing and reduced demands on cognitive resources, and with sufficient use, leading to a high degree of automaticity (e.g., Logan, 1988; see Bybee & McClelland, 2005, for a linguistic perspective).

In chapter 4, we suggested that quasi-regularities in language arise in the same way that a particular pattern of tracks are laid down across a forest, through overlaid traces of endless agents finding the path of local least resistance. Continuing with this analogy, just as a well-trodden forest path becomes more strongly established, so a more frequently used chunk becomes more entrenched, resulting in easier access and faster processing. With sufficiently frequent use, adjacent tracks may blend together, creating somewhat wider paths. For example, the frequent processing of simple transitive sentences, processed individually as multi-word chunks, such as *I want milk, I want candy*, might first lead to a wider track involving the item-based template *I want X*. Repeated use of this template along with others (e.g., *I like X, I see X*) might eventually give rise to a more abstract transitive generalization along the lines of *N V N* (a veritable highway, in terms of our track analogy). Similar proposals for the emergence of basic word-order patterns have been proposed both within emergentist (e.g., O'Grady, 2005, 2013; Tomasello, 2003) and generative perspectives (e.g., Townsend & Bever, 2001). However, just as with generalizations in perception and motor skills, the grammatical abstractions are not explicitly represented, but result from the merging of item-based procedures for chunking. Thus, there is no representation of grammatical structure separate from processing. Learning to process *is* learning the grammar.

To apply the Culicover argument one final time, note that the processing history provides a structure for each sentence, and the operations of the language system determine which processing histories are possible. Moreover, given that every sentence that is produced and comprehended by the human language system has been created by precisely these processes, it is clear that such processing histories, and the constraints defining which histories are possible, must successfully capture the full complexity of natural language. It remains a logical possibility, of course, that in addition there is an entirely separate set of linguistic analyses that can be applied to natural language sentences, and that these might be captured by an entirely different set of rules. But not only would this additional set of rules seem to be entirely redundant, it would also seem to be an astonishing coincidence that a language produced and comprehended by a particular set of processing operations happens, mysteriously, also to be subject to a completely different collection of principles, not embedded in this processing system. It seems much more parsimonious to have a unified account, in which the structure of language arises from the traces of the processing operations that create it.

8.2 The Cumulative Emergence of Linguistic Order

We have argued that the world's natural languages have emerged through cultural evolution operating on successive traces of individual communicative episodes, shaped both by processing and acquisition. But a crucial question remains: why is language so orderly? How have myriad processing episodes created not a chaotic mass of conflicting conventions, but a highly, if partially, structured system linking form and meaning?

Picking up our forest track analogy, notice that on each occasion that an animal navigates through the forest, it is concerned only with reaching its immediate destination as easily as possible. But the cumulative effect of such processing episodes, in breaking down vegetation and gradually creating a network of paths, is by no means chaotic. Indeed, over time, we may expect the pattern of tracks to become increasingly ordered: kinks will be become straightened, paths between ecologically salient locations (e.g., sources of food, shelter, or water) will become more strongly established, and so on. We might similarly suspect that language will become increasingly ordered over long periods of cultural evolution. This appears to be the case, for example, in the transition between pidgins (languages with very limited grammatical and morphological structure that spontaneously emerge when two groups with no common language need to communicate extensively, such as during European colonialism) and the creoles created by subsequent generations from those pidgins—typically with richer and more rigid grammatical structure (Arends, Muysken, & Smith, 1994; see also our discussion of emerging sign languages in chapter 2).

We should anticipate that such order should emerge because the cognitive system does not merely learn lists of words and constructions by rote; it generalizes from past experience to new cases. To the extent that the language is a disordered morass of competing and inconsistent regularities, the language will be difficult to process, and difficult to learn. Thus, the cultural evolution of language, both within individuals and across generations of learners, will impose a strong selection pressure on individual words and constructions to align with each other, forming a system of linguistic regularities (as discussed in chapter 2). Just as stable and orderly forest tracks emerge from the initially arbitrary wanderings of the forest fauna, so an orderly language may emerge from what may, perhaps, have been the rather limited, arbitrary and inconsistent communicative system of early protolanguage. In particular, for example, the need to convey an unlimited number of messages will lead to a drive to recombine linguistic elements in systematic ways, yielding increasingly "compositional" semantics, in which the meaning of a message to a reasonable degree is associated with the meaning of its parts, and the way in which they are composed together (e.g., Kirby, 1999, 2000; Kirby, Cornish, & Smith, 2008).

Yet the tendency toward order is no more than a tendency. Commonly used parts of the language are often wildly irregular (for example, verbs such as *to be*, and *to go* are highly irregular in English, French, and many other languages); frequently used, and formulaic, phrases can have fossilized structures at variance with the rest of the language (e.g., *unaccustomed as I am …*; *hell hath no fury …*; *there be dragons*). And individual words exhibit many semi-regular patterns that seem puzzling, at least if we compare natural language to the iron consistency of artificial languages developed in logic and computer science. To pick a well-known example, for most native English speakers, we have the pattern:

(1) Mary gave a book to the library.
(2) Mary gave the library a book.
(3) Mary donated a book to the library.
(4) *Mary donated the library a book.

The verb *give* permits so-called *dative alternation*, in which (1) can be reformulated as (2). Even though the meaning of *donate* is very close to that of *give,* (3) cannot be rephrased as (4). This type of pattern, and countless other semi-regularities throughout natural languages across all levels of linguistic structure, are puzzling if we assume that language is fundamentally ordered (e.g., by some set of innate universal grammatical principles). Such irregularity is much less puzzling if we view the structure of language as emerging from countless individual processing episodes. Working against the tendency toward order, individual processing episodes are subject to other pressures, such as minimizing the complexity of articulation and making communication as rapid as possible. Moreover, in frequently used parts of the language, where such factors will be most potent, grammaticalization processes of erosion, morphological contraction, shortening of frequent words, and fusing of multiword phrases will introduce further irregularity into the language.

8.3 Return of the Lumpers: Putting the Language Sciences Back Together

One temptation, when struggling with a difficult crossword puzzle, is to consider each clue in isolation. It would seem so much easier to make progress if we didn't have to make answers interlock, and there is nothing more frustrating than discovering that what seemed a perfectly plausible answer to "one down" doesn't fit with an equally plausible proposal for "three across." But while we may *feel* we are making progress by considering each clue in isolation, this progress is, of course, illusory: the proposed solutions won't fit together merely by happy accident. And, of course, if we *do* make the clues fit together successfully, then the last few clues will fall into place straightforwardly; so while progress initially seems harder, it ultimately turns out to be easier, because we can exploit the mutual constraints between the different clues to our advantage.

It seems that many mainstream approaches to language have been lured into an illusory sense of progress, analogous to solving crossword clues in isolation. Of course, sometimes it may be useful to consider clues in isolation just to get some options on the table. And some genuine insights have been reached in this way, even though they do not always fit together into a bigger picture. However, going down this "isolationist route" has its dangers. Once started, such a strategy is difficult to stop because it requires unpicking proposals that may seem defensible when considered in isolation, but turn out to be incompatible when considered together. It is far easier to continue to insist that the clues should be considered in isolation, and not as part of an interlocking puzzle at all.

On a theoretical level, the "splitter" approach, typical of much of the language sciences, is also surprising given the history of scholars drawing parallels between language and biological systems (e.g., Berwick et al., 2013; Boeckx & Piattelli-Palmarini, 2005; Chomsky, 1980b; Lenneberg, 1967; Piattelli-Palmarini, 1989; Pinker, 1999). In fact, in the biological sciences, rich interconnections between explanations at multiple timescales, including evolutionary history, ontogeny, physiology, anatomy, and behavior have been central to creating the spectacular intellectual progress of the last century or more. We advocate a similarly integrated approach, drawing together research on language structure, processing, acquisition, and evolution.

Fortunately, alongside the mainstream generative grammar tradition, a wide range of linguists, computational linguists, psychologists, and other cognitive scientists have, over many decades, been working on the challenge of building an integrated science of language, exploring a variety of relations between language structure, processing, acquisition, and evolution. This book aims to contribute to this goal: to help provide a framework for putting the language sciences back together. As in a crossword puzzle, mutual constraints between different aspects of language can initially appear daunting. But there is no alternative: ignoring the mutual constraints between different aspects of language can only lead to the illusion of progress. Thus, recalling the quotation with which we began this chapter, we must collectively aim to combat "the fragmentation of scientific and humanistic disciplines," in order to create a genuine science of language.

References

Abe, K., & Watanabe, D. (2011). Songbirds possess the spontaneous ability to discriminate syntactic rules. *Nature Neuroscience, 14*, 1067–1074.

Abu-Mostafa, Y. S. (1993). Hints and the VC Dimension. *Neural Computation, 5*, 278–288.

Abzhanov, A., Kuo, W. P., Hartmann, C., Grant, B. R., Grant, P. R., & Tabin, C. J. (2006). The calmodulin pathway and evolution of elongated beak morphology in Darwin's finches. *Nature, 442*, 563–567.

Abzhanov, A., Protas, M., Grant, B. R., Grant, P. R., & Tabin, C. J. (2004). *Bmp4* and morphological variation of beaks in Darwin's finches. *Science, 305*, 1462–1465.

Ackley, D. H., Hinton, G. E., & Sejnowski, T. J. (1985). A learning algorithm for Boltzmann machines. *Cognitive Science*, 9, 147–169.

Adger, D. (2003). *Core syntax*. Oxford: Oxford University Press.

Aho, A. V., Sethi, R., & Ullman, J. D. (1986). *Compilers. Principles, techniques, and tools*. Reading, MA: Addison-Wesley Publishing Company.

Aitchison, J. (2000). *The seeds of speech: Language origin and evolution*. Oxford: Oxford University Press.

Alexander, R. M. (2003). *Principles of animal locomotion*. Princeton, NJ: Princeton University Press.

Allen, J., & Christiansen, M. H. (1996). Integrating multiple cues in word segmentation: A connectionist model using hints. In *Proceedings of the 18th Annual Cognitive Science Society Conference* (pp. 370–375). Mahwah, NJ: Lawrence Erlbaum.

Allopenna, P. D., Magnuson, J. S., & Tanenhaus, M. K. (1998). Tracking the time course of spoken word recognition using eye movements: Evidence for continuous mapping models. *Journal of Memory and Language, 38*, 419–439.

Altmann, G. T. M. (2002). Learning and development in neural networks: the importance of prior experience. *Cognition, 85*, 43–50.

Altmann, G. T. M. (2004). Language-mediated eye movements in the absence of a visual world: The 'blank screen paradigm.' *Cognition, 93*, 79–87.

Altmann, G. T. M., & Kamide, Y. (1999). Incremental interpretation at verbs: Restricting the domain of subsequent reference. *Cognition, 73*, 247–264.

Altmann, G. T. M., & Kamide, Y. (2009). Discourse-mediation of the mapping between language and the visual world: Eye movements and mental representation. *Cognition, 111*, 55–71.

Altmann, G. T. M., & Mirkovic, J. (2009). Incrementality and prediction in human sentence processing. *Cognitive Science, 33*, 583–609.

Altmann, G. T. M., & Steedman, M. J. (1988). Interaction with context during human sentence processing. *Cognition, 30*, 191–238.

Ancel, L. (1999). A quantitative model of the Simpson-Baldwin effect. *Journal of Theoretical Biology, 196*, 197–209.

Andersen, H. (1973). Abductive and deductive change. *Language, 40*, 765–793.

Anderson, J. R. (1993). *Rules of the mind.* Hillsdale, NJ: Lawrence Erlbaum Associates.

Anderson, M. L. (2008). Circuit sharing and the implementation of intelligent systems. *Connection Science, 20*, 239–251.

Anderson, M. L. (2010). Neural reuse: A fundamental organizational principle of the brain. *Behavioral and Brain Sciences, 33*, 245–313.

Andersson, M. B. (1994). *Sexual selection.* Princeton, NJ: Princeton University Press.

Arbib, M. A. (2005). From monkey-like action recognition to human language: An evolutionary framework for neurolinguistics. *Behavioral and Brain Sciences, 28*, 105–124.

Arends, J., Muysken, P., & Smith, N. (Eds.). (1994). *Pidgins and creoles: An introduction* (Vol. 15). Amsterdam: John Benjamins.

Arnon, I. (2010). Re-thinking child difficulty: The effect of NP type on children's processing of relative clauses in Hebrew. *Journal of Child Language, 37*, 27–57.

Arnon, I., & Christiansen, M. H. (submitted). *Multiword units as building blocks for language.*

Arnon, I., & Clark, E. V. (2011). When '*on your feet*' is better than '*feet*': Children's word production is facilitated in familiar sentence-frames. *Language Learning and Development, 7*, 107–129.

Arnon, I., & Cohen Priva, U. (2013). More than words: the effect of multi-word frequency and constituency on phonetic duration. *Language and Speech, 56*, 349–373.

Arnon, I., & Snider, N. (2010). More than words: frequency effects for multiword phrases. *Journal of Memory and Language, 62*, 67–82.

Austin, J. L. (1962). *How to do things with words.* Cambridge, MA: Harvard University Press.

Baayen, R. H., Pipenbrock, R., & Gulikers, L. (1995). *The CELEX lexical database (CD-ROM).* Philadelphia, PA: Linguistic Data Consortium, University of Pennsylvania.

Bach, E., Brown, C., & Marslen-Wilson, W. (1986). Crossed and nested dependencies in German and Dutch: A psycholinguistic study. *Language and Cognitive Processes, 1*, 249–262.

Bahrick, L. E., Lickliter, R., & Flom, R. (2004). Intersensory redundancy guides infants' selective attention, perceptual and cognitive development. *Current Directions in Psychological Science*, *13*, 99–102.

Baker, C. L. (1979). Syntactic theory and the projection problem. *Linguistic Inquiry*, *10*, 533–582.

Baker, C. L., & McCarthy, J. J. (Eds.). (1981). *The logical problem of language acquisition*. Cambridge, MA: MIT Press.

Baker, M. C. (2001). *The atoms of language: The mind's hidden rules of grammar*. New York: Basic Books.

Baker, M. C. (2003). Language differences and language design. *Trends in Cognitive Sciences*, *7*, 349–353.

Baldwin, J. M. (1896). A new factor in evolution. *American Naturalist*, *30*, 441–451.

Balota, D. A., Cortese, M. J., Sergent-Marshall, S. D., Spieler, D. H., & Yapp, M. J. (2004). Visual word recognition for single syllable words. *Journal of Experimental Psychology. General*, *133*, 283–316.

Bannard, C., & Matthews, D. (2008). Stored word sequences in language learning. *Psychological Science*, *19*, 241–248.

Bar, M. (2004). Visual objects in context. *Nature Reviews: Neuroscience*, *5*, 617–629.

Bar, M. (2007). The proactive brain: Using analogies and associations to generate predictions. *Trends in Cognitive Sciences*, *11*, 280–289.

Barkow, J., Cosmides, L., & Tooby, J. (Eds.). (1992). *The adapted mind: Evolutionary psychology and the generation of culture*. New York: Oxford University Press.

Barlow, H. B. (1983). Intelligence, guesswork, language. *Nature*, *304*, 207–209.

Baronchelli, A., Chater, N., Christiansen, M. H., & Pastor-Satorras, R. (2013). Evolution in a changing environment. *PLoS One*, *8*(1), e52742.

Baronchelli, A., Chater, N., Pastor-Satorras, R., & Christiansen, M. H. (2012). The biological origin of linguistic diversity. *PLoS One*, *7*(10), e48029.

Baronchelli, A., Ferrer-i-Cancho, R., Pastor-Satorras, R., Chater, N., & Christiansen, M. H. (2013). Networks in cognitive science. *Trends in Cognitive Sciences*, *17*, 348–360.

Baronchelli, A., Gong, T., Puglisi, A., & Loreto, V. (2010). Modeling the emergence of universality in color naming patterns. *Proceedings of the National Academy of Sciences*, *107*, 2403–2407.

Batali, J. (1994). Innate biases and critical periods: Combining evolution and learning in the acquisition of syntax. In R. Brooks & P. Maes (Eds.), *Artificial Life IV* (pp. 160–171). Cambridge, MA: MIT Press.

Batali, J. (1998). Computational simulations of the emergence of grammar. In J. R. Hurford, M. Studdert Kennedy, & C. Knight (Eds.), *Approaches to the evolution of language: Social and cognitive bases* (pp. 405–426). Cambridge: Cambridge University Press.

Bates, E. (1999). On the nature and nurture of language. In E. Bizzi, P. Calissano, & V. Volterra (Eds.), *Frontiere della biologia: Il cervello di Homo sapiens*. [Frontiers of biology: The brain of homo sapiens] (pp. 241–265). Rome: Giovanni Trecanni.

Bates, E., & MacWhinney, B. (1979). A functionalist approach to the acquisition of grammar. In E. Ochs & B. Schieffelin (Eds.), *Developmental pragmatics* (pp. 167–209). New York: Academic Press.

Bates, E., & MacWhinney, B. (1987). Competition, variation, and language learning. In B. MacWhinney (Ed.), *Mechanisms of language acquisition* (pp. 157–193). Hillsdale, NJ: Erlbaum.

Baumann, T., & Schlangen, D. (2012). INPRO_iSS: A component for just-in-time incremental speech synthesis. In *Proceedings of the ACL 2012 System Demonstrations* (pp. 103–108). Stroudsburg, PA: Association for Computational Linguistics.

Beckers, G. J., Bolhuis, J. J., Okanoya, K., & Berwick, R. C. (2012). Birdsong neurolinguistics: Songbird context-free grammar claim is premature. *Neuroreport*, *23*, 139–145.

Beckner, C., & Bybee, J. (2009). A usage-based account of constituency and reanalysis. *Language Learning*, *59* (Suppl. 1), 27–46.

Beckner, C., Blythe, R., Bybee, J., Christiansen, M. H., Croft, W., Ellis, N., Holland, J., Ke, J., Larsen- Freeman, D., & Schoenemann, T. (2009). Language is a complex adaptive system: Position paper. *Language Learning*, *59* (Suppl. 1), 1–27.

Bedore, L. M., & Leonard, L. B. (1998). Specific language impairment and grammatical morphology: A discriminate function analysis. *Journal of Speech and Hearing Research*, *41*, 1185–1192.

Behrens, H. (2008). Corpora in language acquisition research: History, methods, perspectives. In H. Behrens (Ed.), *Trends in corpus research: Finding structure in data* (pp. xi–xxx) (TILAR Series). Amsterdam: John Benjamins.

Beja-Pereira, A., Luikart, G., England, P. R., Bradley, D. G., Jann, O. C., Bertorelle, G., et al. (2003). Gene-culture coevolution between cattle milk protein genes and human lactase genes. *Nature Genetics*, *35*, 311–313.

Bekoff, M., & Byers, J. A. (Eds.). (1998). *Animal play: Evolutionary, comparative, and ecological perspectives*. Cambridge: Cambridge University Press.

Bellugi, U., & Fischer, S. (1972). A comparison of sign language and spoken language. *Cognition*, *1*, 173–200.

Bergen, B. K. (2004). The psychological reality of phonaesthemes. *Language*, *80*, 290–311.

Bernstein-Ratner, N. (1984). Patterns of vowel modification in motherese. *Journal of Child Language*, *11*, 557–578.

Berwick, R. C. (1985). *The acquisition of syntactic knowledge*. Cambridge, MA: MIT Press.

Berwick, R. C. (2009). What genes cannot learn about language. *Proceedings of the National Academy of Sciences*, *106*, 1685–1686.

Berwick, R. C., Friederici, A. D., Chomsky, N., & Bolhuis, J. J. (2013). Evolution, brain, and the nature of language. *Trends in Cognitive Sciences*, *17*, 89–98.

Berwick, R. C., & Weinberg, A. S. (1984). *The grammatical basis of linguistic performance*. Cambridge, MA: MIT Press.

Bever, T. G. (1970). The cognitive basis for linguistic structures. In J. R. Hayes (Ed.), *Cognition and the development of language* (pp. 279–362). New York: Wiley & Sons.

Bever, T. G., & Langendoen, D. T. (1971). A dynamic model of the evolution of language. *Linguistic Inquiry*, *2*, 433–463.

Bickerton, D. (1984). The language bio-program hypothesis. *Behavioral and Brain Sciences*, *7*, 173–212.

Bickerton, D. (1995). *Language and human behavior*. Seattle, WA: University of Washington Press.

Bickerton, D. (2003). Symbol and structure: a comprehensive framework for language evolution. In M. H. Christiansen & S. Kirby (Eds.), *Language evolution* (pp. 77–93). Oxford: Oxford University Press.

Bijeljac, R., Bertoncini, J., & Mehler, J. (1993). How do 4-day-old infants categorize multisyllabic utterances? *Developmental Psychology*, *29*, 711–721.

Birdsong, D., & Molis, M. (2001). On the evidence for maturational constraints in second-language acquisition. *Journal of Memory and Language*, *44*, 235–249.

Blackburn, S. (1984). *Spreading the word*. Oxford: Oxford University Press.

Blackmore, S. J. (1999). *The meme machine*. Oxford: Oxford University Press.

Blasi, D., Wichmann, S., Stadler, P. F., Hammarström, H. & Christiansen, M. H. (submitted). *Universal sound-meaning associations permeate the world's languages*.

Blaubergs, M. S., & Braine, M. D. S. (1974). Short-term memory limitations on decoding self-embedded sentences. *Journal of Experimental Psychology*, *102*, 745–748.

Bloom, L. (1970). *Language development: Form and function in emerging grammars*. Cambridge, MA: MIT Press.

Bloom, L. (1991). *Language development from two to three*. Cambridge: Cambridge University Press.

Bloom, P. (2001). Précis of how children learn the meanings of words. *Behavioral and Brain Sciences*, *24*, 1095–1103.

Bloomfield, L. (1933). *Language*. New York: Holt, Rinehart and Winston.

Boas, F. (1940). *Race, language, and culture*. Chicago: University of Chicago Press.

Bock, J. K. (1982). Toward a cognitive psychology of syntax: information processing contributions to sentence formulation. *Psychological Review*, *89*, 1–47.

Bock, J. K. (1986). Meaning, sound, and syntax: Lexical priming in sentence production. *Journal of Experimental Psychology: Learning, Memory, and Cognition*, *12*, 575–586.

Bock, J. K., & Loebell, H. (1990). Framing sentences. *Cognition*, *35*, 1–39.

Bock, J. K., & Miller, C. A. (1991). Broken agreement. *Cognitive Psychology*, *23*, 45–93.

Bod, R. (2009). From exemplar to grammar: A probabilistic analogy-based model of language learning. *Cognitive Science*, *33*, 752–793.

Boeckx, C. (2006). *Linguistic minimalism: Origins, concepts, methods, and aims*. New York: Oxford University Press.

Boeckx, C., & Leivada, E. (2013). Entangled parametric hierarchies: Problems for an overspecified universal grammar. *PLoS One*, *8*, e72357.

Boeckx, C., & Piattelli-Palmarini, M. (2005). Language as a natural object, linguistics as a natural science. *Linguistic Review*, *22*, 447–466.

Boland, J. E. (1997). The relationship between syntactic and semantic processes in sentence comprehension. *Language and Cognitive Processes*, *12*, 423–484.

Boland, J. E., Tanenhaus, M. K., & Garnsey, S. M. (1990). Evidence for the immediate use of verb control information in sentence processing. *Journal of Memory and Language*, *29*, 413–432.

Boland, J. E., Tanenhaus, M. K., Garnsey, S. M., & Carlson, G. N. (1995). Verb argument structure in parsing and interpretation: Evidence from *wh*-questions. *Journal of Memory and Language*, *34*, 774–806.

Bolger, D. J., Perfetti, C. A., & Schneider, W. (2005). Cross-cultural effect on the brain revisited: Universal structures plus writing system variation. *Human Brain Mapping*, *25*, 92–104.

Borovsky, A., Elman, J. L., & Fernald, A. (2012). Knowing a lot for one's age: vocabulary skill and not age is associated with anticipatory incremental sentence interpretation in children and adults. *Journal of Experimental Child Psychology*, *112*, 417–436.

Botha, R. P. (2003). *Unravelling the evolution of language*. Amsterdam: Elsevier.

Botvinick, M., & Plaut, D. C. (2004). Doing without schema hierarchies: A recurrent connectionist approach to normal and impaired routine sequential action. *Psychological Review*, *111*, 395–429.

Bowerman, M. (1973). Structural relationships in children's utterances: Semantic or syntactic? In T. Moore (Ed.), *Cognitive development and the acquisition of language* (pp. 197–213). New York: Academic Press.

Bowerman, M. (1974). Learning the structure of causative verbs: A study in the relationship of cognitive, semantic, and syntactic development. *Papers and Reports on Child Language Development*, *8*, 142–178.

Boyd, R., & Richerson, P. J. (2005). *The origin and evolution of cultures*. Oxford: Oxford University Press.

Boyd, R., & Richerson, P. J. (1987). The evolution of ethnic markers. *Cultural Anthropology*, *2*, 65–79.

Boyer, P. (1994). *The naturalness of religious belief: A cognitive theory of religion*. London, CA: University of California Press.

Braine, M. D. S. (1963). The ontogeny of English phrase structure: The first phase. *Language*, *39*, 1–13.

Braine, M. D. S., & Brooks, P. J. (1995). Verb argument structure and the problem of avoiding an overgeneral grammar. In M. Tomasello & W. Merriman (Eds.), *Beyond names for things: Young children's acquisition of verbs* (pp. 353–376). Hillsdale, NJ: Erlbaum.

Brandt, S., Kidd, E., Lieven, E., & Tomasello, M. (2009). The discourse bases of relativization: An investigation of young German and English-speaking children's comprehension of relative clauses. *Cognitive Linguistics*, *20*, 539–570.

Branigan, H., Pickering, M., & Cleland, A. (2000). Syntactic co-ordination in dialogue. *Cognition*, *75*, 13–25.

Brants, T., & Franz, A. (2006). *Web 1T 5-gram Version 1*. Philadelphia, PA: Linguistic Data Consortium.

Bregman, A. S. (1990). *Auditory scene analysis*. Cambridge, MA: MIT Press.

Bremner, A. J., Caparos, S., Davidoff, J., de Fockert, J., Linnell, K. J., & Spence, C. (2013). "Bouba" and "Kiki" in Namibia? A remote culture make similar shape-sound matches, but different shape-taste matches to Westerners. *Cognition*, *126*, 165–172.

Bresnan, J. (1982). *The mental representation of grammatical relations*. Cambridge, MA: MIT Press.

Briscoe, E. J. (2003). Grammatical assimilation. In M. H. Christiansen & S. Kirby (Eds.), *Language evolution* (pp. 295–316). Oxford: Oxford University Press.

Broadbent, D. (1958). *Perception and communication*. London: Pergamon Press.

Brooks, P. J., Braine, M. D. S., Catalano, L., Brody, R. E., & Sudhalter, V. (1993). Acquisition of gender-like noun subclasses in an artificial language: The contribution of phonological markers to learning. *Journal of Memory and Language*, *32*, 79–95.

Brown, G. D. A., Neath, I., & Chater, N. (2007). A temporal ratio model of memory. *Psychological Review*, *114*, 539–576.

Brown, R. (1973). *A first language: The early stages*. Cambridge, MA: Harvard University Press.

Brown, R., & Hanlon, C. (1970). Derivational complexity and order of acquisition in child speech. In J. Hayes (Ed.), *Cognition and the development of language* (pp. 11–53). New York: Wiley.

Brown-Schmidt, S., & Konopka, A. E. (2011). Experimental approaches to referential domains and the on-line processing of referring expressions in unscripted conversation. *Information*, *2*, 302–326.

Brown-Schmidt, S., & Konopka, A. E. (2015). Processes of incremental message planning during conversation. *Psychonomic Bulletin & Review*, *22*, 833–843.

Brown-Schmidt, S., & Tanenhaus, M. K. (2008). Real-time investigation of referential domains in unscripted conversation: A targeted language game approach. *Cognitive Science*, *32*, 643–684.

Burchinal, M., McCartney, K., Steinberg, L., Crosnoe, R., Friedman, S.L., McLoyd, V., Pianta, R., & NICHD Early Child Care Research Network. (2011). Examining the black—white achievement gap among low-income children using the NICHD Study of Early Child Care and Youth Development. *Child Development*, *82*, 1404–1420.

Burling, R. (2005). *The talking ape: How language evolved*. Oxford: Oxford University Press.

Burnard, L., & Aston, G. (1998). *The BNC handbook: Exploring the British National Corpus*. Edinburgh: Edinburgh University Press.

Bybee, J. L. (2002). Sequentiality as the basis of constituent structure. In T. Givón & B. Malle (Eds.), *The evolution of language out of pre-language* (pp. 107–132). Philadelphia, PA: John Benjamins.

Bybee, J. L. (2002). Word frequency and context of use in the lexical diffusion of phonetically conditioned sound change. *Language Variation and Change*, *14*, 261–290.

Bybee, J. L. (2006). From usage to grammar: The mind's response to repetition. *Language*, *82*, 711–733.

Bybee, J. L. (2007). *Frequency of use and the organization of language*. New York: Oxford University Press.

Bybee, J. L. (2009). Language universals and usage-based theory. In M. H. Christiansen, C. Collins, & S. Edelman (Eds.), *Language universals* (pp. 17–39). New York: Oxford University Press.

Bybee, J. L., & Hopper, P. (Eds.). (2001). *Frequency and the emergence of linguistic structure*. Amsterdam: John Benjamins.

Bybee, J. L., & McClelland, J. L. (2005). Alternatives to the combinatorial paradigm of linguistic theory based on general principles of human cognition. *Linguistic Review*, *22*, 381–410.

Bybee, J. L., Perkins, R. D., & Pagliuca, W. (1994). *The evolution of grammar: Tense, aspect and modality in the languages of the world*. Chicago: University of Chicago Press.

Bybee, J. L., & Scheibman, J. (1999). The effect of usage on degrees of constituency: The reduction of *don't* in English. *Linguistics*, *37*, 575–596.

Bybee, J. L., & Slobin, D. I. (1982). Rules and schemas in the development and use of the English past tense. *Language*, *58*, 265–289.

Byrne, R. W., & Byrne, J. M. E. (1993). Complex leaf-gathering skills of mountain gorillas (*Gorilla g. berengei*): Variability and standardization. *American Journal of Primatology*, *31*, 241–261.

Byrne, R. W., & Russon, A. E. (1998). Learning by imitation: A hierarchical approach. *Behavioral and Brain Sciences*, *21*, 667–721.

Calvin, W. H. (1994). The emergence of intelligence. *Scientific American*, *271*, 100–107.

Campbell, D. T. (1965). Variation and selective retention in socio-cultural evolution. In H. R. Barringer, G. I. Blanksten, & R. W. Mack (Eds.), *Social change in developing areas: A reinterpretation of evolutionary theory* (pp. 19–49). Cambridge, MA: Schenkman.

Campbell, L. (2000). What's wrong with grammaticalization? *Language Sciences*, *23*, 113–161.

Campbell, R., & Besner, D. (1981). This and that—Constraints on the pronunciation of new written words. *Quarterly Journal of Experimental Psychology*, *33*, 375–396.

Cann, R., Kempson, R., & Wedgwood, D. (2012). Representationalism and linguistic knowledge. In R. Kempson, T. Fernando, & N. Asher (Eds.), *Philosophy of linguistics* (pp. 357–402). Amsterdam: Elsevier.

Cannon, G. (1991). Jones's "Spring from some common source": 1786–1986. In S. M. Lamb & E. D. Mitchell (Eds.), *Sprung from some common source: Investigations into the pre-history of languages*. Stanford, CA: Stanford University Press.

Carey, S., & Spelke, E. S. (1996). Science and core knowledge. *Philosophy of Science, 63*, 515–533.

Carr, M. F., Jadhav, S. P., & Frank, L. M. (2011). Hippocampal replay in the awake state: A potential substrate for memory consolidation and retrieval. *Nature Neuroscience, 14*, 147–153.

Carroll, S. B. (2001). Chance and necessity: The evolution of morphological complexity and diversity. *Nature, 409*, 1102–1109.

Carroll, S. B. (2005). *Endless forms most beautiful: The new science of evo devo*. New York: W. W. Norton and Co.

Carstairs-McCarthy, A. (1992). *Current morphology*. London: Routledge.

Cartwright, N. (1999). *The dappled world: A study of the boundaries of science*. Cambridge: Cambridge University Press.

Cassidy, K. W., & Kelly, M. H. (1991). Phonological information for grammatical category assignments. *Journal of Memory and Language, 30*, 348–369.

Cassidy, K. W., & Kelly, M. H. (2001). Children's use of phonology to infer grammatical class in vocabulary learning. *Psychonomic Bulletin & Review, 8*, 519–523.

Cavalli-Sforza, L. L., & Feldman, M. W. (2003). The application of molecular genetic approaches to the study of human evolution. *Nature Genetics, 33*, 266–275.

Chang, F., Dell, G. S., & Bock, K. (2006). Becoming syntactic. *Psychological Review, 113*, 234–272.

Chater, N. (2005). Mendelian and Darwinian views of memes and cultural change. In S. Hurley & N. Chater (Eds.), *Perspectives on imitation: From neuroscience to social science* (Vol. 2, pp. 355–362). Cambridge, MA: MIT Press.

Chater, N., & Christiansen, M. H. (2010). Language acquisition meets language evolution. *Cognitive Science, 34*, 1131–1157.

Chater, N., & Christiansen, M. H. (in press). Squeezing through the Now-or-Never bottleneck: Reconnecting language processing, acquisition, change and structure. *Behavioral and Brain Sciences*.

Chater, N., Clark, A., Goldsmith, J. A., & Perfors, A. (2015). *Empiricism and language learnability*. Oxford: Oxford University Press.

Chater, N., McCauley, S.M. & Christiansen, M.H. (in press). Language as skill: Intertwining comprehension and production. *Journal of Memory and Language*.

Chater, N., Reali, F., & Christiansen, M. H. (2009). Restrictions on biological adaptation in language evolution. *Proceedings of the National Academy of Sciences, 106*, 1015–1020.

Chater, N., Tenenbaum, J. B., & Yuille, A. (2006). Probabilistic models of cognition: Conceptual foundations. *Trends in Cognitive Sciences, 10*, 287–291.

Chater, N., & Vitányi, P. (2007). 'Ideal learning' of natural language: Positive results about learning from positive evidence. *Journal of Mathematical Psychology, 51*, 135–163.

Cheng, P. W. (1997). From covariation to causation: A causal power theory. *Psychological Review, 104*, 367.

Cherry, E. C. (1953). Some experiments on the recognition of speech with one and with two ears. *Journal of the Acoustical Society of America, 25*, 975–979.

Childers, J. B., & Tomasello, M. (2006). Are nouns easier to learn than verbs? Three experimental studies. In K. Hirsh-Pasek & R. Golinkoff (Eds.), *Action meets word: How children learn verbs* (pp. 311–335). New York: Oxford University Press.

Chomsky, N. (1956). Three models for the description of language. *I.R.E. Transactions on Information Theory, 2*, 113–124.

Chomsky, N. (1957). *Syntactic structures*. The Hague, Paris: Mouton.

Chomsky, N. (1965). *Aspects of the theory of syntax*. Cambridge, MA: MIT Press.

Chomsky, N. (1970). *Language and Freedom.* Lecture delivered at the University freedom and the Human Sciences Symposium, Loyola University, Chicago, IL.

Chomsky, N. (1972). *Language and mind.* New York: Harcourt, Brace and World (extended edition).

Chomsky, N. (1975). *Reflections on language*. New York: Pantheon Books.

Chomsky, N. (1980 a). Rules and representations. *Behavioral and Brain Sciences, 3*, 1–15.

Chomsky, N. (1980 b). *Rules and representations*. New York: Columbia University Press.

Chomsky, N. (1981). *Lectures on government and binding*. New York: Foris.

Chomsky, N. (1986). *Knowledge of language*. New York: Praeger.

Chomsky, N. (1988). *Language and the problems of knowledge. The Managua Lectures*. Cambridge, MA: MIT Press.

Chomsky, N. (1993). *Language and thought.* Wakefield, RI: Moyer Bell.

Chomsky, N. (1995). *The minimalist program.* Cambridge, MA: MIT Press.

Chomsky, N. (2000). *New horizons in the study of language and mind.* Cambridge: Cambridge University Press.

Chomsky, N. (2005). Three factors in language design. *Linguistic Inquiry, 36*, 1–22.

Chomsky, N. (2010). Some simple evo devo theses: How true might they be for language. In R. K. Larson, V. Déprez, & H. Yamakido (Eds.), *The evolution of language: Biolinguistic perspectives* (pp. 45–62). Cambridge: Cambridge University Press.

Chomsky, N. (2011). Language and other cognitive systems. What is special about language? *Language Learning and Development, 7*, 263–278.

Christiansen, M. H. (1992). The (non) necessity of recursion in natural language processing. In *Proceedings of the 14th Annual Cognitive Science Society Conference* (pp. 665–670). Hillsdale, NJ: Lawrence Erlbaum.

Christiansen, M. H. (1994). *Infinite languages, finite minds: Connectionism, learning and linguistic structure.* Unpublished doctoral dissertation, Centre for Cognitive Science, University of Edinburgh, UK.

Christiansen, M. H. (2000). Using artificial language learning to study language evolution: Exploring the emergence of word universals. J. L. Dessalles & L. Ghadakpour (Eds.), *The Evolution of Language: 3rd International Conference* (pp. 45–48). Paris, France: Ecole Nationale Supérieure des Télécommunications.

Christiansen, M. H., Allen, J., & Seidenberg, M. S. (1998). Learning to segment speech using multiple cues: A connectionist model. *Language and Cognitive Processes, 13*, 221–268.

Christiansen, M. H., & Chater, N. (1994). Generalization and connectionist language learning. *Mind & Language, 9*, 273–287.

Christiansen, M. H., & Chater, N. (1999). Toward a connectionist model of recursion in human linguistic performance. *Cognitive Science, 23*, 157–205.

Christiansen, M. H., & Chater, N. (2008). Language as shaped by the brain. *Behavioral and Brain Sciences, 31*, 489–558.

Christiansen, M. H., & Chater, N. (in press). The Now-or-Never bottleneck: A fundamental constraint on language. *Behavioral and Brain Sciences.*

Christiansen, M. H., Collins, C., & Edelman, S. (Eds.). (2009). *Language universals.* New York: Oxford University Press.

Christiansen, M. H., Conway, C. M., & Onnis, L. (2012). Similar neural correlates for language and sequential learning: Evidence from event-related brain potentials. *Language and Cognitive Processes, 27*, 231–256.

Christiansen, M. H., & Dale, R. (2001). Integrating distributional, prosodic and phonological information in a connectionist model of language acquisition. In *Proceedings of the 23rd Annual Conference of the Cognitive Science Society* (pp. 220–225). Mahwah, NJ: Lawrence Erlbaum.

Christiansen, M. H., & Dale, R. (2004). The role of learning and development in the evolution of language. A connectionist perspective. In D. Kimbrough Oller & U. Griebel (Eds.), *Evolution of communication systems: A comparative approach. The Vienna Series in Theoretical Biology* (pp. 90–109). Cambridge, MA: MIT Press.

Christiansen, M. H., Dale, R., Ellefson, M. R., & Conway, C. M. (2002). The role of sequential learning in language evolution: Computational and experimental studies. In A. Cangelosi & D. Parisi (Eds.), *Simulating the evolution of language* (pp. 165–187). London: Springer-Verlag.

Christiansen, M. H., Dale, R., & Reali, F. (2010). Connectionist explorations of multiple-cue integration in syntax acquisition. In S. P. Johnson (Ed.), *Neoconstructivism: The new science of cognitive development* (pp. 87–108). New York: Oxford University Press.

Christiansen, M. H., & Devlin, J. T. (1997). Recursive inconsistencies are hard to learn: A connectionist perspective on universal word order correlations. *In Proceedings of the 19th Annual Cognitive Science Society Conference* (pp. 113–118). Mahwah, NJ: Lawrence Erlbaum.

Christiansen, M. H., & Ellefson, M. R. (2002). Linguistic adaptation without linguistic constraints: The role of sequential learning in language evolution. In A. Wray (Ed.), *Transitions to language* (pp. 335–358). Oxford: Oxford University Press.

Christiansen, M. H., Kelly, M. L., Shillcock, R. C., & Greenfield, K. (2010). Impaired artificial grammar learning in agrammatism. *Cognition*, *116*, 382–393.

Christiansen, M. H., & Kirby, S. (2003). Language evolution: Consensus and controversies. *Trends in Cognitive Sciences*, *7*, 300–307.

Christiansen, M. H., & MacDonald, M. C. (2009). A usage-based approach to recursion in sentence processing. *Language Learning*, *59* (Suppl. 1), 126–161.

Christiansen, M. H., & Monaghan, P. (2006). Discovering verbs through multiple-cue integration. In K. Hirsh-Pasek & R. M. Golinkoff (Eds.), *Action meets words: How children learn verbs* (pp. 88–107). New York: Oxford University Press.

Christiansen, M. H., Reali, F., & Chater, N. (2011). Biological adaptations for functional features of language in the face of cultural evolution. *Human Biology*, *83*, 247–259.

Church, K. (1982). *On memory limitations in natural language processing*. Bloomington: Indiana University Linguistics Club.

Clark, Alexander. (2011). A learnable representation for syntax using residuated lattices. In P. Groote, M. Egg, & L. Kallmeyer (Eds.), *Formal grammar* (pp. 183–198). Berlin: Springer.

Clark, Alexander, Fox, C., & Lappin, S. (Eds.). (2010). *The handbook of computational linguistics and natural language processing*. Oxford: Blackwell-Wiley.

Clark, Alexander, & Lappin, S. (2010). *Linguistic nativism and the poverty of the stimulus*. Oxford: Blackwell-Wiley.

Clark, Andy. (2013). Whatever next? Predictive brains, situated agents, and the future of cognitive science. *Behavioral and Brain Sciences*, *36*, 181–253.

Clark, H. H. (1975). Bridging. In R. C. Schank & B. L. Nash-Webber (Eds.), *Theoretical issues in natural language processing* (pp. 169–174). New York: Association for Computing Machinery.

Clark, H. H. (1996). *Using language*. Cambridge: Cambridge University Press.

Clark, H. H., & Wilkes-Gibbs, D. (1986). Referring as a collaborative process. *Cognition*, *22*, 1–39.

Clark, J., Yallop, C., & Fletcher, J. (2007). *An introduction to phonetics and phonology* (3rd ed.). Malden, MA: Wiley-Blackwell.

Clément, S., Demany, L., & Semal, C. (1999). Memory for pitch versus memory for loudness. *Journal of the Acoustical Society of America*, *106*, 2805–2811.

Clifton, C., Frazier, L., & Connine, C. (1984). Lexical expectations in sentence comprehension. *Journal of Verbal Learning and Verbal Behavior*, *23*, 696–708.

Coltheart, M. (1980). Iconic memory and visible persistence. *Perception & Psychophysics*, *27*, 183–228.

Comrie, B. (1989). *Language typology and language change*. Oxford: Blackwell.

Conway, C. M., & Christiansen, M. H. (2001). Sequential learning in non-human primates. *Trends in Cognitive Sciences*, *5*, 539–546.

Conway, C. M., & Pisoni, D. B. (2008). Neurocognitive basis of implicit learning of sequential structure and its relation to language processing. *Annals of the New York Academy of Sciences, 1145*, 113–131.

Cook, R. S., Kay, P., & Regier, T. (2005). The world color survey database: History and use. In H. Cohen & C. Lefebvre (Eds.), *Handbook of categorization in cognitive science* (pp. 224–242). Amsterdam: Elsevier.

Cooper, R. P., & Shallice, T. (2006). Hierarchical schemas and goals in the control of sequential behavior. *Psychological Review, 113*, 887–916.

Corballis, M. C. (1992). On the evolution of language and generativity. *Cognition, 44*, 197–226.

Corballis, M. C. (2002). *From hand to mouth: The origins of language*. Princeton, NJ: Princeton University Press.

Corballis, M. C. (2003). From hand to mouth: The gestural origins of language. In M. H. Christiansen & S. Kirby (Eds.), *Language evolution* (pp. 201–218). Oxford: Oxford University Press.

Corballis, M. C. (2007). Recursion, language, and starlings. *Cognitive Science, 31*, 697–704.

Corballis, M. C. (2011). *The recursive mind*. Princeton, NJ: Princeton University Press.

Cornish, H., Dale, R., Kirby, S., & Christiansen, M. H. (submitted). *Sequence memory constraints give rise to linguistic structure through iterated learning*.

Cover, T. M., & Thomas, J. A. (2006). *Elements of information theory* (2nd ed.). Hoboken, NJ: Wiley.

Cowan, N. (2000). The magical number 4 in short-term memory: A reconsideration of mental storage capacity. *Behavioral and Brain Sciences, 24*, 87–114.

Cowie, F. (1999). *What's within? Nativism reconsidered*. New York: Oxford University Press.

Crain, S. (1991). Language acquisition in the absence of experience. *Behavioral and Brain Sciences, 14*, 597–650.

Crain, S., Goro, T., & Thornton, R. (2006). Language acquisition is language change. *Journal of Psycholinguistic Research, 35*, 31–49.

Crain, S., & Lillo-Martin, D. C. (1999). *An introduction to linguistic theory and language acquisition*. Oxford: Blackwell.

Crain, S., & Pietroski, P. (2006). Is Generative Grammar deceptively simple or simply deceptive? *Lingua, 116*, 64–68.

Crick, F., & Mitchison, G. (1983). The function of dream sleep. *Nature, 304*, 111–114.

Cristianini, N., & Shawe-Taylor, J. (2000). *An introduction to support vector machines and other kernel-based learning methods*. Cambridge: Cambridge University Press.

Crocker, M. W., & Corley, S. (2002). Modular architectures and statistical mechanisms. In P. Merlo & S. Stevenson (Eds.), *The lexical basis of sentence processing* (pp. 157–180). Amsterdam: John Benjamins.

Croft, W. (2000). *Explaining language change: An evolutionary approach*. Harlow, Essex: Longman.

Croft, W. (2001). *Radical construction grammar: Syntactic theory in typological perspective*. New York: Oxford University Press.

Croft, W., & Cruise, D. A. (2004). *Cognitive linguistics*. Cambridge: Cambridge University Press.

Crossman, E. R. F. W. (1959). A theory of the acquisition of speed-skill. *Ergonomics, 2*, 153–166.

Crowder, R. G., & Neath, I. (1991). The microscope metaphor in human memory. In W. E. Hockley & S. Lewandowsky (Eds.), *Relating theory and data: Essays on human memory in honour of Bennet B. Murdock. Jr* (pp. 111–125). Hillsdale, NJ: Erlbaum.

Crowley, J., & Katz, L. (1999). Development of ocular dominance columns in the absence of retinal input. *Nature Neuroscience, 2*, 1125–1130.

Culicover, P. W. (1999). *Syntactic nuts: hard cases, syntactic theory, and language acquisition*. Oxford: Oxford University Press.

Culicover, P. W. (2013 a). *Grammar and complexity: Language at the intersection of competence and performance*. Oxford: Oxford University Press.

Culicover, P. W. (2013 b). The role of linear order in the computation of referential dependencies. *Lingua, 136*, 125–144.

Culicover, P. W., & Jackendoff, R. (2005). *Simpler syntax*. Oxford: Oxford University Press.

Culicover, P. W., & Nowak, A. (2003). *Dynamical grammar*. Oxford: Oxford University Press.

Curtiss, S. (1977). *Genie: A psycholinguistic study of a modern-day "wild child"*. New York: Academic Press.

Cutler, A. (Ed.). (1982). *Slips of the tongue and language production*. Berlin: De Gruyter Mouton.

Cutler, A. (1993). Phonological cues to open- and closed-class words in the processing of spoken sentences. *Journal of Psycholinguistic Research, 22*, 109–131.

Cutler, A., & Carter, D. M. (1987). The predominance of strong initial syllables in the English vocabulary. *Computer Speech & Language, 2*, 133–142.

Cutler, A., Hawkins, J. A., & Gilligan, G. (1985). The suffixing preference: A processing explanation. *Linguistics, 23*, 723–758.

Cutler, A., Mehler, J., Norris, D., & Segui, J. (1986). The syllable's differing role in the segmentation of French and English. *Journal of Memory and Language, 25*, 385–400.

Dąbrowska, E. (1997). The LAD goes to school: A cautionary tale for nativists. *Linguistics, 35*, 735–766.

Dąbrowska, E. (2012). Different speakers, different grammars: Individual differences in native language attainment. *Linguistic Approaches to Bilingualism, 2*, 219–253.

Daelemans, W., & Van den Bosch, A. (2005). *Memory-based language processing.* Cambridge: Cambridge University Press.

Dahan, D. (2010). The time course of interpretation in speech comprehension. *Current Directions in Psychological Science, 19,* 121–126.

Dale, P. S., & Cole, K. N. (1991). What's normal? Specific language impairment in an individual differences perspective. *Language, Speech, and Hearing Services in Schools, 22,* 80–83.

Daneman, M., & Carpenter, P. A. (1980). Individual differences in working memory and reading. *Journal of Verbal Learning and Verbal Behavior, 19,* 450–466.

Darwin, C. (1871). *The descent of man, and selection in relation to sex* (Vol. 1). London: John Murray.

Darwin, C. (1877). A biographical sketch of an infant. *Mind, 2,* 285–294.

Davies, A. M. (1987). "Organic" and "Organism" in Franz Bopp. In H. M. Hoenigswald & L. F. Wiener (Eds.), *Biological metaphor and cladistic classification* (pp. 81–107). Philadelphia, PA: University of Pennsylvania Press.

Davies, M. (2008) *The Corpus of Contemporary American English: 450 million words, 1990-present.* Available online at http://corpus.byu.edu/coca/.

Davies, M. (2010). The Corpus of Contemporary American English as the first reliable monitor corpus of English. *Literary and Linguistic Computing, 25,* 447–464.

Dawkins, R. (1976). *The selfish gene.* New York: Oxford University Press.

Dawkins, R. (1986). *The blind watchmaker: Why the evidence of evolution reveals a universe without design.* Harmondsworth, UK: Penguin.

de Leeuw, E., Schmid, M. S., & Mennen, I. (2010). The effects of contact on native language pronunciation in an L2 migrant setting. *Bilingualism: Language and Cognition, 13,* 33–40.

de Ruiter, J. P., & Levinson, S. C. (2008). A biological infrastructure for communication underlies the cultural evolution of language. *Behavioral and Brain Sciences, 31,* 518.

de Saussure, F. (1916). *Course in general linguistics.* New York: McGraw-Hill.

de Vries, M. H., Barth, A. R. C., Maiworm, S., Knecht, S., Zwitserlood, P., & Flöel, A. (2010). Electrical stimulation of Broca's area enhances implicit learning of an artificial grammar. *Journal of Cognitive Neuroscience, 22,* 2427–2436.

de Vries, M. H., Christiansen, M. H., & Petersson, K. M. (2011). Learning recursion: Multiple nested and crossed dependencies. *Biolinguistics, 5,* 10–35.

de Vries, M. H., Geukes, S., Zwitserlood, P., Petersson, K. M., & Christiansen, M. H. (2012). Processing multiple non-adjacent dependencies: Evidence from sequence learning. *Philosophical Transactions of the Royal Society B: Biological Sciences, 367,* 2065–2076.

de Vries, M. H., Monaghan, P., Knecht, S., & Zwitserlood, P. (2008). Syntactic structure and artificial grammar learning: The learnability of embedded hierarchical structures. *Cognition, 107,* 763–774.

Deacon, T. W. (1997). *The symbolic species: The co-evolution of language and the brain.* New York: W.W. Norton.

Dediu, D. (2011). A Bayesian phylogenetic approach to estimating the stability of linguistic features and the genetic biasing of tone. *Proceedings of the Royal Society B: Biological Sciences, 278*, 474–479.

Dediu, D., & Christiansen, M. H. (in press). Language evolution: Constraints and opportunities from modern genetics. *Topics in Cognitive Science.*

Dediu, D., Cysouw, M., Levinson, S. C., Baronchelli, A., Christiansen, M. H., Croft, W., et al. (2013). Cultural evolution of language. In P. J. Richerson & M. H. Christiansen (Eds.), *Cultural evolution: Society, technology, language and religion* (pp. 303–332). Cambridge, MA: MIT Press.

Dediu, D., & Ladd, D. R. (2007). Linguistic tone is related to the population frequency of the adaptive haplogroups of two brain size genes, *ASPM* and *Microcephalin. Proceedings of the National Academy of Sciences, 104*, 10944–10949.

Dehaene, S., & Cohen, L. (2007). Cultural recycling of cortical maps. *Neuron, 56*, 384–398.

Dehaene-Lambertz, G., Dehaene, S., Anton, J. L., Campagne, A., Ciuciu, P., Dehaene, G. P., et al. (2006 a). Functional segregation of cortical language areas by sentence repetition. *Human Brain Mapping, 27*, 360–371.

Dehaene-Lambertz, G., Hertz-Pannier, L., Dubois, J., Meriaux, S., Roche, A., Sigman, M., et al. (2006 b). Functional organization of perisylvian activation during presentation of sentences in preverbal infants. *Proceedings of the National Academy of Sciences, 103*, 14240–14245.

Dell, G. S., Burger, L. K., & Svec, W. R. (1997). Language production and serial order: A functional analysis and a model. *Psychological Review, 104*, 123–147.

Dell, G. S., & Chang, F. (2014). The P-chain: relating sentence production and its disorders to comprehension and acquisition. *Philosophical Transactions of the Royal Society B: Biological Sciences, 369*(1634), 20120394.

DeLong, K. A., Urbach, T. P., & Kutas, M. (2005). Probabilistic word pre-activation during language comprehension inferred from electrical brain activity. *Nature Neuroscience, 8*, 1117–1121.

Dennett, D. C. (1995). *Darwin's dangerous idea: Evolution and the meanings of life.* New York: Simon & Schuster.

Desmet, T., De Baecke, C., Drieghe, D., Brysbaert, M., & Vonk, W. (2006). Relative clause attachment in Dutch: On-line comprehension corresponds to corpus frequencies when lexical variables are taken into account. *Language and Cognitive Processes, 21*, 453–485.

Diamond, J. (1992). *The third chimpanzee: The evolution and future of the human animal.* New York: Harper Collins.

Diamond, J. (1997). *Guns, germs, and steel: The fates of human societies.* New York: Harper Collins.

Dickey, M. W., & Vonk, W. (1997). *Center-embedded structures in Dutch: An on-line study.* Poster presented at the Tenth Annual CUNY Conference on Human Sentence Processing. Santa Monica, CA, March 20–22.

Dickinson, S. (1987). Recursion in development: Support for a biological model of language. *Language and Speech, 30*, 239–249.

Dienes, Z., & McLeod, P. (1993). How to catch a cricket ball. *Perception*, *22*, 1427–1439.

Diessel, H. (2004). *The acquisition of complex sentences*. Cambridge: Cambridge University Press.

Dikker, S., Rabagliati, H., Farmer, T. A., & Pylkkänen, L. (2010). Early occipital sensitivity to syntactic category is based on form typicality. *Psychological Science*, *21*, 629–634.

Dingemanse, M., Blasi, D., Lupyan, G., Christiansen, M.H. & Monaghan, P. (2015). Arbitrariness, iconicity, and systematicity in language. *Trends in Cognitive Sciences*, *19*, 603–615.

Dobzhansky, G. T. (1964). Biology, molecular and organismic. *American Zoologist*, *4*, 443–452.

Donald, M. (1998). Mimesis and the executive suite: Missing links in language evolution. In J. R. Hurford, M. Studdert-Kennedy, & C. Knight (Eds.), *Approaches to the evolution of language* (pp. 44–67). Cambridge: Cambridge University Press.

Dryer, M. S. (1992). The Greenbergian word order correlations. *Language*, *68*, 81–138.

Dunbar, R. I. M. (1998). *Grooming, gossip, and the evolution of language*. Cambridge, MA: Harvard University Press.

Dunbar, R. I. M. (2003). The origin and subsequent evolution of language. In M. H. Christiansen & S. Kirby (Eds.), *Language evolution* (pp. 219–234). New York: Oxford University Press.

Dunn, M., Greenhill, S. J., Levinson, S. C., & Gray, R. D. (2011). Evolved structure of language shows lineage-specific trends in word-order universals. *Nature*, *473*, 79–82.

Dunning, T. (1993). Accurate methods for the statistics of surprise and coincidence. *Computational Linguistics*, *19*, 61–74.

Durrant, P. (2013). Formulaicity in an agglutinating language: the case of Turkish. *Corpus Linguistics and Linguistic Theory*, *9*, 1–38.

Dyer, F. C. (2002). The biology of the dance language. *Annual Review of Entomology*, *47*, 917–949.

Eco, U. (1995). *The search for the perfect language*. London: Blackwell.

Ellefson, M. R., & Christiansen, M. H. (2000). Subjacency constraints without universal grammar: Evidence from artificial language learning and connectionist modeling. In *The Proceedings of the 22nd Annual Conference of the Cognitive Science Society* (pp. 645–650). Mahwah, NJ: Lawrence Erlbaum.

Ellis, A. W., & Young, A. W. (1988). *Human cognitive neuropsychology*. Hillsdale, NJ: Lawrence Erlbaum Associates.

Elliott, L. L. (1962). Backward and forward masking of probe tones of different frequencies. *Journal of the Acoustical Society of America*, *34*, 1116–1117.

Ellis, N. C. (2002). Frequency effects in language processing. *Studies in Second Language Acquisition*, *24*, 143–188.

Elman, J. L. (1990). Finding structure in time. *Cognitive Science*, *14*, 179–211.

Elman, J. L. (1991). Distributed representation, simple recurrent networks, and grammatical structure. *Machine Learning*, *7*, 195–225.

Elman, J. L. (1993). Learning and development in neural networks: The importance of starting small. *Cognition*, *48*, 71–99.

Elman, J. L. (1999). Origins of language: A conspiracy theory. In B. MacWhinney (Ed.), *The emergence of language* (pp. 1–27). Hillsdale, NJ: Lawrence Erlbaum.

Elman, J. L. (2005). Connectionist models of cognitive development: Where next? *Trends in Cognitive Sciences*, *9*, 111–117.

Elman, J. L., Bates, E. A., Johnson, M. H., Karmiloff-Smith, A., Parisi, D., & Plunkett, K. (1996). *Rethinking innateness: A connectionist perspective on development.* Cambridge, MA: MIT Press.

Enard, W., Przeworski, M., Fisher, S. E., Lai, C. S. L., Wiebe, V., Kitano, T., et al. (2002). Molecular evolution of *FOXP2*, a gene involved in speech and language. *Nature*, *418*, 869–872.

Enfield, N. J. (2013). *Relationship thinking: Enchrony, agency, and human sociality.* New York: Oxford University Press.

Engelmann, F., & Vasishth, S. (2009). Processing grammatical and ungrammatical center embeddings in English and German: A computational model. In A. Howes, D. Peebles, & R. Cooper (Eds.), *Proceedings of 9th International Conference on Cognitive Modeling* (pp. 240–245). Manchester, UK.

Erickson, T. D., & Matteson, M. E. (1981). From words to meaning: a semantic illusion. *Journal of Verbal Learning and Verbal Behavior*, *20*, 540–552.

Ericsson, K. A., Chase, W. G., & Faloon, S. (1980). Acquisition of a memory skill. *Science*, *208*, 1181–1182.

Ericsson, K. A., & Kintsch, W. (1995). Long-term working memory. *Psychological Review*, *102*, 211–245.

Evans, J. L., Saffran, J. R., & Robe-Torres, K. (2009). Statistical learning in children with specific language impairment. *Journal of Speech, Language, and Hearing Research: JSLHR*, *52*, 321–335.

Evans, N. (2013). Language diversity as a resource for understanding cultural evolution. In P. J. Richerson & M. H. Christiansen (Eds.), *Cultural evolution: Society, technology, language, and religion* (pp. 233–268). Cambridge, MA: MIT Press.

Evans, N., & Levinson, S. (2009). The myth of language universals: Language diversity and its importance for cognitive science. *Behavioral and Brain Sciences*, *32*, 429–492.

Evans, P. D., Gilbert, S. L., Mekel-Bobrov, N., Vallender, E. J., Anderson, J. R., Vaez-Azizi, L. M., et al. (2005). *Microcephalin*, a gene regulating brain size, continues to evolve adaptively in humans. *Science*, *309*, 1717–1720.

Everett, D. L. (2005). Cultural constraints on grammar and cognition in Pirahã. *Current Anthropology*, *46*, 621–646.

Everett, D. L. (2007). Cultural constraints on grammar in Pirahã: A Reply to Nevins, Pesetsky, and Rodrigues (2007). Available from: http://ling.auf.net/lingBuzz/000427.

Farkas, G., & Beron, K. (2004). The detailed trajectory of oral vocabulary knowledge: differences by class and race. *Social Science Research, 33*, 464–497.

Farmer, T. A., Christiansen, M. H., & Monaghan, P. (2006). Phonological typicality influences on-line sentence comprehension. *Proceedings of the National Academy of Sciences, 103*, 12203–12208.

Farmer, T. A., Fine, A. B., Misyak, J. B., & Christiansen, M. H. (in press). Reading span task performance, linguistic experience, and the processing of unexpected syntactic events. *Quarterly Journal of Experimental Psychology.*

Farmer, T. A., Misyak, J. B., & Christiansen, M. H. (2012). Individual differences in sentence processing. In M. J. Spivey, M. F. Joannisse, & K. McRae (Eds.), *Cambridge Handbook of Psycholinguistics* (pp. 353–364). Cambridge: Cambridge University Press.

Farmer, T. A., Monaghan, P., Misyak, J. B., & Christiansen, M. H. (2011). Phonological typicality influences sentence processing in predictive contexts: A reply to Staub et al. (2009). *Journal of Experimental Psychology: Learning, Memory, and Cognition, 37*, 1318–1325.

Fay, N., Garrod, S., & Roberts, L. (2008). The fitness and functionality of culturally evolved communication systems. *Philosophical Transactions of the Royal Society B: Biological Sciences, 363*, 3553–3561.

Federmeier, K. D. (2007). Thinking ahead: The role and roots of prediction in language comprehension. *Psychophysiology, 44*, 491–505.

Fehér, O., Wang, H., Saar, S., Mitra, P. P., & Tchernichovski, O. (2009). De novo establishment of wild-type song culture in the zebra finch. *Nature, 459*, 564–568.

Fehr, E., & Gächter, S. (2002). Altruistic punishment in humans. *Nature, 415*, 137–140.

Feldman, J. (1997). The structure of perceptual categories. *Journal of Mathematical Psychology, 41*, 145–170.

Fellbaum, C. (2005). WordNet and wordnets. In K. Brown (Ed.), *Encyclopedia of language and linguistics* (2nd ed., pp. 665–670). Oxford: Elsevier.

Fernald, A., Marchman, V. A., & Weisleder, A. (2013). SES differences in language processing skill and vocabulary are evident at 18 months. *Developmental Science, 16*, 234–248.

Ferreira, F., Bailey, K. G., & Ferraro, V. (2002). Good-enough representations in language comprehension. *Current Directions in Psychological Science, 11*, 11–15.

Ferreira, F., & Swets, B. (2002). How incremental is language production? Evidence from the production of utterances requiring the computation of arithmetic sums. *Journal of Memory and Language, 46*, 57–84.

Ferreira, F., & Patson, N. D. (2007). The "good enough" approach to language comprehension. *Language and Linguistics Compass, 1*, 71–83.

Ferreira, V. (2008). Ambiguity, accessibility, and a division of labor for communicative success. *Psychology of Learning and Motivation, 49*, 209–246.

Ferrer-i-Cancho, R. (2004). The Euclidean distance between syntactically linked words. *Physical Review E: Statistical, Nonlinear, and Soft Matter Physics, 70*, 056135.

Ferrer-i-Cancho, R., & Liu, H. (2014). The risks of mixing dependency lengths from sequences of different length. *Glottotheory*, *5*, 143–155.

Field, D. J. (1987). Relations between the statistics of natural images and the response profiles of cortical cells. *Journal of the Optical Society of America. A, Optics and Image Science*, *4*, 2379–2394.

Fillmore, C. J., Kay, P., & O'Connor, M. C. (1988). Regularity and idiomaticity in grammatical constructions: The case of let alone. *Language*, *64*, 501–538.

Fine, A. B., Jaeger, T. F., Farmer, T. A., & Qian, T. (2013). Rapid expectation adaptation during syntactic comprehension. *PLoS One*, *8*(10), e77661.

Finlay, B. L. (2007). Endless minds most beautiful. *Developmental Science*, *10*, 30–34.

Finlay, B. L., & Darlington, R. B. (1995). Linked regularities in the development and evolution of mammalian brains. *Science*, *268*, 1578–1584.

Finlay, B. L., Darlington, R. B., & Nicastro, N. (2001). Developmental structure in brain evolution. *Behavioral and Brain Sciences*, *24*, 263–308.

Fisher, C., & Tokura, H. (1996). Acoustic cues to grammatical structure in infant-directed speech: Cross-linguistic evidence. *Child Development*, *67*, 3192–3218.

Fisher, S. E., & Scharff, C. (2009). FOXP2 as a molecular window into speech and language. *Trends in Genetics*, *25*, 166–177.

Fiske, S. T., & Taylor, S. E. (1984). *Social cognition*. Reading, MA: Addison-Wesley.

Fitneva, S. A., & Spivey, M. J. (2004). Context and language processing: The effect of authorship. In J. C. Trueswell & M. K. Tanenhaus (Eds.), *Approaches to studying world-situated language use: Bridging the language-as-product and language-as-action traditions* (pp. 317–328). Cambridge, MA: MIT Press.

Fitneva, S. A., Christiansen, M. H., & Monaghan, P. (2009). From sound to syntax: Phonological constraints on children's lexical categorization of new words. *Journal of Child Language*, *36*, 967–997.

Fitz, H., Chang, F., & Christiansen, M. H. (2011). A connectionist account of the acquisition and processing of relative clauses. In E. Kidd (Ed.), *The acquisition of relative clauses: Processing, typology and function (TILAR Series)* (pp. 39–60). Amsterdam: John Benjamins.

Flanagan, J. R., & Wing, A. M. (1997). The role of internal models in motor planning and control: evidence from grip force adjustments during movements of hand-held loads. *Journal of Neuroscience*, *17*, 1519–1528.

Fleischman, S. (1982). *The future in thought and language: Diachronic evidence from Romance*. Cambridge, U.K.: Cambridge University Press.

Flöel, A., de Vries, M. H., Scholz, J., Breitenstein, C., & Johansen-Berg, H. (2009). White matter integrity around Broca's area predicts grammar learning success. *NeuroImage*, *4*, 1974–1981.

Fodor, J. A. (1975). *The language of thought*. Cambridge, MA: Harvard University Press.

Fodor, J. D. (1998). Unambiguous triggers. *Linguistic Inquiry*, *29*, 1–36.

Fodor, J. D., Bever, T. G., & Garrett, M. (1974). *The psychology of language*. New York: McGraw Hill.

Ford, M. (1983). A method for obtaining measures of local parsing complexity throughout sentences. *Journal of Verbal Learning and Verbal Behavior, 22*, 203–218.

Forkstam, C., Hagoort, P., Fernández, G., Ingvar, M., & Petersson, K. M. (2006). Neural correlates of artificial syntactic structure classification. *NeuroImage, 32*, 956–967.

Foss, D. J., & Cairns, H. S. (1970). Some effects of memory limitations upon sentence comprehension and recall. *Journal of Verbal Learning and Verbal Behavior, 9*, 541–547.

Fox, B. A., & Thompson, S. A. (1990). A discourse explanation of the grammar of relative clauses in English conversation. *Language, 66*, 297–316.

Frank, R. H. (1988). *Passions within reason: The strategic role of the emotions*. New York: W.W. Norton.

Frank, S. L., & Bod, R. (2011). Insensitivity of the human sentence-processing system to hierarchical structure. *Psychological Science, 22*, 829–834.

Frank, S. L., Bod, R., & Christiansen, M. H. (2012). How hierarchical is language use? *Proceedings of the Royal Society B: Biological Sciences, 279*, 4522–4531.

Frank, S. L., Trompenaars, T., & Vasishth, S. (in press). Cross-linguistic differences in processing double-embedded relative clauses: Working-memory constraints or language statistics? *Cognitive Science*.

Frazier, L., & Fodor, J. D. (1978). The sausage machine: A new two-stage parsing model. *Cognition, 66*, 291–325.

Frean, M. R., & Abraham, E. R. (2004). Adaptation and enslavement in endosymbiont-host associations. *Physical Review E: Statistical, Nonlinear, and Soft Matter Physics, 69*, 051913.

French, R. M. (1999). Catastrophic forgetting in connectionist networks. *Trends in Cognitive Sciences, 3*, 128–135.

Friederici, A. D., Bahlmann, J., Heim, S., Schibotz, R. I., & Anwander, A. (2006). The brain differentiates human and non-human grammars: Functional localization and structural connectivity. *Proceedings of the National Academy of Sciences, 103*, 2458–2463.

Fries, C. C. (1952). *The structure of English: An introduction to the construction of English sentences*. New York: Harcourt, Brace and Co.

Frigo, L., & McDonald, J. L. (1998). Properties of phonological markers that affect the acquisition of gender-like subclasses. *Journal of Memory and Language, 39*, 218–245.

Frishberg, N. (1975). Arbitrariness and iconicity: Historical change in American Sign Language. In D. J. Futuyma & M. Slatkin (Eds.) (1983). *Coevoloution*. Sunderland, MA: Sinauer.

Futuyma, D. J., & Slatkin, M. (Eds.) (1983). *Coevolution*. Sunderland, MA: Sinauer.

Gahl, S., & Garnsey, S. M. (2004). Knowledge of grammar, knowledge of usage: Syntactic probabilities affect pronunciation variation. *Language, 80*, 748–775.

Galef, B. G., & Laland, K. N. (2005). Social learning in animals: Empirical studies and theoretical models. *Bioscience*, *55*, 489–499.

Gallace, A., Tan, H. Z., & Spence, C. (2006). The failure to detect tactile change: A tactile analogue of visual change blindness. *Psychonomic Bulletin & Review*, *13*, 300–303.

Gallese, V. (2008). Mirror neurons and the social nature of language: The neural exploitation hypothesis. *Social Neuroscience*, *3*, 317–333.

Gallistel, C. R., & Gibbon, J. (2000). Time, rate, and conditioning. *Psychological Review*, *107*, 289–344.

Garcia, J., Kimeldorf, D. J., & Koelling, R. A. (1955). Conditioned aversion to saccharin resulting from exposure to gamma radiation. *Science*, *122*, 157–158.

Gasser, M. (2004). The origins of arbitrariness of language. In K. Forbus, D. Gentner, & T. Regier (Eds.), *Proceedings of the 26th Annual Conference of the Cognitive Science Society* (pp. 434–439). Mahwah, NJ: Erlbaum.

Gasser, M., Sethuraman, N., & Hockema, S. (2011). Iconicity in expressives: An empirical investigation. In J. Newman & S. Rice (Eds.), *Empirical and experimental methods in cognitive/functional research* (pp. 163–180). Stanford, CA: Center for the Study of Language and Information.

Gathercole, S. E., & Baddeley, A. D. (1990). Phonological memory deficits in language disordered children—Is there a causal connection? *Journal of Memory and Language*, *29*, 336–360.

Gazdar, G., Klein, E., Pullum, G., & Sag, I. (1985). *Generalized phrase structure grammar*. Cambridge, MA: Harvard University Press.

Geertz, C. (1973). *The interpretation of cultures: Selected essays*. New York, NY: Basic Books.

Gentner, T. Q., Fenn, K. M., Margoliash, D., & Nusbaum, H. C. (2006). Recursive syntactic pattern learning by songbirds. *Nature*, *440*, 1204–1207.

Gerhart, J., & Kirschner, M. (1997). *Cells, embryos and evolution: Toward a cellular and developmental understanding of phenotypic variation and evolutionary adaptability*. Cambridge: Blackwell.

Gibson, E. (1998). Linguistic complexity: Locality of syntactic dependencies. *Cognition*, *68*, 1–76.

Gibson, E., & Thomas, J. (1996). The processing complexity of English center-embedded and self-embedded structures. In C. Schütze (Ed.), *Proceedings of the NELS 26 sentence processing workshop* (pp. 45–71). Cambridge, MA: MIT Press.

Gibson, E., & Thomas, J. (1999). Memory limitations and structural forgetting: The perception of complex ungrammatical sentences as grammatical. *Language and Cognitive Processes*, *14*, 225–248.

Gibson, E., & Wexler, K. (1994). Triggers. *Linguistic Inquiry*, *25*, 407–454.

Gildea, D., & Temperley, D. (2010). Do grammars minimize dependency length? *Cognitive Science*, *34*, 286–310.

Gimenes, M., Rigalleau, F., & Gaonac'h, D. (2009). When a missing verb makes a French sentence more acceptable. *Language and Cognitive Processes*, *24*, 440–449.

Givón, T. (1971) Historical syntax and synchronic morphology: An archaeologist's field trip. *Chicago Linguistics Society*, *7*, 394–415.

Givón, T. (1979). *On understanding grammar*. New York: Academic Press.

Givón, T. (1998). On the co-evolution of language, mind and brain. *Evolution of Communication*, *2*, 45–116.

Givón, T., & Malle, B. F. (Eds.). (2002). *The evolution of language out of pre-language*. Amsterdam: Benjamins.

Gleitman, L., & Wanner, E. (1982). Language acquisition: The state of the state of the art. In E. Wanner & L. Gleitman (Eds.), *Language acquisition: The state of the art* (pp. 3–48). Cambridge: Cambridge University Press.

Gobet, F., Lane, P. C. R., Croker, S., Cheng, P. C. H., Jones, G., Oliver, I., et al. (2001). Chunking mechanisms in human learning. *Trends in Cognitive Sciences*, *5*, 236–243.

Gogate, L. J., & Hollich, G. (2010). Invariance detection within an interactive system: A perceptual gateway to language development. *Psychological Review*, *117*, 496–516.

Gold, E. M. (1967). Language identification in the limit. *Information and Control*, *10*, 447–474.

Goldberg, A. E. (1995). *Constructions: A construction grammar approach to argument structure*. Chicago: University of Chicago Press.

Goldberg, A. E. (2006). *Constructions at work: The nature of generalization in language*. New York: Oxford University Press.

Goldinger, S. D. (1998). Echoes of echoes? An episodic theory of lexical access. *Psychological Review*, *105*, 251–279.

Goldsby, R. A., Kindt, T. K., Osborne, B. A., & Kuby, J. (2003). *Immunology* (5th ed.). New York: W.H. Freeman and Company.

Goldstein, M. H., Waterfall, H. R., Lotem, A., Halpern, J. Y., Schwade, J. A., Onnis, L., et al. (2010). General cognitive principles for learning structure in time and space. *Trends in Cognitive Sciences*, *14*, 249–258.

Golinkoff, R. M., Hirsh-Pasek, K., Bloom, L., Smith, L., Woodward, A., Akhtar, N., et al. (Eds.). (2000). *Becoming a word learner: A debate on lexical acquisition*. New York: Oxford University Press.

Gombrich, E. H. (1960). *Art and illusion: A study in the psychology of pictorial representation*. London: Phaidon Press.

Gómez, R. L. (2002). Variability and detection of invariant structure. *Psychological Science*, *13*, 431–436.

Gómez, R. L., & Gerken, L. A. (1999). Artificial grammar learning by 1-year-olds leads to specific and abstract knowledge. *Cognition*, *70*, 109–135.

Gómez, R. L., & Gerken, L. A. (2000). Infant artificial language learning and language acquisition. *Trends in Cognitive Sciences*, *4*, 178–186.

Goodglass, H. (1993). *Understanding aphasia*. New York: Academic Press.

Goodglass, H., & Kaplan, E. (1983). *The assessment of aphasia and related disorders* (2nd ed.). Philadelphia, PA: Lea & Febiger.

Gopnik, A., Meltzoff, A. N., & Kuhl, P. K. (1999). *The scientist in the crib: Minds, brains, and how children learn.* New York: William Morrow & Co.

Gopnik, M., & Crago, M. B. (1991). Familial aggregation of a developmental language disorder. *Cognition, 39*, 1–50.

Gordon, P. C., Hendrick, R., & Johnson, M. (2001). Memory interference during language processing. *Journal of Experimental Psychology: Learning, Memory, and Cognition, 27*, 1411–1423.

Gorrell, P. (1995). *Syntax and parsing*. Cambridge: Cambridge University Press.

Gould, S. J. (1993). *Eight little piggies: Reflections in natural history*. New York: Norton.

Gould, S. J. (2002). *The structure of evolutionary theory*. Cambridge, MA: Harvard University Press.

Gould, S. J., & Lewontin, R. C. (1979). The spandrels of San Marco and the Panglossian paradigm: A critique of the adaptationist programme. *Proceedings of the Royal Society B: Biological Sciences, 205*, 581–598.

Gould, S. J., & Vrba, E. S. (1982). Exaptation—a missing term in the science of form. *Paleobiology, 8*, 4–15.

Grainger, J., Dufau, S., Montant, M., Ziegler, J. C., & Fagot, J. (2012). Orthographic processing in baboons (Papio papio). *Science, 336*, 245–248.

Gray, R. D., & Atkinson, Q. D. (2003). Language-tree divergence times support the Anatolian theory of Indo-European origin. *Nature, 426*, 435–439.

Graybiel, A. M. (1998). The basal ganglia and chunking of action repertoires. *Neurobiology of Learning and Memory, 70*, 119–136.

Green, A. M., & Angelaki, D. E. (2010). Multisensory integration: Resolving sensory ambiguities to build novel representations. *Current Opinion in Neurobiology, 20*, 353–360.

Green, T. R. G. (1979). Necessity of syntax markers: 2 Experiments with artificial languages. *Journal of Verbal Learning and Verbal Behavior, 18*, 481–496.

Greenfield, P. M. (1991). Language, tools and brain: The ontogeny and phylogeny of hierarchically organized sequential behavior. *Behavioral and Brain Sciences, 14*, 531–595.

Greenfield, P. M., Nelson, K., & Saltzman, E. (1972). The development of rulebound strategies for manipulating seriated cups: A parallel between action and grammar. *Cognitive Psychology, 3*, 291–310.

Greenhill, S. J., Atkinson, Q. D., Meade, A., & Gray, R. D. (2010). The shape and tempo of language evolution. *Proceedings of the Royal Society B: Biological Sciences, 277*, 2443–2450.

Grice, H. P. (1967). *Logic and conversation. William James Lectures. Manuscript.* Harvard University.

Grice, H. P. (1975). Logic and conversation. In D. Davidson & G. Harman (Eds.), *The logic of grammar* (pp. 64–75). Encino, CA: Dickenson.

Griffiths, T. L., & Kalish, M. L. (2005). A Bayesian view of language evolution by iterated learning. In B. Bara, L. Barsalou, & M. Bucciarelli (Eds.) *Proceedings of the*

27th annual conference of the cognitive science society (pp. 827-832). Hillsdale, NJ: Erlbaum.

Gruber, O. (2002). The co-evolution of language and working memory capacity in the human brain. In M. I. Stamenov & V. Gallese (Eds.), *Mirror neurons and the evolution of brain and language* (pp. 77–86). Amsterdam: John Benjamins.

Haber, R. N. (1983). The impending demise of the icon: the role of iconic processes in information processing theories of perception. *Behavioral and Brain Sciences, 6,* 1–55.

Hagoort, P. (2009). Reflections on the neurobiology of syntax. In D. Bickerton & E. Szathmáry (Eds.), *Biological foundations and origin of syntax. Strüngmann Forum Reports* (Vol. 3, pp. 279–296). Cambridge, MA: MIT Press.

Hagstrom, P., & Rhee, R. (1997). The Dependency Locality Theory in Korean. *Journal of Psycholinguistic Research, 26,* 189–206.

Hahn, U., & Nakisa, R. C. (2000). German inflection: Single route or dual route? *Cognitive Psychology, 41,* 313–360.

Haines, R. F. (1991). A breakdown in simultaneous information processing. In G. Obrecht & L. W. Stark (Eds.), *Presbyopia research* (pp. 171–175). New York: Plenum Press.

Hakes, D. T., Evans, J. S., & Brannon, L. L. (1976). Understanding sentences with relative clauses. *Memory & Cognition, 4,* 283–290.

Hakes, D. T., & Foss, D. J. (1970). Decision processes during sentence comprehension: Effects of surface structure reconsidered. *Perception & Psychophysics, 8,* 413–416.

Hakuta, K., Bialystok, E., & Wiley, E. (2003). Critical evidence a test of the critical-period hypothesis for second-language acquisition. *Psychological Science, 14,* 31–38.

Hale, J. (2006). Uncertainty about the rest of the sentence. *Cognitive Science, 30,* 609–642.

Hamilton, E., & Cairns, H. (Eds.). (1961). *Plato: The collected dialogues.* Princeton, NJ: Princeton University Press.

Hamilton, H. W., & Deese, J. (1971). Comprehensibility and subject-verb relations in complex sentences. *Journal of Verbal Learning and Verbal Behavior, 10,* 163–170.

Hamilton, W. D. (1964). The genetical evolution of social behaviour. *Journal of Theoretical Biology, 7,* 1–52.

Hammerton, J., Osborne, M., Armstrong, S., & Daelemans, W. (2002). Introduction to special issue on machine learning approaches to shallow parsing. *Journal of Machine Learning Research, 2,* 551–558.

Hampe, B. (2006) (Ed.). *From perception to meaning: Image schemas in cognitive linguistics.* Berlin: Mouton de Gruyter.

Hare, M., & Elman, J. L. (1995). Learning and morphological change. *Cognition, 56,* 61–98.

Harman, G., & Kulkarni, S. (2007). *Reliable reasoning: Induction and statistical learning theory.* Cambridge, MA: MIT Press.

Harrington, J., Palethorpe, S., & Watson, C. I. (2000). Does the Queen speak the Queen's English? *Nature*, *408*, 927–928.

Harris, Z. S. (1951). *Methods in structural linguistics*. Chicago: University of Chicago Press.

Hart, B., & Risley, T. (1995). *Meaningful differences in the everyday experience of young American children*. Baltimore, MD: Brookes Publishing.

Hasher, L., & Zacks, R. T. (1984). Automatic processing of fundamental information: the case of frequency of occurrence. *American Psychologist*, *39*, 1372–1388.

Haspelmath, M. (1999). Why is grammaticalization irreversible? *Linguistics*, *37*, 1043–1068.

Hasson, U., Yang, E., Vallines, I., Heeger, D. J., & Rubin, N. (2008). A hierarchy of temporal receptive windows in human cortex. *Journal of Neuroscience*, *28*, 2539–2550.

Hauser, M. D. (2006). *Moral minds: How nature designed our universal sense of right and wrong*. New York: Ecco/HarperCollins.

Hauser, M. D., Chomsky, N., & Fitch, W. T. (2002). The faculty of language: What is it, who has it, and how did it evolve? *Science*, *298*, 1569–1579.

Hauser, M. D., & Fitch, W. T. (2003). What are the uniquely human components of the language faculty? In M. H. Christiansen & S. Kirby (Eds.), *Language evolution* (pp. 158–181). Oxford: Oxford University Press.

Hawkins, J. A. (1994). *A performance theory of order and constituency*. Cambridge: Cambridge University Press.

Hawkins, J. A. (2004). *Efficiency and complexity in grammars*. Oxford: Oxford University Press.

Hawkins, J. A. (2009). Language universals and the performance-grammar correspondence hypothesis. In M. H. Christiansen, C. Collins, & S. Edelman (Eds.), *Language universals* (54–78). New York: Oxford University Press.

Hawkins, J. A., & Gilligan, G. (1988). Prefixing and suffixing universals in relation to basic word order. *Lingua*, *74*, 219–259.

Hawks, J. D., Hunley, K., Lee, S.-H., & Wolpoff, M. (2000). Population bottlenecks and Pleistocene human evolution. *Molecular Biology and Evolution*, *17*, 2–22.

Hay, J. (2000). Morphological adjacency constraints: A synthesis. *Proceedings of SCIL 9. MIT Working Papers in Linguistics, 36*, 17–29.

Healy, S. D., & Hurly, T. A. (2004). Spatial learning and memory in birds. *Brain, Behavior and Evolution*, *63*, 211–220.

Healy, S. D., Walsh, P., & Hansell, M. (2008). Nest building in birds. *Current Biology*, *18*, R271–R273.

Heathcote, A., Brown, S., & Mewhort, D. J. K. (2000). The power law repealed: The case for an exponential law of practice. *Psychonomic Bulletin & Review*, *7*, 185–207.

Hecht Orzak, S., & Sober, E. (Eds.). (2001). *Adaptationism and optimality*. Cambridge: Cambridge University Press.

Heider, P., Dery, J., & Roland, D. (2014). The processing of *it* object relative clauses: Evidence against a fine-grained frequency account. *Journal of Memory and Language, 75*, 58–76.

Heimbauer, L. A., Conway, C. M., Christiansen, M. H., Beran, M. J., & Owren, M. J. (2010). Grammar rule-based sequence learning by rhesus macaque (Macaca mulatta). Paper presented at the *33rd Meeting of the American Society of Primatologists*, Louisville, KY. [Abstract in *American Journal of Primatology*, 72, 65].

Heimbauer, L. A., Conway, C. M., Christiansen, M. H., Beran, M. J., & Owren, M. J. (2012). A Serial Reaction Time (SRT) task with symmetrical joystick responding for nonhuman primates. *Behavior Research Methods, 44*, 733–741.

Heimbauer, L.A., Conway, C.M., Christiansen, M.H., Beran, M.J. & Owren, M.J. (submitted). *Artificial grammar learning and generalization by rhesus macaques (Macaca mulatta)*.

Heine, B. (1991). *Grammaticalization*. Chicago: University of Chicago Press.

Heine, B., & Kuteva, T. (2002 a). On the evolution of grammatical forms. In A. Wray (Ed.), *Transitions to language* (pp. 376–397). Oxford, U.K.: Oxford University Press.

Heine, B., & Kuteva, T. (2002 b). *World lexicon of grammaticalization*. Cambridge: Cambridge University Press.

Heine, B., & Kuteva, T. (2007). *The genesis of grammar. A reconstruction*. Oxford: Oxford University Press.

Heine, B., & Kuteva, T. (2012). Grammaticalization theory as a tool for reconstructing language evolution. In M. Tallerman & K. Gibson (Eds.), *Oxford handbook of language evolution* (pp. 512–522). Oxford: Oxford University Press.

Hinojosa, J. A., Moreno, E. M., Casado, P., Munõz, F., & Pozo, M. A. (2005). Syntactic expectancy: An event-related potentials study. *Neuroscience Letters, 378*, 34–39.

Hinton, G. E., & Nowlan, S. J. (1987). How learning can guide evolution. *Complex Systems, 1*, 495–502.

Hintzman, D. L. (1988). Judgments of frequency and recognition memory in a multiple-trace memory model. *Psychological Review, 95*, 528–551.

Hirsh-Pasek, K., & Golinkoff, R. (Eds.). (1996). *Action meets word: How children learn verbs*. New York: Oxford University Press.

Hockett, C. F. (1954). Two models of grammatical description. *Word, 10*, 210–234.

Hockett, C. F. (1960). The origin of speech. *Scientific American, 203*, 89–96.

Hoen, M., Golembiowski, M., Guyot, E., Deprez, V., Caplan, D., & Dominey, P. F. (2003). Training with cognitive sequences improves syntactic comprehension in agrammatic aphasics. *Neuroreport, 14*, 495–499.

Hoey, M. (2005). *Lexical priming: A new theory of words and language*. Hove, UK: Psychology Press.

Hoff, E. (2006). How social contexts support and shape language development. *Developmental Review, 26*, 55–88.

Hofmeister, P., & Sag, I. A. (2010). Cognitive constraints and island effects. *Language, 86*, 366–415.

Holden, C., & Mace, R. (1997). Phylogenetic analysis of the evolution of lactose digestion in adults. *Human Biology*, *69*, 605–628.

Holmes, V. M., & O'Regan, J. K. (1981). Eye fixation patterns during the reading of relative clause sentences. *Journal of Verbal Learning and Verbal Behavior*, *20*, 417–430.

Holmes, W. G., & Sherman, P. W. (1982). The ontogeny of kin recognition in two species of ground squirrels. *American Zoologist*, *22*, 491–517.

Honey, C. J., Thesen, T., Donner, T. H., Silbert, L. J., Carlson, C. E., Devinsky, O., et al. (2012). Slow cortical dynamics and the accumulation of information over long timescales. *Neuron*, *76*, 423–434.

Hoover, M. L. (1992). Sentence processing strategies in Spanish and English. *Journal of Psycholinguistic Research*, *21*, 275–299.

Hopper, P., & Traugott, E. (2003). *Grammaticalization* (2nd ed.). Cambridge: Cambridge University Press.

Hornstein, N. (2001). *Move! A minimalist approach to construal*. Oxford: Blackwell.

Hornstein, N., & Boeckx, C. (2009). Universals in light of the varying aims of linguistic theory. In M. H. Christiansen, C. Collins, & S. Edelman (Eds.), *Language universals* (pp. 79–98). New York: Oxford University Press.

Hornstein, N., & Lightfoot, D. (Eds.). (1981). *Explanations in linguistics: The logical problem of language acquisition*. London: Longman.

Hruschka, D., Christiansen, M. H., Blythe, R. A., Croft, W., Heggarty, P., Mufwene, S. S., et al. (2009). Building social cognitive models of language change. *Trends in Cognitive Sciences*, *13*, 464–469.

Hsiao, F., & Gibson, E. (2003). Processing relative clauses in Chinese. *Cognition*, *90*, 3–27.

Hsu, A., Chater, N., & Vitányi, P. (2011). The probabilistic analysis of language acquisition: Theoretical, computational, and experimental analysis. *Cognition*, *120*, 380–390.

Hsu, A., Chater, N., & Vitányi, P. (2013). Language learning from positive evidence, reconsidered: A simplicity-based approach. *Topics in Cognitive Science*, *5*, 35–55.

Hsu, H. J., Tomblin, J. B., & Christiansen, M. H. (2014). Impaired statistical learning of non-adjacent dependencies in adolescents with specific language impairment. *Frontiers in Psychology*, *5*, 175. doi:.10.3389/fpsyg.2014.00175

Huang, Y. (2000). *Anaphora: A cross-linguistic study*. Oxford: Oxford University Press.

Hubel, D. H., & Wiesel, T. N. (1970). The period of susceptibility to the physiological effects of unilateral eye closure in kittens. *Journal of Physiology*, *206*, 419–436.

Hudson Kam, C. L., & Newport, E. L. (2005). Regularizing unpredictable variation: The roles of adult and child learners in language formation and change. *Language Learning and Development*, *1*, 151–195.

Huettig, F., Quinlan, P., McDonald, S., & Altmann, G. T. M. (2006). Word co-occurrence statistics predict language-mediated eye movements in the visual world. *Acta Psychologica*, *121*, 65–80.

Hupp, J. M., Sloutsky, V. M., & Culicover, P. W. (2009). Evidence for a domain general mechanism underlying the suffixation preference in language. *Language and Cognitive Processes, 24*, 876–909.

Hurford, J. R. (1990). Nativist and functional explanations in language acquisition. In I. M. Roca (Ed.), *Logical issues in language acquisition* (pp. 85–136). Dordrecht: Foris.

Hurford, J. R. (1991). The evolution of the critical period for language learning. *Cognition, 40*, 159–201.

Hurford, J. R. (1999). The evolution of language and languages. In R. Dunbar, C. Knight, & C. Power (Eds.), *The evolution of culture* (pp. 173–193). Edinburgh, UK: Edinburgh University Press.

Hurford, J. R. (2003). The language mosaic and its evolution. In M. H. Christiansen & S. Kirby (Eds.), *Language evolution* (pp. 38–57). Oxford: Oxford University Press.

Hurford, J. R., & Kirby, S. (1999). Co-evolution of language size and the critical period. In D. Birdsong (Ed.), *Second language acquisition and the critical period hypothesis* (pp. 39–63). Mahwah, NJ: Erlbaum.

Hurley, S. (2008). The shared circuits model (SCM): How control, mirroring, and simulation can enable imitation, deliberation, and mindreading. *Behavioral and Brain Sciences, 31*, 1–58.

Hurley, S., & Chater, N. (2005). (Eds.). *Perspectives on imitation: From neuroscience to social science. Volume 1. Mechanisms of imitation and imitation in animals*. Cambridge, MA: MIT Press.

Hurtado, N., Marchman, V. A., & Fernald, A. (2008). Does input influence uptake? Links between maternal talk, processing speed and vocabulary size in Spanish- learning children. *Developmental Science, 11*, F31–F39.

Huttenlocher, J., Vasilyeva, M., Cymerman, E., & Levine, S. (2002). Language input and child syntax. *Cognitive Psychology, 45*, 337–374.

Hutton, J., & Kidd, E. (2011). Structural priming in comprehension of relative clauses: In search of a frequency by regularity interaction. In E. Kidd (Ed.), *The acquisition of relative clauses: typology, processing, and function* (pp. 227–242). Amsterdam: John Benjamins.

Hüllen, W. (2003). *A history of Roget's Thesaurus: Origins, development, and design*. New York: Oxford University Press.

Ide, N., & Suderman, K. (2004). The American National Corpus First Release. In *Proceedings of the Fourth Language Resources and Evaluation Conference* (pp. 1681–1684). Lisbon, Portugal.

Imai, M., & Kita, S. (2014). The sound symbolism bootstrapping hypothesis for language acquisition and language evolution. *Philosophical Transactions of the Royal Society B: Biological Sciences, 369*, 20130298.

Imai, M., Kita, S., Nagumo, M., & Okada, H. (2008). Sound symbolism facilitates early verb learning. *Cognition, 109*, 54–65.

Ingram, D. (1989). *First language acquisition: Method, description, and explanation*. Cambridge: Cambridge University Press.

Jablonka, E., & Lamb, M. J. (1989). The inheritance of acquired epigenetic variations. *Journal of Theoretical Biology*, *139*, 69–83.

Jackendoff, R. (1999). Possible stages in the evolution of the language capacity. *Trends in Cognitive Sciences*, *3*, 272–279.

Jackendoff, R. (2002). *Foundations of language: Brain, meaning, grammar, evolution.* New York: Oxford University Press.

Jackendoff, R. (2007). A parallel architecture perspective on language processing. *Brain Research*, *1146*, 2–22.

Jackendoff, R. (2010). *Meaning and the lexicon: The parallel architecture 1975–2010.* New York: Oxford University Press.

Jacob, F. (1977). Evolution and tinkering. *Science*, *196*, 1161–1166.

Jacoby, L. L., Baker, J. G., & Brooks, L. R. (1989). The priority of the specific: Episodic effects in picture identification. *Journal of Experimental Psychology: Learning, Memory, and Cognition*, *15*, 275–281.

Jaeger, T. F. (2006). *Redundancy and syntactic reduction in spontaneous speech.* Unpublished Dissertation, Stanford University, Stanford, CA.

Jaeger, T. F. (2010). Redundancy and reduction: Speakers manage syntactic information density. *Cognitive Psychology*, *61*, 23–62.

Jaeger, T. F., & Tily, H. (2011). On language 'utility': Processing complexity and communicative efficiency. *Wiley Interdisciplinary Reviews: Cognitive Science*, *2*, 323–335.

Jäger, G., & Rogers, J. (2012). Formal language theory: Refining the Chomsky hierarchy. *Philosophical Transactions of the Royal Society B: Biological Sciences*, *367*, 1956–1970.

James, W. (1890). *Principles of psychology* (Vol. 1). New York: Holt.

Jenkins, L. (2000). *Biolinguistics: Exploring the biology of language*. Cambridge: Cambridge University Press.

Jensen, M. S., Yao, R., Street, W. N., & Simons, D. J. (2011). Change blindness and inattentional blindness. *Wiley Interdisciplinary Reviews: Cognitive Science*, *2*, 529–546.

Jerne, N. K. (1985). The generative grammar of the immune system. *Bioscience Reports*, *5*, 439–451.

Jescheniak, J., & Levelt, W. J. M. (1994). Word frequency effects in production. *Journal of Experimental Psychology: Learning, Memory, and Cognition*, *20*, 824–843.

Johansson, S. (2006). Working backwards from modern language to proto-grammar. In A. Cangelosi, A. D. M. Smith, & K. Smith (Eds.), *The evolution of language* (pp. 160–167). Singapore: World Scientific.

Johnson, J. S., & Newport, E. L. (1989). Critical period effects in second language learning: The influence of maturational state on the acquisition of English as a second language. *Cognitive Psychology*, *21*, 60–99.

Johnson, J. S., & Newport, E. L. (1991). Critical period effects on universal properties of language: The status of subjacency in the acquisition of a second language. *Cognition*, *39*, 215–258.

Johnson, S. P. (2010). How infants learn about the visual world. *Cognitive Science*, *34*, 1158–1184.

Johnson-Laird, P. N. (1983). *Mental models: Towards a cognitive science of language, inference, and consciousness*. Cambridge, MA: Harvard University Press.

Johnson-Pynn, J., Fragaszy, D. M., Hirsch, M. H., Brakke, K. E., & Greenfield, P. M. (1999). Strategies used to combine seriated cups by chimpanzees (*Pantroglodytes*), bonobos (*Pan paniscus*), and capuchins (*Cebus apella*). *Journal of Comparative Psychology*, *113*, 137–148.

Jolsvai, H., McCauley, S. M., & Christiansen, M. H. (2013). Meaning overrides frequency in idiomatic and compositional multiword chunks. In M. Knauff, M. Pauen, N. Sebanz, & I. Wachsmuth (Eds.), *Proceedings of the 35th Annual Conference of the Cognitive Science Society* (pp. 692–697). Austin, TX: Cognitive Science Society.

Jones, G. (2012). Why chunking should be considered as an explanation for developmental change before short-term memory capacity and processing speed. *Frontiers in Psychology*, *3*, 167. doi:.10.3389/fpsyg.2012.00167

Jones, G., Gobet, F., Freudenthal, D., Watson, S. E., & Pine, J. M. (2014). Why computational models are better than verbal theories: the case of nonword repetition. *Developmental Science*, *17*, 298–310.

Joshi, A. K. (1990). Processing crossed and nested dependencies: An automaton perspective on the psycholinguistic results. *Language and Cognitive Processes*, *5*, 1–27.

Joshi, A. K., & Schabes, Y. (1997). Tree-adjoining grammars. In G. Rosenberg & A. Salomaa (Eds.), *Handbook of formal languages* (pp. 69–123). Springer Berlin Heidelberg.

Juliano, C., & Tanenhaus, M. K. (1994). A constraint-based lexicalist account of the subject/object attachment preference. *Journal of Psycholinguistic Research*, *23*, 459–471.

Jurafsky, D. (1996). A probabilistic model of lexical and syntactic access and disambiguation. *Cognitive Science*, *20*, 137–194.

Jurafsky, D., Bell, A., Gregory, M. L., & Raymond, W. D. (2001). Probabilistic relations between words: Evidence from reduction in lexical production. In J. L. Bybee & P. Hopper (Eds.), *Frequency and the emergence of linguistic structure* (pp. 229–254). Amsterdam: John Benjamins.

Jurafsky, D., Martin, J. H., Kehler, A., Vander Linden, K., & Ward, N. (2000). *Speech and language processing: An introduction to natural language processing, computational linguistics, and speech recognition*. Upper Saddle River: Prentice Hall.

Jusczyk, P. W. (1997). *The discovery of spoken language*. Cambridge, MA: MIT Press.

Jusczyk, P. W. (1999). How infants begin to extract words from speech. *Trends in Cognitive Sciences*, *3*, 323–328.

Just, M. A., & Carpenter, P. A. (1992). A capacity theory of comprehension: Individual differences in working memory. *Psychological Review*, *99*, 122–149.

Just, M. A., Carpenter, P. A., & Woolley, J. D. (1982). Paradigms and processes in reading comprehension. *Journal of Experimental Psychology. General*, *3*, 228–238.

Kading, J. (1897). *Häufigkeitswörterbuch der deutschen Sprache*. Steglitz: Privately published.

Kamide, Y. (2008). Anticipatory processes in sentence processing. *Language and Linguistics Compass*, *2*, 647–670.

Kamide, Y., Altmann, G. T. M., & Haywood, S. (2003). The time-course of prediction in incremental sentence processing: Evidence from anticipatory eye-movements. *Journal of Memory and Language*, *49*, 133–159.

Kamin, L. J. (1969). Predictability, surprise, attention and conditioning. In B. A. Campbell & R. M. Church (Eds.), *Punishment and aversive behavior* (pp. 279–296). New York: Appleton-Century-Crofts.

Karlsson, F. (2007). Constraints on multiple center-embedding of clauses. *Journal of Linguistics*, *43*, 365–392.

Karlsson, F. (2010). Syntactic recursion and iteration. In H. van der Hulst (Ed.), *Recursion and human language* (pp. 43–67). Berlin: Mouton de Gruyter.

Karmiloff-Smith, A., & Inhelder, B. (1975). If you want to get ahead, get a theory. *Cognition*, *3*, 195–212.

Kaschak, M. P., & Glenberg, A. M. (2004). This construction needs learned. *Journal of Experimental Psychology. General*, *133*, 450–467.

Katz, J. J., & Postal, P. M. (1964). *An integrated theory of linguistic description*. Cambridge, MA: MIT Press.

Kauffman, S. A. (1995). *The origins of order: Self-organization and selection in evolution*. Oxford: Oxford University Press.

Keller, R. (1994). *On language change: The invisible hand in language*. London: Routledge.

Kelly, M. H. (1988). Phonological biases in grammatical category shifts. *Journal of Memory and Language*, *27*, 343–358.

Kelly, M. H. (1992). Using sound to solve syntactic problems: The role of phonology in grammatical category assignments. *Psychological Review*, *99*, 349–364.

Kelly, M. H., & Bock, J. (1988). Stress in time. *Journal of Experimental Psychology: Human Perception and Performance*, *14*, 389–403.

Kempson, R., Meyer-Viol, W., & Gabbay, D. (2001). *Dynamic syntax: The flow of language understanding*. Malden, MA: Blackwell.

Kidd, E. (Ed.). (2011). *The acquisition of relative clauses: Processing, typology and function*. Amsterdam: John Benjamins.

Kidd, E., & Bavin, E. L. (2002). English-speaking children's comprehension of relative clauses: Evidence for general-cognitive and language-specific constraints on development. *Journal of Psycholinguistic Research*, *31*, 599–617.

Kidd, E., Brandt, S., Lieven, E., & Tomasello, M. (2007). Object relatives made easy: A cross-linguistic comparison of the constraints influencing young children's processing of relative clauses. *Language and Cognitive Processes*, *22*, 860–897.

Kimball, J. (1973). Seven principles of surface structure parsing in natural language. *Cognition*, *2*, 15–47.

King, J., & Just, M. A. (1991). Individual differences in syntactic processing: The role of working memory. *Journal of Memory and Language, 30*, 580–602.

Kirby, S. (1998). Fitness and the selective adaptation of language. In J. R. Hurford, M. Studdert-Kennedy, & C. Knight (Eds.), *Approaches to the evolution of language: Social and cognitive bases* (pp. 359–383). New York: Cambridge University Press.

Kirby, S. (1999). *Function, selection and innateness: The emergence of language universals*. Oxford: Oxford University Press.

Kirby, S. (2000). Syntax without natural selection: How compositionality emerges from vocabulary in a population of learners. In C. Knight (Ed.), *The evolutionary emergence of language: social function and the origins of linguistic form* (pp. 303–323). Cambridge: Cambridge University Press.

Kirby, S., Cornish, H., & Smith, K. (2008). Cumulative cultural evolution in the laboratory: An experimental approach to the origins of structure in human language. *Proceedings of the National Academy of Sciences, 105*, 10681–10685.

Kirby, S., Dowman, M., & Griffiths, T. (2007). Innateness and culture in the evolution of language. *Proceedings of the National Academy of Sciences, 104*, 5241–5245.

Kirby, S., & Hurford, J. R. (1997). Learning, culture and evolution in the origin of linguistic constraints. In P. Husbands & I. Harvey (Eds.), *ECAL97* (pp. 493–502). Cambridge, MA: MIT Press.

Kirby, S., & Hurford, J. R. (2002). The emergence of linguistic structure: An overview of the iterated learning model. In A. Cangelosi & D. Parisi (Eds.), *Simulating the evolution of language* (pp. 121–148). London: Springer Verlag.

Klein, E. (1966). *A comprehensive etymological dictionary of the English language: Dealing with the origin of words and their sense development thus illustrating the history of civilization and culture*. Amsterdam: Elsevier.

Kluender, R., & Kutas, M. (1993). Subjacency as a processing phenomenon. *Language and Cognitive Processes, 8*, 573–633.

Köhler, W. (1929). *Gestalt psychology*. New York: Liveright.

Konopka, A. E. (2012). Planning ahead: How recent experience with structures and words changes the scope of linguistic planning. *Journal of Memory and Language, 66*, 143–162.

Kovic, V., Plunkett, K., & Westermann, G. (2010). The shape of words in the brain. *Cognition, 114*, 19–28.

Kraljic, T., & Brennan, S. (2005). Prosodic disambiguation of syntactic structure: For the speaker or for the addressee? *Cognitive Psychology, 50*, 194–231.

Kucera, H., & Francis, W. N. (1967). *Computational analysis of present-day American English*. Providence, RI: Brown University Press.

Kuhl, P. K. (1987). The special mechanisms debate in speech research: Categorization tests on animals and infants. In S. Harnad (Ed.), *Categorical perception: The groundwork of cognition* (pp. 355–386). Cambridge: Cambridge University Press.

Kuhl, P. K. (1999). Speech, language, and the brain: Innate preparation for learning. In M. Konishi & M. Hauser (Eds.), *Neural mechanisms of communication* (pp. 419–450). Cambridge, MA: MIT Press.

Kuperman, V., Stadthagen-Gonzalez, H., & Brysbaert, M. (2012). Age-of-acquisition ratings for 30,000 English words. *Behavior Research Methods*, *44*, 978–990.

Kutas, M., Federmeier, K. D., & Urbach, T. P. (2014). The "negatives" and "positives" of prediction in language. In M. S. Gazzaniga & G. R. Mangun (Eds.), *The Cognitive Neurosciences V* (pp. 649–656). Cambridge, MA: MIT Press.

Kvasnicka, V., & Pospichal, J. (1999). An emergence of coordinated communication in populations of agents. *Artificial Life*, *5*, 318–342.

Laertius, D. (1853). *The lives and opinions of eminent philosophers* (C. D. Yonge, Trans.). London: H. G. Bohn.

Lai, C. S. L., Fisher, S. E., Hurst, J. A., Vargha-Khadem, F., & Monaco, A. P. (2001). A forkhead-domain gene is mutated in a severe speech and language disorder. *Nature*, *413*, 519–523.

Lai, C. S. L., Gerrelli, D., Monaco, A. P., Fisher, S. E., & Copp, A. J. (2003). FOXP2 expression during brain development coincides with adult sites of pathology in a severe speech and language disorder. *Brain*, *126*, 2455–2462.

Lakoff, G., & Johnson, M. (1980). *Metaphors we live by*. Chicago: University of Chicago Press.

Lanyon, S. J. (2006). A saltationist approach for the evolution of human cognition and language. In A. Cangelosi, A. D. M. Smith, & K. Smith (Eds.), *The Evolution of language* (pp. 176–183). Singapore: World Scientific.

Larkin, W., & Burns, D. (1977). Sentence comprehension and memory for embedded structure. *Memory & Cognition*, *5*, 17–22.

Lashley, K. S. (1951). The problem of serial order in behavior. In L. A. Jeffress (Ed.), *Cerebral mechanisms in behavior* (pp. 112–146). New York: Wiley.

Laubichler, M. D., & Maienschein, J. (Eds.). (2007). *From embryology to evo-devo: A history of developmental evolution*. Cambridge, MA: MIT Press.

Lee, Y. S. (1997). Learning and awareness in the serial reaction time task. In *Proceedings of the 19th Annual Conference of the Cognitive Science Society*. (pp. 119–124). Hillsdale, NJ: Lawrence Erlbaum Associates.

Leech, R., Aydelott, J., Symons, G., Carnevale, J., & Dick, F. (2007). The development of sentence interpretation: Effects of perceptual, attentional and semantic interference. *Developmental Science*, *10*, 794–813.

Lenneberg, E. H. (1967). *Biological foundations of language*. New York: Wiley.

Leonard, L. (1987). Is specific language impairment a useful construct? In S. Rosenberg (Ed.), *Advances in applied psycholinguistics: (Vol. 1) Disorders of first-language development* (pp. 1–39). New York: Cambridge University Press.

Lerner, Y., Honey, C. J., Silbert, L. J., & Hasson, U. (2011). Topographic mapping of a hierarchy of temporal receptive windows using a narrated story. *Journal of Neuroscience*, *31*, 2906–2915.

Levelt, W. J. M. (2001). Spoken word production: A theory of lexical access. *Proceedings of the National Academy of Sciences*, *98*, 13464–13471.

Levelt, W. J. M. (2012). *A history of psycholinguistics: The pre-Chomskyan era*. Oxford: Oxford University Press.

Levinson, S. C. (1987). Pragmatics and the grammar of anaphora: A partial pragmatic reduction of binding and control phenomena. *Journal of Linguistics*, *23*, 379–434.

Levinson, S. C. (2000). *Presumptive meanings: The theory of generalized conversational implicature*. Cambridge, MA: MIT Press.

Levinson, S. C. (2013). Recursion in pragmatics. *Language*, *89*, 149–162.

Levinson, S. C., & Dediu, D. (2013). The interplay of genetic and cultural factors in ongoing language evolution. In P. J. Richerson & M. H. Christiansen (Eds.), *Cultural evolution: Society, technology, language & religion* (pp. 219–232). Cambridge, MA: MIT Press.

Levy, R. (2008). Expectation-based syntactic comprehension. *Cognition*, *106*, 1126–1177.

Lewis, R. L. (1996). A theory of grammatical but unacceptable embeddings. *Journal of Psycholinguistic Research*, *25*, 93–116.

Lewis, R. L., & Vasishth, S. (2005). An activation-based model of sentence processing as skilled memory retrieval. *Cognitive Science*, *29*, 375–419.

Lewis, R. L., Vasishth, S., & Van Dyke, J. A. (2006). Computational principles of working memory in sentence comprehension. *Trends in Cognitive Sciences*, *10*, 447–454.

Lewontin, R. C. (1998). The evolution of cognition: Questions we will never answer. In D. Scarborough & S. Sternberg (Eds.), *An invitation to cognitive science, Volume 4: Methods, models, and conceptual issues.* Cambridge, MA: MIT Press.

Li, M., & Vitányi, P. (1997). *An introduction to Kolmogorov complexity theory and its applications* (2nd ed.). Berlin: Springer.

Lieberman, M. D., Chang, G. Y., Chiao, J., Bookheimer, S. Y., & Knowlton, B. J. (2004). An event-related fMRI study of artificial grammar learning in a balanced chunk strength design. *Journal of Cognitive Neuroscience*, *16*, 427–438.

Lieberman, P. (1984). *The biology and evolution of language*. Cambridge, MA: Harvard University Press.

Lieberman, P. (1991). Speech and brain evolution. *Behavioral and Brain Sciences*, *14*, 566–568.

Lieberman, P. (2000). *Human language and our reptilian brain*. Cambridge, MA: Harvard University Press.

Lieberman, P. (2003). Motor control, speech, and the evolution of human language. In M. H. Christiansen & S. Kirby (Eds.), *Language evolution* (pp. 255–271). New York: Oxford University Press.

Lightfoot, D. (1989). The child's trigger experience: Degree-0 learnability. *Behavioral and Brain Sciences*, *12*, 321–334.

Lightfoot, D. (2000). The spandrels of the linguistic genotype. In C. Knight, M. Studdert-Kennedy, & J. R. Hurford (Eds.), *The evolutionary emergence of language: Social function and the origins of linguistic form* (pp. 231–247). Cambridge: Cambridge University Press.

Lively, S. E., Pisoni, D. B., & Goldinger, S. D. (1994). Spoken word recognition. In M. A. Gernsbacher (Ed.), *Handbook of psycholinguistics* (pp. 265–318). San Diego, CA: Academic Press.

Livingstone, D., & Fyfe, C. (2000). Modelling language-physiology coevolution. In C. Knight, M., Studdert-Kennedy, and J. R. Hurford (Eds.), *The emergence of language: Social function and the origins of linguistic form* (pp. 199–215). Cambridge: Cambridge University Press.

Lobina, D. J. (2014). What linguists are talking about when talking about…. *Language Sciences, 45*, 56–70.

Locke, J. L., & Bogin, B. (2006). Language and life history: A new perspective on the development and evolution of human language. *Behavioral and Brain Sciences, 29*, 259–280.

Logan, G. D. (1988). Toward an instance theory of automatization. *Psychological Review, 95*, 492–527.

Logan, G. D. (2002). An instance theory of attention and memory. *Psychological Review, 109*, 376–400.

Long, F., Yang, Z., & Purves, D. (2006). Spectral statistics in natural scenes predict hue, saturation, and brightness. *Proceedings of the National Academy of Sciences, 103*, 6013–6018.

Loritz, D. (1999). *How the brain evolved language*. Oxford: Oxford University Press.

Louwerse, M. M., Dale, R., Bard, E. G., & Jeuniaux, P. (2012). Behavior matching in multimodal communication is synchronized. *Cognitive Science, 36*, 1404–1426.

Lum, J. A. G., Conti-Ramsden, G. M., Morgan, A. T., & Ullman, M. T. (2014). Procedural learning deficits in Specific Language Impairment (SLI): A meta-analysis of serial reaction time task performance. *Cortex, 51*, 1–10.

Lum, J. A. G., Conti-Ramsden, G. M., Page, D., & Ullman, M. T. (2012). Working, declarative and procedural memory in specific language impairment. *Cortex, 48*, 1138–1154.

Lupyan, G., & Christiansen, M. H. (2002). Case, word order, and language learnability: Insights from connectionist modeling. In *Proceedings of the 24th Annual Conference of the Cognitive Science Society* (pp. 596–601). Mahwah, NJ: Lawrence Erlbaum Associates.

MacDermot, K. D., Bonora, E., Sykes, N., Coupe, A. M., Lai, C. S. L., Vernes, S. C., et al. (2005). Identification of *FOXP2* truncation as a novel cause of developmental speech and language deficits. *American Journal of Human Genetics, 76*, 1074–1080.

MacDonald, M. C. (1994). Probabilistic constraints and syntactic ambiguity resolution. *Language and Cognitive Processes, 9*, 157–201.

MacDonald, M. C. (2013). How language production shapes language form and comprehension. *Frontiers in Psychology, 4*, 226. doi:.10.3389/fpsyg.2013.00226

MacDonald, M. C., & Christiansen, M. H. (2002). Reassessing working memory: A comment on Just & Carpenter (1992) and Waters & Caplan (1996). *Psychological Review, 109*, 35–54.

MacDonald, M. C., Just, M. A., & Carpenter, P. A. (1992). Working memory constraints on the processing of syntactic ambiguity. *Cognitive Psychology, 24*, 56–98.

MacDonald, M. C., Pearlmutter, N. J., & Seidenberg, M. S. (1994). The lexical nature of syntactic ambiguity resolution. *Psychological Review, 101*, 676–703.

MacKay, D. J. C. (2003). *Information theory, inference, and learning algorithms.* Cambridge: Cambridge University Press.

MacNeilage, P. F. (1998). The frame/content theory of evolution of speech production. *Behavioral and Brain Sciences, 21*, 499–511.

MacWhinney, B. (2000). *The CHILDES project: Tools for analyzing talk* (3rd ed.). Mahwah, NJ: Lawrence Erlbaum Associates.

MacWhinney, B., & Snow, C. (1985). The child language exchange system. *Journal of Child Language, 12*, 271–296.

Maess, B., Koelsch, S., Gunter, T. C., & Friederici, A. D. (2001). Musical syntax is processed in Broca's area: An MEG study. *Nature Neuroscience, 4*, 540–545.

Magyari, L., & de Ruiter, J. P. (2012). Prediction of turn-ends based on anticipation of upcoming words. *Frontiers in Psychology, 3*, 376. doi:.10.3389/fpsyg.2012.00376

Mak, W. M., Vonk, W., & Schriefers, H. (2008). Discourse structure and relative clause processing. *Memory & Cognition, 36*, 170–181.

Malle, B. F. (2002). The relation between language and theory of mind in development and evolution. In T. Givón & B. Malle (Eds.), *The evolution of language out of pre-language* (pp. 265–284). Philadelphia, PA: John Benjamins.

Mani, N., & Huettig, F. (2012). Prediction during language processing is a piece of cake—But only for skilled producers. *Journal of Experimental Psychology: Human Perception and Performance, 38*, 843–847.

Manning, C. D., & Schütze, H. (1999). *Foundations of statistical natural language processing*. Cambridge, MA: MIT Press.

Mantel, N. (1967). The detection of disease clustering and a generalized regression approach. *Cancer Research, 27*, 209–220.

Maratsos, M. P., & Chalkley, M. A. (1980). The internal language of children's syntax: The ontogenesis and representation of syntactic categories. In K. E. Nelson (Ed.), *Children's language* (Vol. 2, pp. 127–214). New York: Gardner Press.

Marchand, H. (1969). *The categories and types of present-day English word-formation* (2nd ed.). Munich: C.H. Beck'sche Verlagsbuchhandlung.

Marchman, V. A., & Fernald, A. (2008). Speed of word recognition and vocabulary knowledge in infancy predict cognitive and language outcomes in later childhood. *Developmental Science, 11*, F9–F16.

Marcus, G. F., Vouloumanos, A., & Sag, I. A. (2003). Does Broca's play by the rules? *Nature Neuroscience, 6*, 651–652.

Marcus, M. P. (1980). *Theory of syntactic recognition for natural languages*. Cambridge, MA: MIT Press.

Marks, L. E. (1968). Scaling of grammaticalness of self-embedded English sentences. *Journal of Verbal Learning and Verbal Behavior, 7*, 965–967.

Marler, P., & Slabbekoorn, H. (Eds.). (2004). *Nature's music: The science of birdsong.* San Diego, CA: Elsevier.

Marr, D. (1976). Early processing of visual information. *Philosophical Transactions of the Royal Society B: Biological Sciences, 275*, 483–519.

Marr, D. (1982). *Vision.* San Francisco: W. H. Freeman.

Marslen-Wilson, W. D. (1975). Sentence perception as an interactive parallel process. *Science, 189*, 226–228.

Marslen-Wilson, W. D. (1987). Functional parallelism in spoken word-recognition. *Cognition, 25*, 71–102.

Marslen-Wilson, W. D., Tyler, L. K., & Koster, C. (1993). Integrative processes in utterance resolution. *Journal of Memory and Language, 32*, 647–666.

Martin, R. C. (1995). Working memory doesn't work: A critique of Miyake et al.'s capacity theory of aphasic comprehension deficits. *Cognitive Neuropsychology, 12*, 623–636.

Martin, R. C. (2006). The neuropsychology of sentence processing: Where do we stand? *Cognitive Neuropsychology, 23*, 74–95.

Martin, R. C., & He, T. (2004). Semantic STM and its role in sentence processing: A replication. *Brain and Language, 89*, 76–82.

Martin, R. C., Shelton, J. R., & Yaffee, L. S. (1994). Language processing and working memory: Neuropsychological evidence for separate phonological and semantic capacities. *Journal of Memory and Language, 33*, 83–111.

Martins, M. D. (2012). Distinctive signatures of recursion. *Philosophical Transactions of the Royal Society B: Biological Sciences, 367*, 2055–2064.

Mattys, S. L., Jusczyk, P. W., Luce, P. A., & Morgan, J. L. (1999). Phonotactic and prosodic effects on word segmentation in infants. *Cognitive Psychology, 38*, 465–494.

Mattys, S. L., White, L., & Melhorn, J. F. (2005). Integration of multiple speech segmentation cues: A hierarchical framework. *Journal of Experimental Psychology. General, 134*, 477–500.

Maurer, D., Pathman, T., & Mondloch, C. J. (2006). The shape of boubas: Sound—shape correspondences in toddlers and adults. *Developmental Science, 9*, 316–322.

McCauley, S. M., & Christiansen, M. H. (2011). Learning simple statistics for language comprehension and production: The CAPPUCCINO model. In L. Carlson, C. Hölscher, & T. Shipley (Eds.), *Proceedings of the 33rd Annual Conference of the Cognitive Science Society* (pp. 1619–1624). Austin, TX: Cognitive Science Society.

McCauley, S. M., & Christiansen, M. H. (2014 a). Acquiring formulaic language: A computational model. *Mental Lexicon, 9*, 419–436.

McCauley, S. M., & Christiansen, M. H. (2014 b). Prospects for usage-based computational models of grammatical development: Argument structure and semantic roles. *Wiley Interdisciplinary Reviews: Cognitive Science, 5*, 489–499.

McCauley, S. M., & Christiansen, M. H. (2014 c). Reappraising lexical specificity in children's early syntactic combinations. In *Proceedings of the 36th Annual Conference of the Cognitive Science Society* (pp. 1000–1005). Austin, TX: Cognitive Science Society.

McCauley, S. M., & Christiansen, M. H. (2015). *Individual differences in chunking ability predict on-line sentence processing*. In D. C. Noelle & R. Dale (Eds.), *Proceedings of the 37th Annual Conference of the Cognitive Science Society* (pp. 1553–1558). Austin, TX: Cognitive Science Society.

McCloskey, M., & Cohen, N. J. (1989). Catastrophic interference in connectionist networks: the sequential learning problem. *Psychology of Learning and Motivation, 24*, 109–165.

McEnery, T., & Wilson, A. (1996). *Corpus linguistics*. Edinburgh: Edinburgh University Press.

McMahon, A. M. S. (1994). *Understanding language change*. Cambridge: Cambridge University Press.

Meier, R. P., & Bower, G. H. (1986). Semantic reference and phrasal grouping in the acquisition of a miniature phrase structure language. *Journal of Memory and Language, 25*, 492–505.

Mekel-Bobrov, N., Gilbert, S. L., Evans, P. D., Vallender, E. J., Anderson, J. R., Hudson, R. R., et al. (2005). Ongoing adaptive evolution of *ASPM*, a brain size determinant in Homo sapiens. *Science, 309*, 1720–1722.

Mermillod, M., Bugaïska, A., & Bonin, P. (2013). The Stability-Plasticity Dilemma: Investigating the continuum from catastrophic forgetting to age-limited learning effects. *Frontiers in Psychology, 4*, 504. doi:.10.3389/fpsyg.2013.00504

Mesoudi, A., Whiten, A., & Laland, K. (2006). Toward a unified science of cultural evolution. *Behavioral and Brain Sciences, 29*, 329–383.

Meyer, D. E., & Schvaneveldt, R. W. (1971). Facilitation in recognizing pairs of words: Evidence of a dependence between retrieval operations. *Journal of Experimental Psychology, 90*, 227–234.

Michel, J. B., Shen, Y. K., Aiden, A. P., Veres, A., Gray, M. K., Pickett, J. P., et al. (2011). Quantitative analysis of culture using millions of digitized books. *Science, 331*, 176–182.

Miller, C., & Swift, K. (1980). *The handbook of nonsexist writing*. New York: Harper & Row.

Miller, G. A. (1956). The magical number seven, plus or minus two: Some limits on our capacity for processing information. *Psychological Review, 63*, 81–97.

Miller, G. A. (1962). Some psychological studies of grammar. *American Psychologist, 17*, 748–762.

Miller, G. A. (2013). "WordNet—About us." WordNet. Retrieved from http://wordnet.princeton.edu.

Miller, G. A., & Chomsky, N. (1963). Finitary models of language users. In R. D. Luce, R. R. Bush, & E. Galanter (Eds.), *Handbook of mathematical psychology* (Vol. 2, pp. 419–492). New York: Wiley.

Miller, G. A., Galanter, E., & Pribram, K. H. (1960). *Plans and the structure of behavior*. New York: Holt, Rinehart & Winston.

Miller, G. A., & Isard, S. (1964). Free recall of self-embedded English sentences. *Information and Control, 7*, 292–303.

Miller, G. A., & McKean, K. O. (1964). A chronometric study of some relations between sentences. *Quarterly Journal of Experimental Psychology, 16*, 297–308.

Miller, G. A., & Taylor, W. G. (1948). The perception of repeated bursts of noise. *Journal of the Acoustical Society of America, 20*, 171–182.

Mintz, T. H., Newport, E. L., & Bever, T. G. (2002). The distributional structure of grammatical categories in speech to young children. *Cognitive Science, 26*, 393–424.

Misyak, J. B., & Christiansen, M. H. (2010). When "more" in statistical learning means "less" in language: Individual differences in predictive processing of adjacent dependencies. In R. Catrambone, & S. Ohlsson (Eds.), *Proceedings of the 32nd Annual Cognitive Science Society Conference* (pp. 2686–2691). Austin, TX: Cognitive Science Society.

Misyak, J. B., & Christiansen, M. H. (2012). Statistical learning and language: An individual differences study. *Language Learning, 62*, 302–331.

Misyak, J. B., Christiansen, M. H., & Tomblin, J. B. (2010). Sequential expectations: The role of prediction-based learning in language. *Topics in Cognitive Science, 2*, 138–153.

Misyak, J. B., Goldstein, M. H., & Christiansen, M. H. (2012). Statistical-sequential learning in development. In P. Rebuschat, & J. Williams (Eds.), *Statistical learning and language acquisition* (pp. 13–54). Berlin: Mouton de Gruyter.

Mitchell, D. C., & Cuetos, F. (1991). The origins of parsing strategies. In C. Smith (Ed.), *Current issues in natural language processing* (pp. 1–12). Austin, TX: Center for Cognitive Science, University of Austin.

Mitchell, D. C., Cuetos, F., Corley, M. M. B., & Brysbaert, M. (1995). Exposure-based models of human parsing: Evidence for the use of coarse- grained (non-lexical) statistical records. *Journal of Psycholinguistic Research, 24*, 469–488.

Mithen, S. J. (1996). *The prehistory of the mind: A search for the origins of art, science and religion*. London: Thames & Hudson.

Mithun, M. (2010). The fluidity of recursion and its implications. In H. van der Hulst (Ed.), *Recursion and human language* (pp. 17–41). Berlin: Mouton de Gruyter.

Miyamoto, E., & Nakamura, M. (2003). Subject/Object asymmetries in the processing of relative clauses in Japanese. In G. Garding, & M. Tsujimura (Eds.), *Proceedings of WCCFL, 22*, 342–355.

Miyazaki, M., Hidaka, S., Imai, M., Yeung, H. H., Kantartzis, K., Okada, H., et al. (2013). The facilitatory role of sound symbolism in infant word learning. In M. Knauff, M. Pauen, N. Sebanz & I. Wachsmuth (Eds.), *Proceedings of the 35th Annual Conference of the Cognitive Science Society* (pp. 3080-3085). Austin, TX: Cognitive Science Society.

Monaghan, P., Chater, N., & Christiansen, M. H. (2003). Inequality between the classes: Phonological and distributional typicality as predictors of lexical processing. In *Proceedings of the 25th Annual Conference of the Cognitive Science Society* (pp. 810–815). Mahwah, NJ: Lawrence Erlbaum.

Monaghan, P., Chater, N., & Christiansen, M. H. (2005). The differential role of phonological and distributional cues in grammatical categorisation. *Cognition, 96*, 143–182.

Monaghan, P., & Christiansen, M. H. (2008). Integration of multiple probabilistic cues in syntax acquisition. In H. Behrens (Ed.), *Trends in corpus research: Finding structure in data (TILAR Series)* (pp. 139–163). Amsterdam: John Benjamins.

Monaghan, P., & Christiansen, M. H. (2014). Multiple cues in language acquisition. In P. Brooks & V. Kempe (Eds.), *Encyclopedia of language development* (pp. 389–392). Thousand Oaks, CA: Sage Publications.

Monaghan, P., Christiansen, M. H., & Chater, N. (2007). The Phonological-Distributional Coherence Hypothesis: Cross-linguistic evidence in language acquisition. *Cognitive Psychology*, *55*, 259–305.

Monaghan, P., Christiansen, M. H., & Fitneva, S. A. (2011). The arbitrariness of the sign: Learning advantages from the structure of the vocabulary. *Journal of Experimental Psychology. General*, *140*, 325–347.

Monaghan, P., Christiansen, M. H., Farmer, T. A., & Fitneva, S. A. (2010). Measures of phonological typicality: Robust coherence and psychological validity. *Mental Lexicon*, *5*, 281–299.

Monaghan, P., Shillcock, R. C., Christiansen, M. H., & Kirby, S. (2014). How arbitrary is language? *Philosophical Transactions of the Royal Society B: Biological Sciences*, *369*, 20130299.

Montag, J. L., & MacDonald, M. C. (2015). Text exposure predicts spoken production of complex sentences in 8- and 12-year-old children and adults. *Journal of Experimental Psychology. General, 144,* 447-468.

Morgan, J. L., & Demuth, K. (Eds.). (1996). *Signal to syntax: Bootstrapping from speech to grammar in early acquisition*. Mahwah, NJ: Lawrence Erlbaum.

Morgan, J. L., Meier, R. P., & Newport, E. L. (1987). Structural packaging in the input to language learning: Contributions of prosodic and morphological marking of phrases to the acquisition of language. *Cognitive Psychology*, *19*, 498–550.

Morgan, J. L., & Saffran, J. R. (1995). Emerging integration of sequential and suprasegmental information in preverbal speech segmentation. *Child Development*, *66*, 911–936.

Morgan, J. L., Shi, R., & Allopenna, P. (1996). Perceptual bases of grammatical categories. In J. L. Morgan & K. Demuth (Eds.), *Signal to syntax: Bootstrapping from speech to grammar in early acquisition* (pp. 263–283). Mahwah, NJ: Lawrence Erlbaum Associates.

Morowitz, H. (1979). *The wine of life and other essays on society, energy and living things*. New York: St. Martin's Press.

Morton, J. (1969). The interaction of information in word recognition. *Psychological Review*, *76*, 165–178.

Mufwene, S. (2008). *Language evolution: Contact, competition and change*. London: Continuum International Publishing Group.

Müller, M. (1870). Darwinism tested by the science of language. *Nature*, *1*, 256–259.

Munakata, Y., McClelland, J. L., Johnson, M. H., & Siegler, R. S. (1997). Rethinking infant knowledge: toward an adaptive process account of successes and failures in object permanence tasks. *Psychological Review*, *104*, 686–713.

Munroe, S., & Cangelosi, A. (2002). Learning and the evolution of language: the role of cultural variation and learning cost in the Baldwin Effect. *Artificial Life*, *8*, 311–339.

Murdock, B. B., Jr. (1968). Serial order effects in short-term memory. *Journal of Experimental Psychology Monograph*, *76*(Supplement), 1–15.

Murphy, G. L. (2002). *The big book of concepts*. Cambridge, MA: MIT Press.

Musso, M., Moro, A., Glauche, V., Rijntjes, M., Reichenbach, J., Buchel, C., et al. (2003). Broca's area and the language instinct. *Nature Neuroscience*, *6*, 774–781.

Nadig, A. S., & Sedivy, J. C. (2002). Evidence of perspective-taking constraints in children's on-line reference resolution. *Psychological Science*, *13*, 329–336.

Navon, D., & Miller, J. O. (2002). Queuing or sharing? A critical evaluation of the single-bottleneck notion. *Cognitive Psychology*, *44*, 193–251.

Neal, R. M., & Hinton, G. E. (1998). A view of the EM algorithm that justifies incremental, sparse, and other variants. In M. I. Jordan (Ed.), *Learning in graphical models* (pp. 355–368). Dordrecht, Netherlands: Kluwer.

Nerlich, B. (1989). The evolution of the concept of 'linguistic evolution' in the 19th and 20th century. *Lingua*, *77*, 101–112.

Nettle, D. (1999). Is the rate of linguistic change constant? *Lingua*, *108*, 119–136.

Nettle, D., & Dunbar, R. I. M. (1997). Social markers and the evolution of reciprocal exchange. *Current Anthropology*, *38*, 93–99.

Nevins, A. (2010). *Locality in vowel harmony*. Cambridge, MA: MIT Press.

Nevins, A., Pesetsky, D., & Rodrigues, C. (2007). *Pirahã exceptionality: A reassessment* [On-line]. Available from: http://ling.auf.net/lingBuzz/000411.

Newell, A., & Rosenbloom, P. S. (1981). Mechanisms of skill acquisition and the law of practice. In J. R. Anderson (Ed.), *Cognitive skills and their acquisition* (pp. 1–55). Hillsdale, NJ: Erlbaum.

Newmeyer, F. J. (1991). Functional explanation in linguistics and the origins of language. *Language & Communication*, *11*, 3–28.

Newmeyer, F. J. (2003). What can the field of linguistics tell us about the origin of language? In M. H. Christiansen & S. Kirby (Eds.), *Language evolution* (pp. 58–76). New York: Oxford University Press.

Nicol, J. L., Forster, K. I., & Veres, C. (1997). Subject-verb agreement processes in comprehension. *Journal of Memory and Language*, *36*, 569–587.

Nissen, M. J., & Bullemer, P. (1987). Attentional requirements of learning: Evidence from performance measures. *Cognitive Psychology*, *19*, 1–32.

Niyogi, P. (2006). *The computational nature of language learning and evolution*. Cambridge, MA: MIT Press.

Niyogi, P., & Berwick, R. C. (1996). A language learning model for finite parameter spaces. *Cognition*, *61*, 161–193.

Norman, D. A., & Shallice, T. (1986). Attention to action: Willed and automatic control of behavior. In R. J. Davidson, G. E. Schwartz, & D. Shapiro (Eds.), *Consciousness and self-regulation* (pp. 1–18). New York: Plenum Press.

Nosofsky, R. M. (1986). Attention, similarity, and the identification—categorization relationship. *Journal of Experimental Psychology. General*, *115*, 39–57.

Nottebohm, F. (1969). The "critical period" for song learning. *Ibis*, *111*, 386–387.

Noveck, I. A., & Reboul, A. (2008). Experimental pragmatics: A Gricean turn in the study of language. *Trends in Cognitive Sciences, 12*, 425–431.

Novick, J. M., Trueswell, J. C., & Thompson-Schill, S. L. (2005). Cognitive control and parsing: Re-examining the role of Broca's area in sentence comprehension. *Cognitive, Affective & Behavioral Neuroscience, 5*, 263–281.

Novogrodsky, R., & Friedmann, N. (2006). The production of relative clauses in syntactic SLI: A window to the nature of the impairment. *Advances in Speech Language Pathology, 8*, 364–375.

Nowak, M. A., Komarova, N. L., & Niyogi, P. (2001). Evolution of universal grammar. *Science, 291*, 114–118.

Nygaard, L. C., Cook, A. E., & Namy, L. L. (2009). Sound to meaning correspondences facilitate word learning. *Cognition, 112*, 181–186.

Nygaard, L. C., Sommers, M. S., & Pisoni, D. B. (1994). Speech perception as a talker-contingent process. *Psychological Science, 5*, 42–46.

O'Grady, W. (1997). *Syntactic development.* Chicago, IL: University of Chicago Press.

O'Grady, W. (2005). *Syntactic carpentry: An emergentist approach to syntax.* Mahwah, NJ: Erlbaum.

O'Grady, W. (2011). Relative clauses: Processing and acquisition. In E. Kidd (Ed.), *The acquisition of relative clauses: Processing, typology and function* (pp. 13–38). Amsterdam: John Benjamins.

O'Grady, W. (2013). The illusion of language acquisition. *Approaches to Bilingualism, 3*, 253–285.

O'Grady, W. (2015). Anaphora and the case for emergentism. In B. MacWhinney & W. O'Grady (Eds.), *The handbook of language emergence* (pp. 100–122). Hoboken, NJ: Wiley-Blackwell.

O'Regan, J. K., & Noë, A. (2001). A sensorimotor account of vision and visual consciousness. *Behavioral and Brain Sciences, 24*, 939–973.

Odling-Smee, F. J., Laland, K. N., & Feldman, M. W. (2003). *Niche construction: The neglected process in evolution.* Princeton, NJ: Princeton University Press.

OED Online (2013). Oxford: Oxford University Press (accessed July 5, 2013). http://www.oed.com.

Oh, Y., Pellegrino, F., Marsico, E., & Coupé, C. (2013) A quantitative and typological approach to correlating linguistic complexity. In T. Wielfaert, K. Heylen, & D. Speelman (Eds.), *Proceedings of the 5th Conference on Quantitative Investigations in Theoretical Linguistics* (pp. 71–75). Leuven, Belgium: University of Leuven.

Olson, K. R., & Spelke, E. S. (2008). Foundations of cooperation in young children. *Cognition, 108*, 222–231.

Onnis, L., & Christiansen, M. H. (2008). Lexical categories at the edge of the word. *Cognitive Science, 32*, 184–221.

Onnis, L., Monaghan, P., Richmond, K., & Chater, N. (2005). Phonology impacts segmentation in online speech processing. *Journal of Memory and Language, 53*, 225–237.

Orr, D. B., Friedman, H. L., & Williams, J. C. C. (1965). Trainability of listening comprehension of speeded discourse. *Journal of Educational Psychology*, *56*, 148–156.

Otake, T., Hatano, G., Cutler, A., & Mehler, J. (1993). Mora or syllable? Speech segmentation in Japanese. *Journal of Memory and Language*, *32*, 258–278.

Ozturk, O., Krehm, M., & Vouloumanos, A. (2013). Sound symbolism in infancy: evidence for sound—shape cross-modal correspondences in 4-month-olds. *Journal of Experimental Child Psychology*, *114*, 173–186.

Packard, M., & Knowlton, B. (2002). Learning and memory functions of the basal ganglia. *Annual Review of Neuroscience*, *25*, 563–593.

Pagel, M., Atkinson, Q. D., & Meade, A. (2007). Frequency of word-use predicts rates of lexical evolution throughout Indo-European history. *Nature*, *449*, 717–721.

Pallier, C., Christophe, A., & Mehler, J. (1997). Language-specific listening. *Trends in Cognitive Sciences*, *1*, 129–132.

Pani, J. R. (2000). Cognitive description and change blindness. *Visual Cognition*, *7*, 107–126.

Parker, A. R. (2006). Evolving the narrow language faculty: Was recursion the pivotal step? In A. Cangelosi, A. Smith, & K. Smith (Eds.), *Proceedings of the 6th International Conference on the Evolution of Language* (pp. 239–246). London: World Scientific Publishing.

Pashler, H. (1988). Familiarity and visual change detection. *Perception & Psychophysics*, *44*, 369–378.

Patel, A. D., Gibson, E., Ratner, J., Besson, M., & Holcomb, P. J. (1998). Processing syntactic relations in language and music: An event-related potential study. *Journal of Cognitive Neuroscience*, *10*, 717–733.

Patel, A. D., Iversen, J. R., Wassenaar, M., & Hagoort, P. (2008). Musical syntactic processing in agrammatic Broca's aphasia. *Aphasiology*, *22*, 776–789.

Pavani, F., & Turatto, M. (2008). Change perception in complex auditory scenes. *Perception & Psychophysics*, *70*, 619–629.

Pearlmutter, N. J., Garnsey, S. M., & Bock, K. (1999). Agreement processes in sentence comprehension. *Journal of Memory and Language*, *41*, 427–456.

Pearlmutter, N. J., & MacDonald, M. C. (1995). Individual differences and probabilistic constraints in syntactic ambiguity resolution. *Journal of Memory and Language*, *34*, 521–542.

Pellegrino, F., Coupé, C., & Marsico, E. (2011). A cross-language perspective on speech information rate. *Language*, *87*, 539–558.

Pelucchi, B., Hay, J. F., & Saffran, J. R. (2009). Statistical learning in a natural language by 8- month odl infants. *Child Development*, *80*, 674–685.

Peña, M., Bonnatti, L., Nespor, M., & Mehler, J. (2002). Signal-driven computations in speech processing. *Science*, *298*, 604–607.

Pennisi, E. (2004). The first language? *Science*, *303*, 1319–1320.

Percival, W. K. (1987). Biological analogy in the study of languages before the advent of comparative grammar. In H. M. Hoenigswald & L. F. Wiener (Eds.), *Biological metaphor and cladistic classification* (pp. 3–38). Philadelphia, PA: University of Pennsylvania Press.

Pereira, P., & Schabes, Y. (1992). Inside—outside reestimation from partially bracketed corpora. *Proceedings of the 30th annual meeting on Association for Computational Linguistics* (pp. 128–135). Stroudsburg, PA: Association for Computational Linguistics.

Perniss, P., Thompson, R. L., & Vigliocco, G. (2010). Iconicity as a general property of language: evidence from spoken and signed languages. *Frontiers in Psychology, 1*, 227. doi:.10.3389/fpsyg.2010.00227

Perry, G., Dominy, N., Claw, K., Lee, A., Fiegler, H., Redon, R., et al. (2007). Diet and the evolution of human amylase gene copy number variation. *Nature Genetics, 39*, 1256–1260.

Peterfalvi, J. M., & Locatelli, F (1971). L'acceptabilité des phrases. [The acceptability of sentences]. *L'Année Psychologique, 71*, 417–427.

Petersson, K. M. (2005). On the relevance of the neurobiological analogue of the finite state architecture. *Neurocomputing, 65–66*, 825–832.

Petersson, K. M., Folia, V., & Hagoort, P. (2012). What artificial grammar learning reveals about the neurobiology of syntax. *Brain and Language, 120*, 83–95.

Petersson, K. M., Forkstam, C., & Ingvar, M. (2004). Artificial syntactic violations activate Broca's region. *Cognitive Science, 28*, 383–407.

Phillips, B. S. (2006). *Word frequency and lexical diffusion*. New York: Palgrave Macmillan.

Phillips, C. (1996). Merge right: An approach to constituency conflicts. In J. Camacho, L. Choueiri & M. Watanabe (Eds.) *Proceedings of West Coast Conference on Formal Linguistics, Volume 15* (pp. 381–395), Chicago, IL: University of Chicago Press.

Phillips, C. (2003). Linear order and constituency. *Linguistic Inquiry, 34*, 37–90.

Piantadosi, S., Tily, H., & Gibson, E. (2012). The communicative function of ambiguity in language. *Cognition, 122*, 280–291.

Piattelli-Palmarini, M. (1989). Evolution, selection and cognition: From "learning" to parameter setting in biology and in the study of language. *Cognition, 31*, 1–44.

Piattelli-Palmarini, M. (1994). Ever since language and learning: Afterthoughts on the Piaget-Chomsky debate. *Cognition, 50*, 315–346.

Piattelli-Palmarini, M., Hancock, R., & Bever, T. G. (2008). Language as ergonomic perfection. *Behavioral and Brain Sciences, 31*, 530–531.

Piattelli-Palmarini, M., & Uriagereka, J. (2004). The immune syntax: The evolution of the language virus. In L. Jenkins (Ed.), *Variations and universals in biolinguistics* (pp. 341–377). Amsterdam: Elsevier.

Pickering, M. J., & Branigan, H. P. (1998). The representation of verbs: Evidence from syntactic priming in language production. *Journal of Memory and Language, 39*, 633–651.

Pickering, M. J., & Garrod, S. (2004). Toward a mechanistic psychology of dialogue. *Behavioral and Brain Sciences, 27*, 169–226.

Pickering, M. J., & Garrod, S. (2007). Do people use language production to make predictions during comprehension? *Trends in Cognitive Sciences, 11*, 105–110.

Pickering, M. J., & Garrod, S. (2013). An integrated theory of language production and comprehension. *Behavioral and Brain Sciences, 36*, 1–19.

Pierrehumbert, J. (2002). *Word-specific phonetics*. Berlin: Mouton de Gruyter.

Pine, J. M., Freudenthal, D., Krajewski, G., & Gobet, F. (2013). Do young children have adult-like syntactic categories? Zipf's law and the case of the determiner. *Cognition, 127*, 345–360.

Pinker, S. (1979). Formal models of language learning. *Cognition, 7*, 217–283.

Pinker, S. (1984). *Language learnability and language development*. Cambridge, MA: Harvard University Press.

Pinker, S. (1994). *The language instinct: How the mind creates language*. New York, NY: William Morrow and Company.

Pinker, S. (1997). *How the mind works*. New York: W. W. Norton.

Pinker, S. (1999). *Words and rules*. New York: Basic Books.

Pinker, S. (2003 a). *The blank slate: The modern denial of human nature*. New York: Penguin.

Pinker, S. (2003 b). Language as an adaptation to the cognitive niche. In M. H. Christiansen, & S. Kirby (Eds.), *Language evolution* (pp. 16–37). Oxford: Oxford University Press.

Pinker, S., & Bloom, P. (1990). Natural language and natural selection. *Behavioral and Brain Sciences, 13*, 707–727.

Pinker, S., & Jackendoff, R. (2005). The faculty of language: What's special about it? *Cognition, 95*, 201–236.

Pinker, S., & Jackendoff, R. (2009). The components of language: What's specific to language, and what's specific to humans? In M. H. Christiansen, C. Collins, & S. Edelman (Eds.), *Language universals* (pp. 126–151). New York: Oxford University Press.

Pinker, S., & Prince, A. (1988). On language and connectionism: Analysis of a parallel distributed processing model of language acquisition. *Cognition, 28*, 73–193.

Pomerantz, J. R., & Kubovy, M. (1986). Theoretical approaches to perceptual organization: Simplicity and likelihood principles. In K. R. Boff, L. Kaufman, & J. P. Thomas (Eds.) *Handbook of perception and human performance. Volume 2: Cognitive processes and performance* (pp. 36-1–36-46). New York: Wiley.

Potter, M. C., & Lombardi, L. (1998). Syntactic priming in immediate recall of sentences. *Journal of Memory and Language, 38*, 265–282.

Poulet, J. F. A., & Hedwig, B. (2006). New insights into corollary discharges mediated by identified neural pathways. *Trends in Neurosciences, 30*, 14–21.

Powell, A., & Peters, R. G. (1973). Semantic clues in comprehension of novel sentences. *Psychological Reports, 32*, 1307–1310.

Pratt-Hartmann, I. (2010). Computational complexity in natural language. In A. Clark, C. Fox, & S. Lappin (Eds.), *The handbook of computational linguistics and natural language processing* (pp. 43–73). Oxford: Blackwell-Wiley.

Premack, D. (1985). 'Gavagai!' or the future history of the animal language controversy. *Cognition, 19*, 207–296.

Pulman, S. G. (1985). A parser that doesn't. *Proceedings of the 2nd European Meeting of the Association for Computational Linguistics* (pp. 128–135). Geneva: Association for Computational Linguistics.

Pulman, S. G. (1986). Grammars, parsers, and memory limitations. *Language and Cognitive Processes, 2*, 197–225.

Pylyshyn, Z. W. (1973). The role of competence theories in cognitive psychology. *Journal of Psycholinguistic Research, 2*, 21–50.

Quine, W. V. O. (1960). *Word and object.* Cambridge, MA: MIT Press.

Quirk, R. (1960). Towards a description of English Usage. *Transactions of the Philological Society, 59*, 40–61.

Ramachandran, V. S., & Hubbard, E. M. (2001). Synaesthesia—A window into perception, thought and language. *Journal of Consciousness Studies, 8*, 3–34.

Ratcliff, R. (1990). Connectionist models of recognition memory: constraints imposed by learning and forgetting functions. *Psychological Review, 97*, 285–308.

Reali, F. (2014). Frequency affects object relative clause processing: Some evidence in favor of usage-based accounts. *Language Learning, 64*, 685–714.

Reali, F., & Christiansen, M. H. (2005). Uncovering the richness of the stimulus: Structure dependence and indirect statistical evidence. *Cognitive Science, 29*, 1007–1028.

Reali, F., & Christiansen, M. H. (2007 a). Processing of relative clauses is made easier by frequency of occurrence. *Journal of Memory and Language, 57*, 1–23.

Reali, F., & Christiansen, M. H. (2007 b). Word chunk frequencies affect the processing of pronominal object-relative clauses. *Quarterly Journal of Experimental Psychology, 60*, 161–170.

Reali, F., & Christiansen, M. H. (2009). Sequential learning and the interaction between biological and linguistic adaptation in language evolution. *Interaction Studies, 10*, 5–13.

Reali, F., Chater, N., & Christiansen, M. H. (2014). The paradox of linguistic complexity and community size. In E. A. Cartmill, S. Roberts, H. Lyn, & H. Cornish (Eds.), *Proceedings of the 10th International Conference on the Evolution of Language* (pp. 270–277) Singapore: World Scientific.

Reali, F., Christiansen, M. H., & Monaghan, P. (2003). Phonological and distributional cues in syntax acquisition: Scaling up the connectionist approach to multiple-cue integration. In *Proceedings of the 25th Annual Conference of the Cognitive Science Society* (pp. 970–975). Mahwah, NJ: Lawrence Erlbaum.

Reber, A. (1967). Implicit learning of artificial grammars. *Journal of Verbal Learning and Verbal Behavior, 6*, 855–863.

Redington, M., Chater, N., & Finch, S. (1998). Distributional information: A powerful cue for acquiring syntactic categories. *Cognitive Science, 22*, 425–469.

Reeder, P. A. (2004). *Language learnability and the evolution of word order universals: Insights from artificial grammar learning*. Honors thesis, Department of Psychology, Cornell University, Ithaca, NY.

Reich, P. (1969). The finiteness of natural language. *Language*, *45*, 831–843.

Regier, T., Kay, P., & Khetarpal, N. (2007). Color naming reflects optimal partitions of color space. *Proceedings of the National Academy of Sciences*, *104*, 1436–1441.

Reilly, J., & Kean, J. (2007). Formal distinctiveness of high- and low-imageability nouns: Analyses and theoretical implications. *Cognitive Science*, *31*, 1–12.

Reilly, J., Westbury, C., Kean, J., & Peelle, J. E. (2012). Arbitrary symbolism in natural language revisited: When word forms carry meaning. *PLoS One*, *7*(8), e42286.

Reimers-Kipping, S., Hevers, W., Pääbo, S., & Enard, W. (2011). Humanized Foxp2 specifically affects cortico-basal ganglia circuits. *Neuroscience*, *175*, 75–84.

Reinhart, T. (1983). *Anaphora and semantic interpretation*. Chicago: Chicago University Press.

Remez, R. E., Fellowes, J. M., & Rubin, P. E. (1997). Talker identification based on phonetic information. *Journal of Experimental Psychology: Human Perception and Performance*, *23*, 651–666.

Remez, R. E., Ferro, D. F., Dubowski, K. R., Meer, J., Broder, R. S., & Davids, M. L. (2010). Is desynchrony tolerance adaptable in the perceptual organization of speech? *Attention, Perception & Psychophysics*, *72*, 2054–2058.

Rensink, R. A. (2005). Change blindness. In L. Itti, G. Rees, & J. K. Tsotsos (Eds.), *Neurobiology of attention* (pp. 76–81). San Diego, CA: Elsevier Academic Press.

Rensink, R. A., O'Regan, J. K., & Clark, J. J. (1997). To see or not to see: The need for attention to perceive changes in scenes. *Psychological Science*, *8*, 368–373.

Reuland, E. J. (2008). Why neo-adaptationism fails. *Behavioral and Brain Sciences*, *31*, 531–532.

Reuland, E. J. (2011). *Anaphora and language design*. Cambridge, MA: MIT Press.

Richerson, P. J., & Boyd, R. (2005). *Not by genes alone: How culture transformed human evolution*. Chicago: Chicago University Press.

Richerson, P. J., & Christiansen, M. H. (Eds.). (2013). *Cultural evolution: Society, technology, language and religion*. Cambridge, MA: MIT Press.

Riordan, B., & Jones, M. N. (2011). Redundancy in perceptual and linguistic experience: Comparing feature-based and distributional models of semantic representation. *Topics in Cognitive Science*, *3*, 303–345.

Ritt, N. (2004). *Selfish sounds and linguistic evolution: A Darwinian approach to language change*. Cambridge: Cambridge University Press.

Roland, D., Elman, J., & Ferreira, V. (2006). Why is that? Structural prediction and ambiguity resolution in a very large corpus of English sentences. *Cognition*, *98*, 245–272.

Roland, D., Mauner, G., O'Meara, C., & Yun, H. (2012). Discourse expectations and relative clause processing. *Journal of Memory and Language*, *66*, 479–508.

Rossel, S., Corlija, J., & Schuster, S. (2002). Predicting three-dimensional target motion: How archer fish determine where to catch their dislodged prey. *Journal of Experimental Biology*, *205*, 3321–3326.

Roth, F. P. (1984). Accelerating language learning in young children. *Child Language*, *11*, 89–107.

Rumelhart, D. E., & McClelland, J. L. (1986). On learning the past tenses of English verbs. In J. L. McClelland & D. E. Rumelhart, & the PDP Research Group (Eds.), *Parallel distributed processing: Explorations in the microstructure of cognition. Volume2: Psychological and biological models* (pp. 216–271). Cambridge, MA: MIT Press.

Saad, D. (Ed.). (1998). *On-line learning in neural networks.* Cambridge: Cambridge University Press.

Saffran, J. R. (2001). The use of predictive dependencies in language learning. *Journal of Memory and Language*, *44*, 493–515.

Saffran, J. R. (2002). Constraints on statistical language learning. *Journal of Memory and Language*, *47*, 172–196.

Saffran, J. R. (2003). Statistical language learning: Mechanisms and constraints. *Current Directions in Psychological Science*, *12*, 110–114.

Saffran, J. R., Aslin, R. N., & Newport, E. L. (1996). Statistical learning by 8-month-old infants. *Science*, *274*, 1926–1928.

Saffran, J. R., Hauser, M., Seibel, R., Kapfhamer, J., Tsao, F., & Cushman, F. (2008). Grammatical pattern learning by human infants and cotton-top tamarin monkeys. *Cognition*, *107*, 479–500.

Sag, I. A., & Pollard, C. J. (1987). *Head-driven phrase structure grammar: An informal synopsis.* CSLI Report 87–79. Stanford, CA: Stanford University.

Sagarra, N., & Herschensohn, J. (2010). The role of proficiency and working memory in gender and number processing in L1 and L2 Spanish. *Lingua*, *120*, 2022–2039.

Sahin, N. T., Pinker, S., Cash, S. S., Schomer, D., & Halgren, E. (2009). Sequential processing of lexical, grammatical, and articulatory information within Broca's area. *Science*, *326*, 445.

Sandler, W. (2012). Dedicated gestures and the emergence of sign language. *Gesture*, *12*, 265–307.

Sandler, W., Meir, I., Padden, C., & Aronoff, M. (2005). The emergence of grammar: Systematic structure in a new language. *Proceedings of the National Academy of Sciences*, *102*, 2661–2665.

Sanford, A. J., & Sturt, P. (2002). Depth of processing in language comprehension: Not noticing the evidence. *Trends in Cognitive Sciences*, *6*, 382–386.

Schelling, T. C. (1960). *The strategy of conflict.* Cambridge, MA: Harvard University Press.

Schleicher, A. (1863). *Die Darwinsche Theorie und die Sprachwissenschaft.* Weimar: Böhlau.

Schlesinger, I. M. (1975). Why a sentence in which a sentence is embedded is difficult. *Linguistics*, *153*, 53–66.

Schlosser, G., & Wagner, G. P. (Eds.). (2004). *Modularity in development and evolution.* Chicago, IL: University of Chicago Press.

Schmidt, R. A., & Wrisberg, C. A. (2004). *Motor learning and performance* (3rd ed.). Champaign, IL: Human Kinetics.

Schmidtke, D. S., Conrad, M., & Jacobs, A. M. (2014). Phonological iconicity. *Frontiers in Psychology*, *5*. doi:.10.3389/fpsyg.2014.00080

Schoenemann, P. T. (1999). Syntax as an emergent characteristic of the evolution of semantic complexity. *Minds and Machines*, *9*, 309–346.

Schoenemann, P. T. (2009). Evolution of brain and language. *Language Learning*, *59* (Suppl. 1), 162–186.

Schreiweis, C., Bornschein, U., Burguière, E., Kerimoglu, C., Schreiter, S., Dannemann, M., et al. (2014). Humanized Foxp2 accelerates learning by enhancing transitions from declarative to procedural performance. *Proceedings of the National Academy of Sciences*, *111*, 14253–14258.

Schultz, W., Dayan, P., & Montague, P. R. (1997). A neural substrate of prediction and reward. *Science*, *275*, 1593–1599.

Scott-Brown, K. C., Baker, M. R., & Orbach, H. S. (2000). Comparison blindness. *Visual Cognition*, *7*, 253–267.

Searcy, W. A., & Nowicki, S. (2001). *The evolution of animal communication: Reliability and deception in signaling systems*. Princeton, NJ: Princeton University Press.

Sedlmeier, P. E., & Betsch, T. E. (Eds.). (2002). *Etc. Frequency processing and cognition*. New York: Oxford University Press.

Seidenberg, M. S. (1985). The time course of phonological code activation in two writing systems. *Cognition*, *19*, 1–30.

Seidenberg, M. S. (1997). Language acquisition and use: Learning and applying probabilistic constraints. *Science*, *275*, 1599–1603.

Seidenberg, M. S., & MacDonald, M. (2001). Constraint-satisfaction in language acquisition. In M. H. Christiansen & N. Chater (Eds.), *Connectionist psycholinguistics* (pp. 281–318). Westport, CT: Ablex.

Senghas, A., Kita, S., & Özyürek, A. (2004). Children creating core properties of language: Evidence from an emerging sign language in Nicaragua. *Science*, *305*, 1779–1782.

Sereno, J. A., & Jongman, A. (1990). Phonological and form-class relations in the lexicon. *Journal of Psycholinguistic Research*, *19*, 387–404.

Sereno, J. A., & Jongman, A. (1995). Acoustic correlates of grammatical class. *Language and Speech*, *38*, 57–76.

Sereno, M. I. (1991). Four analogies between biological and cultural/linguistic evolution. *Journal of Theoretical Biology*, *151*, 467–507.

Servan-Schreiber, D., Cleeremans, A., & McClelland, J. L. (1991). Graded state machines: The representation of temporal dependencies in simple recurrent networks. *Machine Learning*, *7*, 161–193.

Shadmehr, R., & Wise, S. P. (2005). *The computational neurobiology of reaching and pointing: A foundation for motor learning*. Cambridge, MA: MIT Press.

Shafer, V. L., Shucard, D. W., Shucard, J. L., & Gerken, L. A. (1998). An electrophysiological study of infants' sensitivity to the sound patterns of English speech. *Journal of Speech, Language, and Hearing Research: JSLHR*, *41*, 874–886.

Shannon, C. (1948). A mathematical theory of communication. *Bell System Technical Journal*, *27*, 623–656.

Shi, R., Morgan, J., & Allopenna, P. (1998). Phonological and acoustic bases for earliest grammatical category assignment: A cross-linguistic perspective. *Journal of Child Language*, *25*, 169–201.

Shi, R., Werker, J. F., & Morgan, J. L. (1999). Newborn infants' sensitivity to perceptual cues to lexical and grammatical words. *Cognition*, *72*, B11–B21.

Shieber, S. (1985). Evidence against the context-freeness of natural language. *Linguistics and Philosophy*, *8*, 333–343.

Siegel, D. (1978). The adjacency constraint and the theory of morphology. *North East Linguistics Society*, *8*, 189–197.

Sigman, M., & Dehaene, S. (2005). Parsing a cognitive task: A characterization of the mind's bottleneck. *PLoS Biology*, *3*(2), e37. doi:.10.1371/journal.pbio.0030037

Silbert, L. J., Honey, C. J., Simony, E., Poeppel, D., & Hasson, U. (2014). Coupled neural systems underlie the production and comprehension of naturalistic narrative speech. *Proceedings of the National Academy of Sciences*, *111*, E4687–E4696.

Simon, H. A. (1956). Rational choice and the structure of the environment. *Psychological Review*, *63*, 129–138.

Simoncelli, E. P., & Olshausen, B. A. (2001). Natural image statistics as neural representation. *Annual Review of Neuroscience*, *24*, 1193–1215.

Simons, D. J., & Chabris, C. F. (1999). Gorillas in our midst: Sustained inattentional blindness for dynamic events. *Perception*, *28*, 1059–1074.

Simons, D. J., & Levin, D. T. (1998). Failure to detect changes to people during a real-world interaction. *Psychonomic Bulletin & Review*, *5*, 644–649.

Simons, D. J., & Rensink, R. A. (2005). Change blindness: Past, present, and future. *Trends in Cognitive Sciences*, *9*, 16–20.

Siyanova-Chanturia, A., Conklin, K., & Van Heuven, W. J. B. (2011). Seeing a phrase "time and again" matters: The role of phrasal frequency in the processing of multiword sequences. *Journal of Experimental Psychology: Learning, Memory, and Cognition*, *37*, 776–784.

Slobin, D. I. (1973). Cognitive prerequisites for the development of grammar. In C. A. Ferguson & D. I. Slobin (Eds.), *Studies of child language development* (pp. 175–208). New York: Holt, Rinehart & Winston.

Slobin, D. I., & Bever, T. G. (1982). Children use canonical sentence schemas: A crosslinguistic study of word order and inflections. *Cognition*, *12*, 229–265.

Sloboda, J. A. (1985). *The musical mind: The cognitive psychology of music*. Oxford: Oxford University Press.

Sloutsky, V. M. (2010). From perceptual categories to concepts: What develops? *Cognitive Science*, *34*, 1244–1286.

Smith, A. D. M. (2014). Models of language evolution and change. *Wiley Interdisciplinary Reviews: Cognitive Science*, *5*, 281–293.

Smith, K. (2002). Natural selection and cultural selection in the evolution of communication. *Adaptive Behavior*, *10*, 25–44.

Smith, K. (2004). The evolution of vocabulary. *Journal of Theoretical Biology, 228*, 127–142.

Smith, K., Brighton, H., & Kirby, S. (2003). Complex systems in language evolution: the cultural emergence of compositional structure. *Advances in Complex Systems, 6*, 537–558.

Smith, K., & Kirby, S. (2008). Cultural evolution: implications for understanding the human language faculty and its evolution. *Philosophical Transactions of the Royal Society B: Biological Sciences, 363*, 3591–3603.

Smith, K., & Wonnacott, E. (2010). Eliminating unpredictable variation through iterated learning. *Cognition, 116*, 444–449.

Smith, L. B., Colunga, E., & Yoshida, H. (2010). Knowledge as process: Contextually-cued attention and early word learning. *Cognitive Science, 34*, 1287–1314.

Smith, M., & Wheeldon, L. (1999). High-level processing scope in spoken sentence production. *Cognition, 73*, 205–246.

Snedeker, J., & Trueswell, J. (2003). Using prosody to avoid ambiguity: Effects of speaker awareness and referential context. *Journal of Memory and Language, 48*, 103–130.

Solan, Z., Horn, D., Ruppin, E., & Edelman, S. (2005). Unsupervised learning of natural languages. *Proceedings of the National Academy of Sciences, 102*, 11629–11634.

Spector, F., & Maurer, D. (2009). Synesthesia: A new approach to understanding the development of perception. *Developmental Psychology, 45*, 175–189.

Sperber, D., & Wilson, D. (1986). *Relevance*. Oxford: Blackwell.

Sperling, G. (1960). The information available in brief visual presentations. *Psychological Monographs, 74* (Whole No. 11).

Spieler, D. H., & Balota, D. A. (1997). Bringing computational models of word naming down to the item level. *Psychological Science, 8*, 411–416.

St. Clair, M. C., Monaghan, P., & Christiansen, M. H. (2010). Learning grammatical categories from distributional cues: Flexible frames for language acquisition. *Cognition, 116*, 341–360.

St. Clair, M. C., Monaghan, P., & Ramscar, M. (2009). Relationships between language structure and language learning: The suffixing preference and grammatical categorization. *Cognitive Science, 33*, 1317–1329.

Stabler, E. P. (1994). The finite connectivity of linguistic structure. In C. Clifton, L. Frazier, & K. Rayner (Eds.), *Perspectives on sentence processing* (pp. 303–336). Hillsdale, NJ: Lawrence Erlbaum Associates.

Stabler, E. P. (2009). Computational models of language universals: Expressiveness, learnability and consequences. In M. H. Christiansen, C. Collins, & S. Edelman (Eds.), *Language universals* (pp. 200–223). New York: Oxford University Press.

Stallings, L., MacDonald, M., & O'Seaghdha, P. (1998). Phrasal ordering constraints in sentence production: phrase length and verb disposition in heavy-NP shift. *Journal of Memory and Language, 39*, 392–417.

Stanovich, K. E., & Cunningham, A. E. (1992). Studying the consequences of literacy within a literate society: The cognitive correlates of print exposure. *Memory & Cognition, 20*, 51–68.

Stanovich, K. E., & Cunningham, A. E. (1993). Where does knowledge come from? Specific associations between print exposure and information acquisition. *Journal of Educational Psychology, 85*, 211–229.

Staub, A., & Clifton, C., Jr. (2006). Syntactic prediction in language comprehension: Evidence from either … or. *Journal of Experimental Psychology: Learning, Memory, and Cognition, 32*, 425–436.

Steedman, M. (1987). Combinatory grammars and parasitic gaps. *Natural Language and Linguistic Theory, 5*, 403–439.

Steedman, M. (2000). *The syntactic process.* Cambridge, MA: MIT Press.

Steels, L. (2003). Evolving grounded communication for robots. *Trends in Cognitive Sciences, 7*, 308–312.

Stephens, D. W., Brown, J. S., & Ydenberg, R. C. (2007). *Foraging: Behavior and ecology.* Chicago: University of Chicago Press.

Stephens, G. J., Honey, C. J., & Hasson, U. (2013). A place for time: the spatiotemporal structure of neural dynamics during natural audition. *Journal of Neurophysiology, 110*, 2019–2026.

Stephens, G. J., Silbert, L. J., & Hasson, U. (2010). Speaker—listener neural coupling underlies successful communication. *Proceedings of the National Academy of Sciences, 107*, 14425–14430.

Stephenson, A. G., Mulville, D. R., Bauer, F. H., Dukeman, G. A., Norvig, P., LaPiana, L. S., et al. (1999). *Mars Climate Orbiter Mishap Investigation Board Phase I Report.* Washington, DC: NASA.

Stevick, R. D. (1963). The biological model and historical linguistics. *Language, 39*, 159–169.

Stivers, T., Enfield, N. J., Brown, P., Englert, C., Hayashi, M., Heinemann, T., (2009). Universals and cultural variation in turn-taking in conversation. *Proceedings of the National Academy of Sciences, 106*, 10587–10592.

Stolz, W. S. (1967). A study of the ability to decode grammatically novel sentences. *Journal of Verbal Learning and Verbal Behavior, 6*, 867–873.

Storkel, H. L. (2001). Learning new words: Phonotactic probability in language development. *Journal of Speech, Language, and Hearing Research: JSLHR, 44*, 1321–1337.

Storkel, H. L. (2003). Learning new words II: Phonotactic probability in verb learning. *Journal of Speech, Language, and Hearing Research: JSLHR, 46*, 1312–1323.

Street, J., & Dąbrowska, E. (2010). More individual differences in language attainment: How much do adult native speakers of English know about passives and quantifiers? *Lingua, 120*, 2080–2094.

Studdert-Kennedy, M., & Goldstein, L. (2003). Launching language: The gestural origin of discrete infinity. In M. H. Christiansen & S. Kirby (Eds.), *Language evolution* (pp. 235–254). New York: Oxford University Press.

Studdert-Kennedy, M. (1986). Some developments in research on language behavior. In N. J. Smelser & D. R. Gerstein (Eds.), *Behavioral and social science: Fifty years of discovery: In commemoration of the fiftieth anniversary of the "Ogburn Report: Recent Social Trends in the United States"* (pp. 208–248). Washington, DC: National Academy Press.

Sturt, P., & Crocker, M. W. (1996). Monotonic syntactic processing: A cross-linguistic study of attachment and reanalysis. *Language and Cognitive Processes*, *11*, 449–494.

Swaab, T., Brown, C. M., & Hagoort, P. (2003). Understanding words in sentence contexts: The time course of ambiguity resolution. *Brain and Language*, *86*, 326–343.

Tallal, P., Miller, S., Bedi, G., Byma, G., Wang, X., Nagarajan, S., et al. (1996). Language comprehension in language-learning impaired children improved with acoustically modified speech. *Science*, *271*, 77–81.

Tallerman, M., Newmeyer, F., Bickerton, D., Nouchard, D., Kann, D., & Rizzi, L. (2009). What kinds of syntactic phenomena must biologists, neurobiologists, and computer scientists try to explain and replicate. In D. Bickerton & E. Szathmáry (Eds.), *Biological foundations and origin of syntax* (pp. 135–157). Cambridge, MA: MIT Press.

Tanenhaus, M. K., Spivey-Knowlton, M. J., Eberhard, K. M., & Sedivy, J. C. (1995). Integration of visual and linguistic information in spoken language comprehension. *Science*, *268*, 1632–1634.

Tanenhaus, M. K., & Trueswell, J. C. (1995). Sentence comprehension. In J. Miller & P. Eimas (Eds.), *Handbook of cognition and perception* (pp. 217–262). San Diego, CA: Academic Press.

Tenenbaum, J. B. (1999). Bayesian modeling of human concept learning. In M. Kearns, M. S. Solla, & D. Cohn (Ed.), *Advances in Neural Information Processing Systems* (Vol. 11) (pp. 59–65). Cambridge, MA: MIT Press.

Tenenbaum, J. B., Kemp, C., & Shafto, P. (2007). Theory-based Bayesian models of inductive reasoning. In A. Feeney & E. Heit (Eds.), *Inductive reasoning* (pp. 167–204).Cambridge: Cambridge University Press.

Theakston, A. L. (2004). The role of entrenchment in children's and adults' performance on grammaticality judgment tasks. *Cognitive Development*, *19*, 15–34.

Thornton, R., MacDonald, M. C., & Gil, M. (1999). Pragmatic constraint on the interpretation of complex noun phrases in Spanish and English. *Journal of Experimental Psychology: Learning, Memory, and Cognition*, *25*, 1347–1365.

Tomalin, M. (2011). Syntactic structures and recursive devices: A legacy of imprecision. *Journal of Logic Language and Information*, *20*, 297–315.

Tomasello, M. (1992). *First verbs: A case study of early grammatical development*. Cambridge: Cambridge University Press.

Tomasello, M. (2000). Do you children have adult syntactic competence? *Cognition*, *74*, 209–253.

Tomasello, M. (2003). *Constructing a language: A usage-based theory of language acquisition*. Cambridge, MA: Harvard University Press.

Tomasello, M. (2004). What kind of evidence could refute the UG hypothesis? *Studies in Language*, *28*, 642–644.

Tomasello, M., Carpenter, M., Call, J., Behne, T., & Moll, H. (2005). Understanding and sharing intentions: The origins of cultural cognition. *Behavioral and Brain Sciences*, *28*, 675–691.

Tomblin, J. B., Mainela-Arnold, E., & Zhang, X. (2007). Procedural learning in adolescents with and without specific language impairment. *Language Learning and Development*, *3*, 269–293.

Tomblin, J. B., Records, N. L., & Zhang, X. (1996). A system for the diagnosis of specific language impairment in kindergarten children. *Journal of Speech and Hearing Research*, *39*, 1284–1294.

Tomblin, J. B., Shriberg, L., Murray, J., Patil, S., & Williams, C. (2004). Speech and language characteristics associated with a 7/13 translocation involving *FOXP2. American Journal of Medical Genetics*, *130B*, 97.

Tomblin, J. B., & Zhang, X. (1999). Are children with SLI a unique group of language learners? In H. Tager-Flusberg (Ed.), *Neurodevelopmental disorders: Contributions to a new framework from the cognitive neurosciences* (pp. 361–382). Cambridge, MA: The MIT Press.

Tooby, J., & Cosmides, L. (2005). Conceptual foundations of evolutionary psychology. In A. Rosenberg & R. Arp (Eds.), *Philosophy of biology: An anthology* (pp. 375–386). Hoboken, NJ: Wiley-Blackwell.

Townsend, D. J., & Bever, T. G. (2001). *Sentence comprehension: The integration of habits and rules*. Cambridge, MA: MIT Press.

Treisman, A. (1964). Selective attention in man. *British Medical Bulletin*, *20*, 12–16.

Treisman, A., & Schmidt, H. (1982). Illusory conjunctions in the perception of objects. *Cognitive Psychology*, *14*, 107–141.

Tremblay, A., & Baayen, H. (2010). Holistic processing of regular four-word sequences: A behavioral and ERP study of the effects of structure, frequency, and probability on immediate free recall. In D. Wood (Ed.), *Perspectives on formulaic language* (pp. 151–167). New York: Continuum International Publishing.

Tremblay, A., Derwing, B., Libben, G., & Westbury, C. (2011). Processing advantages of lexical bundles: Evidence from self-paced reading and sentence recall tasks. *Language Learning*, *61*, 569–613.

Trivers, R. L. (1971). The evolution of reciprocal altruism. *Quarterly Review of Biology*, *46*, 35–57.

Trotzke, A., Bader, M., & Frazier, L. (2013). Third factors and the performance interface in language design. *Biolinguistics*, *7*, 1–34.

Trudgill, P. (2011). *Sociolinguistic typology: Social determinants of linguistic complexity*. New York: Oxford University Press.

Trueswell, J. C., Sekerina, I., Hill, N. M., & Logrip, M. L. (1999). The kindergarten path effect: Studying on-line sentence processing in young children. *Cognition*, *73*, 89–134.

Trueswell, J. C., & Tanenhaus, M. K. (1994). Towards a lexicalist framework of constraint-based syntactic ambiguity resolution. In C. Clifton, L. Frazier, & K. Rayner (Eds.), *Perspectives on sentence processing* ((pp. 155–179). Hillsdale, NJ: Erlbaum.

Trueswell, J. C., Tanenhaus, M. K., & Garnsey, S. M. (1994). Semantic influences on parsing: Use of thematic role information in syntactic ambiguity resolution. *Journal of Memory and Language*, *33*, 285–318.

Trueswell, J. C., Tanenhaus, M. K., & Kello, C. (1993). Verb-specific constraints in sentence processing: Separating effects of lexical preference from garden-paths. *Journal of Experimental Psychology: Learning, Memory, and Cognition*, *19*, 528–553.

Tsutsui, K., Taira, M., & Sakata, H. (2005). Neural mechanisms of three-dimensional vision. *Neuroscience Research*, *51*, 221–229.

Tversky, A., & Kahneman, D. (1973). Availability: A heuristic for judging frequency and probability. *Cognitive Psychology*, *5*, 207–233.

Twyman, A. D., & Newcombe, N. S. (2010). Five reasons to doubt the existence of a geometric module. *Cognitive Science*, *34*, 1315–1356.

Uddén, J., Folia, V., Forkstam, C., Ingvar, M., Fernández, G., Overeem, S., et al. (2008). The inferior frontal cortex in artificial syntax processing: An rTMS study. *Brain Research*, *1224*, 69–78.

Uehara, K., & Bradley, D. (1996). The effect of *-ga* sequences on processing Japanese multiply center-embedded sentences. In *11th Pacific-Asia conference on language, information, and computation* (pp. 187–196). Seoul: Kyung Hee University.

Ullman, M. T. (2004). Contributions of neural memory circuits to language: The declarative/procedural model. *Cognition*, *92*, 231–270.

Ullman, M. T., & Pierpont, E. I. (2005). Specific language impairment is not specific to language: The procedural deficit hypothesis. *Cortex*, *41*, 399–433.

Ullman, S. (1979). *The interpretation of visual motion*. Cambridge, MA: MIT Press.

Valian, V., & Coulson, S. (1988). Anchor points in language learning: The role of marker frequency. *Journal of Memory and Language*, *27*, 71–86.

Valian, V., Solt, S., & Stewart, J. (2009). Abstract categories or limited-scope formulae? The case of children's determiners. *Journal of Child Language*, *36*, 743–778.

Van Berkum, J. J., Brown, C. M., Zwitserlood, P., Kooijman, V., & Hagoort, P. (2005). Anticipating upcoming words in discourse: evidence from ERPs and reading times. *Journal of Experimental Psychology: Learning, Memory, and Cognition*, *31*, 443–467.

van den Brink, D., Brown, C. M., & Hagoort, P. (2001). Electrophysiological evidence for early contextual influences during spoken-word recognition: N200 versus N400 effects. *Journal of Cognitive Neuroscience*, *13*, 967–985.

Van der Lely, H. K. J., & Battell, J. (2003). WH-movement in children with grammatical SLI: A test of the RDDR hypothesis. *Language*, *79*, 153–181.

Van der Lely, H. K. J., & Pinker, S. (2014). The biological basis of language: Insight from developmental grammatical impairments. *Trends in Cognitive Sciences*, *18*, 586–595.

Van Dyke, J. A., & Johns, C. L. (2012). Memory Interference as a determinant of language comprehension. *Language and Linguistics Compass*, *6*, 193–211.

Van Everbroeck, E. (1999). Language type frequency and learnability: A connectionist appraisal. In *Proceedings of the 21st Annual Conference of the Cognitive Science Society* (pp. 755–760). Mahwah, NJ: Lawrence Erlbaum Associates.

Van Everbroeck, E. (2003). Language type frequency and learnability from a connectionist perspective. *Linguistic Typology*, *7*, 1–50.

van Gompel, R. P., & Liversedge, S. P. (2003). The influence of morphological information on cataphoric pronoun assignment. *Journal of Experimental Psychology: Learning, Memory, and Cognition*, *29*, 128–139.

Vapnik, V. N., & Chervonenkis, A. (1971). On the uniform convergence of relative frequencies of events to their probabilities. *Theory of Probability and Its Applications*, *16*, 264–280.

Vasishth, S., Chen, Z., Li, Q., & Guo, G. (2013). Processing Chinese relative clauses: Evidence for the subject-relative advantage. *PLoS One*, *8*(10), e77006.

Vasishth, S., Suckow, K., Lewis, R. L., & Kern, S. (2010). Short-term forgetting in sentence comprehension: Crosslinguistic evidence from verb-final structures. *Language and Cognitive Processes*, *25*, 533–567.

Vicari, G., & Adenzato, M. (2014). Is recursion language-specific? Evidence of recursive mechanisms in the structure of intentional action. *Consciousness and Cognition*, *26*, 169–188.

Voight, B. F., Kudaravalli, S., Wen, X., & Pritchard, J. K. (2006). A map of recent positive selection in the human genome. *PLoS Biology*, *4*, e72.

von Humboldt, W. (1999). *On language: On the diversity of human language construction and its influence on the mental development of the human species*. Cambridge: Cambridge University Press. (1836).

Vouloumanos, A., & Werker, J. F. (2007). Listening to language at birth: Evidence for a bias for speech in neonates. *Developmental Science*, *10*, 159–164.

Waddington, C. H. (1942). Canalization of development and the inheritance of acquired characters. *Nature*, *150*, 563–565.

Walker, P., Bremner, J. G., Mason, U., Spring, J., Mattock, K., Slater, A., et al. (2010). Preverbal infants' sensitivity to synaesthetic cross-modality correspondences. *Psychological Science*, *21*, 21–25.

Wallace, R. S. (2005). *Be your own botmaster* (2nd Ed.). Oakland, CA: ALICE A.I. Foundation.

Wallentin, M., & Frith, C. D. (2008). Language is shaped for social interactions, as well as by the brain. *Behavioral and Brain Sciences*, *31*, 536–537.

Wang, D., & Li, H. (2007). Nonverbal language in cross-cultural communication. *Sino-US English Teaching*, *4*, 66–70.

Wang, M. D. (1970). The role of syntactic complexity as a determiner of comprehensibility. *Journal of Verbal Learning and Verbal Behavior*, *9*, 398–404.

Wang, W. S.-Y. (1969). Competing changes as a cause of residue. *Language*, *45*, 9–25.

Wang, W. S.-Y. (Ed.). (1977). *The lexicon in phonological change*. The Hague: Mouton.

Wang, W. S.-Y. (1978). The three scales of diachrony. In B. B. Kachru (Ed.), *Linguistics in the seventies: Directions and prospects* (pp. 63–75). Urbana, IL: Department of Linguistics, University of Illinois.

Wang, W. S.-Y., & Cheng, C.-C. (1977). Implementation of phonological change: The Shaung-feng Chinese case. In W. S.-Y. Wang (Ed.), *The lexicon in phonological change* (pp. 86–100). The Hague: Mouton.

Warren, P., & Marslen-Wilson, W. (1987). Continuous uptake of acoustic cues in spoken word recognition. *Perception & Psychophysics*, *41*, 262–275.

Warren, R. M., Obusek, C. J., Farmer, R. M., & Warren, R. P. (1969). Auditory sequence: Confusion of patterns other than speech or music. *Science*, *164*, 586–587.

Warren, T., & Gibson, E. (2002). The influence of referential processing on sentence complexity. *Cognition*, *85*, 79–112.

Wasow, T., & Arnold, J. (2003). Post-verbal constituent ordering in English. In G. Rohdenburg & B. Mondorf (Eds.), *Determinants of grammatical variation in English* (pp. 119–154). Berlin: Mouton de Gruyter.

Weber, B. H., & Depew, D. J. (Eds.). (2003). *Evolution and learning: The Baldwin effect reconsidered*. Cambridge, MA: MIT Press.

Weissenborn, J., & Höhle, B. (Eds.). (2001). *Approaches to bootstrapping: Phonological, lexical, syntactic and neurophysiological aspects of early language acquisition*. Philadelphia, PA: John Benjamins.

Wells, J. B., Christiansen, M. H., Race, D. S., Acheson, D. J., & MacDonald, M. C. (2009). Experience and sentence processing: Statistical learning and relative clause comprehension. *Cognitive Psychology*, *58*, 250–271.

Werker, J. F., & Tees, R. C. (1999). Influences on infant speech processing: Toward a new synthesis. *Annual Review of Psychology*, *50*, 509–535.

Wexler, K., & Culicover, P. W. (1980). *Formal principles of language acquisition*. Cambridge, MA: MIT Press.

Wheeldon, L., & Lahiri, A. (1997). Prosodic units in speech production. *Journal of Memory and Language*, *37*, 356–381.

Wicha, N. Y. Y., Moreno, E. M., & Kutas, M. (2004). Anticipating words and their gender: an event-related brain potential study of semantic integration, gender expectancy, and gender agreement in Spanish sentence reading. *Journal of Cognitive Neuroscience*, *16*, 1272–1288.

Wilbur, R. B., & Nolkn, S. B. (1986). The duration of syllables in American Sign Language. *Language and Speech*, *29*, 263–280.

Wilkins, J. (1668). *An essay towards a real character and a philosophical language*. London: Gellibrand.

Wilkins, W. K., & Wakefield, J. (1995). Brain evolution and neurolinguistic pre-conditions. *Behavioral and Brain Sciences*, *18*, 161–182.

Wilson, B., Slater, H., Kikuchi, Y., Milne, A., Marslen-Wilson, W., Smith, K., et al. (2013). Auditory artificial grammar learning in macaque and marmoset monkeys. *Journal of Neuroscience, 33*, 18825–18835.

Wilson, E. O. (1971). *The insect societies*. Cambridge, MA: Harvard University Press.

Wilson, M., & Emmorey, K. (2006). Comparing sign language and speech reveals a universal limit on short-term memory capacity. *Psychological Science, 17*, 682–683.

Winograd, T. (1972). Understanding natural language. *Cognitive Psychology, 3*, 1–191.

Wittgenstein, L. (1953). *Philosophical investigations*. Oxford: Blackwell.

Wolpert, D. M., Diedrichsen, J., & Flanagan, J. R. (2011). Principles of sensorimotor learning. *Nature Reviews: Neuroscience, 12*, 739–751.

Wynne, T., & Coolidge, F. L. (2008). A stone-age meeting of minds. *American Scientist, 96*, 44–51.

Yamashita, H., & Chang, F. (2001). "Long before short" preference in the production of a head-final language. *Cognition, 81*, B45–B55.

Yamauchi, H. (2001). The difficulty of the Baldwinian account of linguistic innateness. In J. Kelemen & P. Sosík (Eds.), *ECAL01* (pp. 391–400). Prague: Springer.

Yang, C. (2002). *Knowledge and learning in natural language*. New York: Oxford University Press.

Yang, C. (2006). *The infinite gift: How children learn and unlearn the languages of the world*. New York: Simon and Schuster.

Yang, C. (2013). Ontogeny and phylogeny of language. *Proceedings of the National Academy of Sciences, 110*, 6324–6327.

Yopak, K. E., Lisney, T. J., Darlington, R. B., Collin, S. P., Montgomery, J. C., & Finlay, B. L. (2010). A conserved pattern of brain scaling from sharks to primates. *Proceedings of the National Academy of Sciences, 107*, 12946–12951.

Zajonc, R. B. (1980). Feeling and thinking: Preferences need no inferences. *American Psychologist, 35*, 151–175.

Zeevat, H. (2006). Grammaticalisation and evolution. In A. Cangelosi, A. D. M. Smith, & K. Smith (Eds.), *The Evolution of Language* (pp. 372–378). Singapore: World Scientific.

Zhu, L., Chen, Y., Torrable, A., Freeman, W., & Yuille, A. L. (2010). Part and appearance sharing: Recursive compositional models for multi-view multi-object detection. *IEEE Computer Society Conference on Computer Vision and Pattern Recognition*.

Zipf, G. K. (1949). *Human behavior and the principle of least-effort*. Cambridge, MA: Addison-Wesley.

Zuidema, W. (2003). How the poverty of the stimulus solves the poverty of the stimulus. In S. Becker, S. Thrun, & K. Obermayer (Eds.), *Advances in Neural Information Processing Systems 15* (pp. 51–58). Cambridge, MA: MIT Press.

Name Index

Page numbers are annotated with the letters *f*, *t*, *b*, and *n* to indicate references that occur in a figure, table, box, or footnote.

Subject Index

Page numbers are annotated with the letters *f*, *t*, *b*, and *n* to indicate references that occur in a figure, table, box, or footnote.